Monitoring Forest Biodiversity

Monitoring Forest Biodiversity

Improving Conservation through Ecologically Responsible Management

Toby Gardner

publishing for a sustainable future

London • New York

First published in 2010 by Earthscan

Earthscan
2 Park Square, Milton Park, Abingdon, Oxon OX14 4RN
Simultaneously published in the USA and Canada by Earthscan
711 Third Avenue, New York, NY 10017
Earthscan is an imprint of the Taylor & Francis Group, an informa business

First issued in paperback 2011

ISBN: 978-1-84407-654-3 hardback
ISBN: 978-0-415-50715-8 paperback

Typeset by FiSH Books, Enfield
Cover design by Susanne Harris

A catalogue record for this book is available from the British Library
Library of Congress Cataloging-in-Publication Data

Gardner, Toby.
  Monitoring forest biodiversity : improving conservation through ecologically-responsible management / Toby Gardner.
     p. cm.
  Includes bibliographical references and index.
  ISBN 978-1-84407-654-3 (hardback)
  1. Sustainable forestry. 2. Biodiversity conservation. 3. Forest biodiversity. 4. Forest management. I. Title.
  SD387.S87G37 2010
  333.75'16—dc22

                                                                    2010000823

Contents

Purpose of this book: How can monitoring contribute to forest
biodiversity conservation?

Structure and scope of the book

PART I: THE CONTEXT OF MONITORING FOREST BIODIVERSITY

Biodiversity in logged forests

Biodiversity in regenerating forests

Biodiversity in agroforestry systems

Biodiversity in tree plantations

An ecosystem approach to forest conservation

The origins of sustainable forest management (SFM)

Sustainable forest management as a guiding vision versus
a measurable standard

Criteria and indicators in forest management

Scientific uncertainty and biodiversity conservation
in human-modified forest ecosystems

The purpose of biodiversity monitoring as a guide to management

List of Figures, Tables and Boxes

FIGURES

TABLES

BOXES

Foreword

This book expertly combines three key topics, which have also been my passions for more than 25 years – forest biodiversity, forest management and ecological monitoring. I am passionate about these subjects because it is essential they are done well – for the sake of forest biodiversity conservation and to improve forest management so that it can progress far more quickly along the path towards true ecological sustainability. These passions are clearly shared by Toby Gardner.

Although there have been many volumes on forest biodiversity, on forest management and on monitoring, this book is a unique contribution because it is the first to integrate these important topics. Such integration is a major step forward. It unites three vast and complex arenas of scientific literature and then eloquently re-presents key findings in a way that is accessible to a broad readership. However, this book is far more than an outstanding review of the literature; it includes many new insights that have the potential to revolutionize forest monitoring programmes. Indeed, right from the outset, this publication highlights two key reasons that motivated the writing of it and, in turn, this is why I believe this book is so valuable:

- The record of monitoring has been truly appalling to date and there is clearly an urgent need for major reform to improve that record.
- Better monitoring is pivotal to better forest management and this is, in turn, absolutely critical for the conservation of forests and forest biodiversity.

As Toby Gardner correctly points out, the stakes are high if monitoring and forest management continues to be done badly – a large amount of the world's biodiversity occurs in forests most of which are not formally protected but rather located in places affected by human use. As a counter to this, Toby Gardner's excellent book contains many valuable lessons and recommendations on ways to improve forest monitoring, promote far better and more ecologically sustainable forest management, and improve biodiversity conservation programmes.

This book discusses an array of valuable topics that are central to all aspects of monitoring, from the reasons why it is important and its purpose to conceptual frameworks to underpin monitoring programmes, what should be monitored, how to best interpret monitoring data and why the success of monitoring programmes, broad engagement from a wide range of stakeholders and depends on an adaptive approach to monitoring programmes is pivotal.

Researchers, policy makers and forest managers need to read this book. And when they have finished, they should keep a copy close by as a valuable reference text.

David Lindenmayer
Fenner School of Environment and Society
The Australian National University, Canberra
Australia

Acknowledgements

This book would be a shadow of its current self without the help I received from many colleagues and friends along the way. Discussions about many of the ideas and issues presented here, together with constructive criticism I received on earlier versions of the text, have been invaluable in improving both its structure and content. I am particularly indebted to Jos Barlow, Renata Pardini and Joern Fischer, who each reviewed more than half of the book and provided a constant source of perceptive comments and recommendations, as well as welcome support and encouragement to finish. Many other people provided valuable reviews of individual chapters or sets of chapters, and I am extremely grateful to Cris Banks, Corey Bradshaw, Sharon Brooks, Tim Caro, Finn Danielsen, Robert Ewers, Emily Fitzherbert, Joern Fischer Valerie Kapos, Alexander Lees, Julio Louzada, William Magnusson, Liz Nichols, Luke Parry, Jeff Sayer, Navjot Sodhi and Brendan Wintle.

I am thankful to my editors Jeff Sayer and Tim Hardwick for their encouragement and support in contributing to the Earthscan Forestry Library Series, and particularly to Tim for his unwavering patience in the face of repeatedly failed deadlines. Many of the ideas presented in the book were developed while I was a visiting scholar at the University of Queensland and a postdoctoral associate at the Federal University of Lavras and I am indebted to both Hugh Possingham and Julio Louzada for their hospitality and provision of such a fertile research environment. Much of the writing of the first sections was done while on retreat in Yorkshire and I am extremely thankful for the support and company of my mother, Josephine Gardner during that time. The remainder of the book was completed in Cambridge, and I am very appreciative of my friends and colleagues for their moral support throughout. I am also thankful to the Fundação de Amparo à Pesquisa do Estado de Minas Gerais and the Natural Environmental Research Council for funding my work while this book was written.

Writing this book frequently meant that I had to sacrifice time away from those who are dearest to me. To Emily and Liz, and my family Josephine, David, Sam, Emma, Angus, Aulay, Natalie, Ishbel, Eilidh, India, Cora and Aurelia, thank you. This book would never have been possible without your support.

List of Acronyms and Abbreviations

AIC	Aikaike's information criterion
AM	adaptive management
C&I	criteria and indicators
CBD	Convention on Biological Diversity
CIFOR	Center for International Forestry Research
dbh	diameter at breast height
DLMP	dynamic landscape metapopulation
EFI	European Forest Institute
FAO	Food and Agriculture Organization
FMU	forest management unit
FSC	Forest Stewardship Council
GDM	generalized dissimilarity modelling
GFIS	Global Forest Information Service
GLM	general linear model
HH	habitat hectares
HSI	habitat suitability indices
ICDP	integrated conservation and development project
IT	information-theoretic
ITTO	International Tropical Timber Organization
IUCN	International Union for Conservation of Nature
LDC	least disturbed condition
LiDAR	light detection and ranging
MCPFE	Ministerial Conference on the Protection of Forests in Europe
MDC	minimally disturbed condition
NDFI	normalized difference fraction index
NHT	null hypothesis testing
P&C	principles and criteria
PC&I	principles, criteria and indicators
PCA	principal component analysis
PSR	pressure–state–response
RIL	reduced impact logging
SAD	species–abundance distribution
SAN	Sustainable Agriculture Network
SFM	sustainable forest management
SMA	spectral mixture analysis
VHRS	Variable Retention Harvest System
WCFSD	World Commission on Forests and Sustainable Development
ZOI	Zone of Interaction

Introduction

In writing this book, my goal is to appeal to all those involved in the study and management of forest ecosystems – academic and professional ecologists, forest owners, inhabitants and managers, government agencies and third-party certification authorities – to think about ways in which monitoring biodiversity can help achieve more responsible approaches to forest management.

I was motivated in particular by three broadly supported observations:

1   Recognition of the urgent need to develop improved strategies for conserving forest biodiversity in areas outside reserves. For much of the world, the only forests that remain are those that have been managed for some form of human use.
2   A widespread sense of dissatisfaction in the contribution made by biodiversity monitoring programmes to forest management thus far. Many monitoring programmes represent little more than opportunistic surveillance exercises. They employ a reactive philosophy to measure a piecemeal collection of indicators that are disconnected from the actual management process and therefore of little value in guiding any conservation action.
3   A belief that with a bit of work, biodiversity monitoring has the potential to make a much more meaningful contribution to forest conservation and the development of ecologically responsible approaches to management.

While major advances have been made at the level of international forest policy to harmonize the goals and objectives of responsible forest management, comparatively little progress has been achieved in translating these agreements into on-the-ground management standards. A key ingredient in promoting more responsible management approaches is the development of monitoring programmes that can guide progress towards long-term conservation goals. In this book, I draw upon research findings and practical experience in forestry, forest ecology and conservation biology to analyse some of the challenges that have faced monitoring programmes thus far, and then build upon this understanding to identify some of the ways in which biodiversity monitoring can make a positive contribution to forest conservation.

In essence, this book simultaneously represents a cry of protest, an advocacy of reform and an expression of hope. First, a cry of protest against the continued implementation of poorly conceived biodiversity monitoring programmes that contribute very little, if anything, to the conservation management of natural resources. Second, an advocacy of reform of the status quo to encourage the development of future monitoring programmes that draw effectively upon lessons learnt to date. Finally, an expression of hope in reflection of the real opportunities that exist to draw upon support within broader society, including local communities, international and national policy forums, management authorities and the scientific community, to implement more ecologically responsible systems of forest management.

THE PURPOSE OF THIS BOOK: HOW CAN MONITORING CONTRIBUTE TO THE CONSERVATION OF FOREST BIODIVERSITY?

The goal of ecologically responsible forest management is to minimize, wherever possible, the impact of human activity on forest biodiversity and associated ecological processes. This is achieved through the development of management strategies that help ensure progress towards long-term conservation goals. The central motivation behind this book is the identification of ways in which biodiversity monitoring can be effective in facilitating and guiding this process.

In very general terms, there are two reasons why we may wish to monitor biodiversity as an aid to conservation (i.e. ignoring pure research). The first is to provide a surveillance record of change over time in some component of biodiversity that is of conservation interest. The second is to learn about how different human activities impact biodiversity and thus identify ways in which we develop more responsible management practices. This book is concerned with the second.

There are various benefits to be gained from a simple 'surveillance style' record of biodiversity change over time. Information on population or species declines can be used to kick-start conservation action, both in the form of a regulatory mechanism (e.g. as is done commonly in the management of fish stocks) and as a way of raising public and political awareness about environmental issues. Long-term monitoring of biodiversity across a network of sites can also help in developing an improved understanding of background levels of variability in natural systems, as well as capture information on the importance of poorly understood and unpredictable human-related threats in areas that are seemingly isolated from any direct disturbance (e.g. the impacts of climate change and disease on amphibians; Pounds et al, 2006). Surveillance-style monitoring can also be an effective way to engage non-scientists in conservation. Good examples of this are long-term, nationwide bird surveys conducted in Britain and North America that involve thousands of volunteers while also feeding information into national indicators of biodiversity loss (e.g. as developed by the British Trust for Ornithology on behalf of the UK government www.bto.org/research/indicators/index.htm).

However, despite the benefits that can be gained from a surveillance-style approach, there are some serious limitations regarding its utility as a practical aid for conservation management (Nichols and Williams, 2006). The main shortcoming is that *it is disconnected from the management process.* By this I mean that the design of the monitoring programme is not centred on assessing the impact of particular human activities or conservation strategies. Any information that is generated can only be used to guide a reactive approach to management – it presumes that a clear and workable plan of action is already available, and that such a plan can be launched into place once the warning bells start ringing. Unfortunately, such plans do not exist for the majority of conservation problems. Moreover, by only supporting a reactive approach to management there is an optimistic assumption in the use of surveillance monitoring that the maintenance of the status quo represents an adequate conservation goal.

In this book I argue that, in addition to any other benefit, monitoring should be viewed as an explicit mechanism for learning about how to improve opportunities for conservation. To what extent are current management strategies adequate and where can cost-effective improvements be made? I take the position that monitoring and management should be viewed as highly interconnected activities where neither exercise makes proper sense unless it is accompanied by the other.

Biodiversity is the variety of life, in all of its many manifestations. It is a broad unifying concept, encompassing all forms, levels and combinations of natural variation, at all levels of biological organization (Gaston and Spicer, 2004). The focus of most of this book is on biodiversity as a multi-species concept. Much has been written on the management and monitoring of individual species (see Williams et al, 2001; and MacKenzie et al, 2006 for reviews), while far less practical guidance is available for the arguably more complex task of evaluating change across entire species assemblages. In order to understand the drivers of observed changes in biodiversity, it is necessary to monitor not only changes in species but also changes in management activities themselves, as well as the changes to forest structure and function that are a direct result of management. Consequently, attention is given throughout the book to the relationships between different types of indicators, including biological, structural and management indicators, and the ways in which they contribute complementary information to the management process.

THE STRUCTURE AND SCOPE OF THE BOOK

To address the challenges outlined above, the book is structured into four main sections.

In the first section, I set the scene by describing the importance of human-modified and managed forests for biodiversity conservation (Chapter 1), the history and current status of forest management and the process of developing more ecologically responsible management standards, including criteria and indicators for monitoring and evaluation (Chapter 2), the need for forest biodiversity monitoring (Chapter 3), and a general typology of monitoring approaches and indicators for biodiversity monitoring (Chapter 4).

In the second section, I provide a critical appraisal of some of the key challenges of purpose (Chapter 5), design (Chapter 6) and reality (Chapter 7) that currently face the implementation of successful forest biodiversity monitoring programmes.

This analysis serves to highlight some of the major issues and controversies that have characterized many past attempts to monitor biodiversity, and helps provide a useful foundation for developing more effective programmes in the future. Particular attention is given to the significant problems that have hindered the selection and interpretation of species-based indicators (Chapter 6).

The third and core section of the book builds upon the foundation provided by the previous sections, and draws widely from lessons learnt in forestry, ecology and conservation science, to develop an operational framework for monitoring that has the potential to improve conservation opportunities in managed forests. Chapter 8 sets the groundwork for thinking about the overarching purpose of monitoring and considers ways in which biodiversity monitoring interacts with the management process as a whole. Thinking about the end-use of different forms of monitoring information is essential for setting appropriate goals and objectives and guiding the whole monitoring design process.

Discussion and analysis of the monitoring process itself encompasses two principal and interconnected stages that help to structure the organization of individual chapters: a scoping stage to define monitoring goals, objectives and indicators (Chapters 9–12), and a coupled design-implementation stage to determine how different types of indicators interrelate within the overall monitoring-management system (Chapter 13), how to identify appropriate strategies for sampling design and data collection (Chapter 14), and data analysis and interpretation (Chapter 15). In Chapter 16, I highlight some of challenges of making biodiversity monitoring work effectively in the real world, including the role of people in monitoring and the often complementary contributions made by scientists and local communities, and the problem of ensuring the long-term success and viability of monitoring.

Early consideration is given in Chapter 2 as to the origins and meaning of the term 'sustainable forest management'. In recognition of its visionary yet ultimately non-verifiable nature, many commentators feel that the term is inappropriate and prefer phrases such as 'responsible forest management' or 'good forest stewardship'. However, use of the term sustainable is now firmly established within all major stakeholder groups and at various levels of policy and management (Higman et al, 2005). For most of this book, I refer to responsible forest management when discussing actual management practices and the development of standards and guidelines, but occasionally mention the concept of sustainable forest management when relating to questions of long-term vision and overall policy goals.

Biodiversity monitoring is most useful when conducted in an area that is directly or indirectly impacted by some form of human-induced threat, disturbance or management intervention that can, in theory, be curtailed, mitigated or otherwise adjusted in order to enhance the long-term survival of native species and maintenance of critical ecological processes. The management-orientated approach to monitoring that I advocate in this book follows many of the basic principles of applied research, but is distinct in so far as it maintains a constant engagement with the management system, seeking to drive a process of continuous improvement and refinement as part of the adaptive management cycle.

Conservation opportunities exist in landscapes that are actively managed for production, as well as those that have undergone some form of human use in the past. The scope of this book therefore, encompasses both the monitoring of existing forest management plans that have been developed with conservation concerns in mind (whether legal or voluntary) and the unplanned human activities or threatening processes where – provided there is sufficient political and socio-economic incentive to do so – the introduction of new practices (or the mitigation of existing economic activity) could generate substantial benefits for biodiversity in the future.

While responsible management systems for different types of production forest are currently under different stages of development (see Chapter 2), the principles discussed here are equally relevant to all types of human-modified forest, including native logging ones, plantations, agroforestry systems and regenerating secondary forests. Furthermore, and despite the focus on forest systems, the principles of good practice that lie behind meaningful biodiversity monitoring programmes are generic and could apply equally to other biomes around the world.

In writing this book, my intention has been to present a broad analysis of the key issues relating to forest biodiversity monitoring. One of the main barriers to meaningful biodiversity monitoring is the organizational insularity that currently exists between researchers, managers and bureaucrats involved in natural resource management (Field et al, 2007). Typically, there is very little transfer of knowledge and understanding about forest management and biodiversity conservation between these groups, a situation that effectively stifles the potential contribution that any monitoring programme can make to the management process. The core readership of this book is expected to consist of applied ecologists and conservation scientists interested in developing a broader appreciation of the factors that contribute towards the design of meaningful biodiversity monitoring programmes. However, it should also appeal to managers and forest planners who are responsible for enhancing biodiversity in human-modified forest landscapes that are managed primarily for economic production (whether community or privately owned), as well as individuals involved in forest policy, standard development and the certification process.

In many ways, this book is more concerned with the *why* than the *how* of monitoring. Although coverage is reasonably comprehensive, I have purposefully avoided going into excessive technical detail on specific issues, and as such the book is not intended to serve as a detailed manual on exactly how to design and implement all of the components of the biodiversity monitoring process. Excellent specialized texts already exist for some technical subjects, such as the statistical analysis of ecological data, and no attempt has been made to repeat this information in detail here. What has been given far less treatment in the scientific literature is the question of *why we should monitor in the first place*. A clear understanding of purpose, goals and objectives is vital in determining the focus of monitoring work (i.e. choice of indicators and target species, sampling design, data analysis), and the way in which different elements of the overall monitoring framework interact with each other. One of the central arguments I make in the book is that to make monitoring more meaningful we need first to step back from much of the technical detail that surrounds issues of sampling methodology, data collection and analysis, and question more closely how biodiversity information can best be integrated into the management process. A superbly designed monitoring programme that succeeds in collecting statistically robust data in a cost-effective manner is of no use if it addresses objectives that are of limited relevance to management, or if it fails to be sustainable in the long term.

It is not my intention to be overly prescriptive about the ways in which biodiversity evaluation and monitoring can make a meaningful contribution to management. Vast differences in the ecological and social context of different landscapes preclude the notion that any single 'one-size-fits-all' recipe book exists. However, while recognizing the importance of context, neither is it useful to merely list different alternatives from which to choose. We have learnt a considerable amount about the general principles of good monitoring practice, including which stages are essential to ensuring success, and how certain monitoring approaches may work better under certain circumstances. Therefore, in some instances I have proposed a particular approach as being more defensible or likely to deliver greater and more reliable rewards, while also attempting to highlight important alternatives and counter-arguments where they exist. A number of reviewers of the book questioned the distinction between a short-term 'biodiversity assessment or evaluation' and a long-term 'biodiversity monitoring programme', and the practical relevance of any such distinction. In the simplest terms, a monitoring programme needs to include at least two repeat measurements to be deserving of the name but beyond that I would say that the distinction is subjective. Studies of biodiversity take place over a myriad of timescales as determined by the requirements of a given research objective (e.g. the timescales over which a particular attribute of biodiversity or threatening process varies) and the availability of funding and people to support the work. Because human-modified landscapes are highly dynamic and are constantly being faced with both new threats (e.g. climate change) and new conservation opportunities (e.g. investment from an ecosystem service scheme), it is *always* worth planning for the long term, even if there is no guarantee that the necessary funds are available at the outset. The task of designing and managing a landscape to support biodiversity conservation is ultimately a continuous process of enactment where information needs are constantly changing in response to new challenges and opportunities (Haila, 2007; Gardner et al, 2009).

One of the key words running throughout the entire book is *adaptive*. The philosophy and process of adaptive management is central to addressing the enormous uncertainty that characterizes our understanding of the sustainable management of natural resources (whether for forests or any other ecosystem), and provides a guiding framework by which monitoring can engage with the management process. Because we live in such an uncertain world, it is also important that the monitoring process itself is flexible and adaptive, and able to respond appropriately to both ecological and social surprises (Lindenmayer and Likens, 2009). Biodiversity monitoring, when done well, is not a trivial exercise. In keeping with this adaptive theme, it is my hope that the ideas presented in this book will, more than anything else, provide a catalyst for forest managers, researchers, regulation authorities, civil society groups and conservation organizations to think more keenly and critically about how biodiversity monitoring can make a more useful contribution to forest conservation. I very much welcome constructive feedback on any of the issues discussed here, and especially in light of practical experiences in implementing biodiversity monitoring programmes in forests around the world.

Toby Gardner
Cambridge, November 2009
tobyagardner@gmail.com

PART I

The Context of Monitoring Forest Biodiversity

CHAPTER 1

Biodiversity Conservation in Human-modified and Managed Forests

Synopsis

- Forest ecosystems support about two-thirds of the world's terrestrial biodiversity.
- Few forests left on Earth have escaped some form of human impact.
- Logged and degraded forests, as well as agroforests, plantations and regenerating secondary forest can provide valuable habitat for species that are unable to survive in agricultural and urban areas and improve connectivity among remaining areas of undisturbed forest.
- To be successful, the conservation of forest biodiversity therefore needs to be integrated with other human activities. An ecosystem approach is needed that encompasses not only primary forest reserves but also a host of off-reserve conservation measures, including the maintenance of landscape connectivity, buffer zones around reserves, landscape heterogeneity, sensitive management of production forest stands and natural disturbance regimes, and the protection of aquatic ecosystems.

Efforts to improve human welfare have led to the domestication of landscapes and ecosystems worldwide to enhance food supplies and reduce exposure to natural dangers (Kareiva et al, 2007). As a consequence, there are few places left on Earth that have escaped some form of human impact. Changes to the extent and condition of many habitats and ecosystems are, with few exceptions, negative, anthropogenic in origin and showing little evidence of abatement (Sala, 2005). Indeed, it is now accepted that human activity is driving the sixth major extinction crisis the planet has ever faced, with widespread losses in both habitats and species (Jenkins et al, 2003; Balmford and Bond, 2005; Mace et al, 2005).

The threat posed to biodiversity by human activity is perhaps most conspicuous in the case of forest ecosystems – they cover a vast area of the Earth's surface, are host to a significant proportion of the world's terrestrial biota, and suffer from widespread and ongoing clearance and degradation. Modification of forests could be considered one of the defining features of human societies (Lindenmayer and Franklin, 2003). In the UN's Food and Agriculture Organization's (FAO) 2005

Global Forest Assessment, the total forest area was estimated to be just less than 4 billion hectares (ha) or an average of 0.62ha per capita (FAO, 2006). However, deforestation, mainly through the conversion of forests to agricultural land, continues at an alarmingly high rate – about 13 million hectares per year. Only about one third of the global forest estate remains as primary forest; although new data illustrating widespread historical impacts and more cryptic forms of present-day degradation cast doubt on the existence of truly 'virgin' or 'pristine' forests (Box 1.1). The remaining two-thirds (approximately 2.7 billion hectares) of the total forest area exists in a varyingly human-modified condition (from logging and other extraction, fragmentation and burning), while just over 10 per cent of the total is in the form of plantation and agroforests (FAO, 2006).

Apart from being somewhat daunting and depressing, this evidence of widespread and ongoing human modification of the world's forests has at least two profound implications for the conservation of biodiversity.

First, forests support approximately 65 per cent of the world's terrestrial taxa (World Commission on Forests and Sustainable Development (WCFSD), 1999) and the sheer extent to which we have dominated the biosphere means that we have no choice but to integrate conservation efforts with other human activities, and seek to enhance the opportunities for species and populations to persist in human-modified forest systems (Gardner et al, 2009). It is broadly accepted that strictly protected areas provide a necessary yet grossly inadequate component of a broader strategy to safeguard the future of the world's biota. There are increasingly few situations where forest wildlife is effectively safeguarded simply by erecting a fence around a reserve. Gap analyses show that approximately one quarter of the world's threatened species live outside protected areas (Rodrigues et al, 2004), and that most of the world's terrestrial eco-regions fall significantly short of the 10 per cent protection target proposed by the International Union for Conservation of Nature (IUCN) (Schmitt et al, 2008). For much of the world, this shortfall in protected areas is not explained by a simple lack of political will and financial resources, but by the fact that many regions no longer contain large areas of undisturbed forest. This is true in both temperate forest ecosystems, like Western Europe and much of the US, which lost their last areas of forested wilderness hundreds of years ago, but also tropical forest ecosystems that are host to such a significant proportion of the world's biodiversity yet continue to suffer unprecedented rates of deforestation and degradation (Box 1.2; Bradshaw et al, 2008; Hansen et al, 2008).

Second, forest landscapes, even when comprising degraded, semi-natural or artificial forest habitat can provide the last refuges for species that are unable to survive in agricultural land. Consequently, the retention or management of structurally and floristically complex habitats, including logged and degraded forests, agroforests, plantation and secondary forests, can often ensure the persistence of forest species in managed landscapes (Lamb et al, 2005; Barlow et al, 2007a; Chazdon, 2008; Chazdon et al. 2009; Gardner et al. 2009; Scales and Marsden, 2008). In addition to providing conservation benefits through the provision of species habitat per se, modified forest systems can provide an important conservation service through enhancing landscape connectivity and enabling populations of forest species to move between remnant areas of undisturbed forest (Fischer et al, 2006; Pardini et

Box 1.1 What is pristine forest?

Set against the burgeoning spread of human activities and globalization, it is becoming increasingly difficult to identify areas of forest that can be truly considered 'pristine' or 'intact'. Some scientists have gone so far as to consider that the traditional concept of an intact ecosystem is obsolete, and instead propose a classification system based on global patterns of human interaction with ecosystems, demonstrating that much of the world currently exists in the form of different 'anthropogenic biomes' (Ellis and Ramankutty, 2008).

Evidence that challenges the concept of pristine forest ecosystems comes in a number of forms. First, for many regions, large areas of intact vegetation simply no longer exist, as is the case of the Atlantic forest hotspot of Brazil, which has been reduced, except for a few conservation units, to a fragmented network of very small remnants (< 100ha), mainly composed of secondary forest, and immersed in agricultural or urban matrices (Ribeiro et al, 2009).

Second, in vast areas of seemingly intact and contiguous forest, such as the Amazon basin, more cryptic forms of disturbance, for instance the extraction of timber and non-timber forest products, are increasingly evident across much of the area. For example Broadbent et al (2008) calculated that in 2002 53 per cent of the Brazilian Amazon was less than 2 kilometres (km) from an edge; while Peres and Lake (2003) found that only about 1 per cent of the Brazilian Amazon is both strictly protected on paper and reasonably safe from extractive activities targeted to game vertebrates and other valuable non-timber forest products.

Third, it is becoming increasingly evident that global climate change is influencing the structure and function of forest ecosystems, even in areas that suffer from no discernible direct human impact. For example, one of the consequences of increased drought frequency in tropical forests is an increased prevalence of wildfire, as seen in the Amazon in 2005 (Aragão et al, 2007). More cryptically, some studies have demonstrated possible climate-change linked drivers of large-scale increases in the rate of vegetation dynamics in remote parts of the Brazilian Amazon (e.g. Phillips et al, 2004).

Finally, even when accounting for cryptic forms of present-day disturbance, archaeological and paleoecological studies over the last two decades suggest that many contemporary pristine habitats have in fact undergone some form of human disturbance in the past (e.g. Willis et al, 2004; Willis and Birks, 2006). For example, the Upper Xingu region of Brazil comprises one of the largest contiguous tracts of tropical rainforest in the Amazon today. Emerging archaeological evidence suggests that parts of this region had been densely populated by pre-European human settlements (circa ~1250 to ~1600 AD), and that extensive forests underwent large-scale transformation to agricultural areas and urbanized centres (Heckenberger et al, 2003; Willis et al, 2004). Many further examples of extensive pre-European disturbance have been found in areas that conservationists today frequently describe as 'pristine' or 'intact', including Southeast Asia, Papua New Guinea and Central America (Willis et al, 2004). In most of these cases, forest regeneration followed the abandonment of human settlements and agricultural activities, resulting in the old-growth stands that are regarded as pristine today.

The implications for biodiversity conservation of such widespread current and historic human impacts on forest ecosystems are unclear. However, the addition of such a layer of complexity and uncertainty in management for regions that were otherwise considered intact highlights the growing need for biodiversity monitoring programmes outside of those areas that are directly managed for human use.

Box 1.2 The tropical deforestation crisis

A recent review of the conservation prospects for tropical biodiversity concluded that we are squarely in the middle of a tropical biodiversity tragedy and on a trajectory toward disaster (Bradshaw et al, 2008). While other researchers hold a more optimistic view of the future of tropical species (e.g. Wright and Muller-Landau, 2006) there is no escaping the fact that the majority of the world's biodiversity hotspots, regions that are simultaneously very diverse and highly threatened, occur in the tropical biome (see www.biodiversityhotspots.org).

Despite increased awareness of the importance of tropical forests, deforestation rates have not slowed in recent years. A comprehensive analysis (FAO, 2006) showed that annual tropical deforestation rates increased by 8.5 per cent during 2000–2005 when compared to the 1990s. Importantly, forest loss has not been consistent across all regions of the Tropics with some areas experiencing much greater proportional losses than others. The greatest losses have occurred in Southeast Asia, where present day maps of forest cover bear no comparison to their historical counterparts (Figure 1.1).

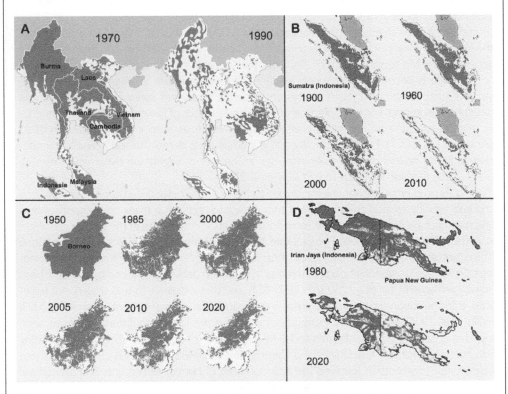

Figure 1.1 Forest disappearance across Southeast Asia during the last century

Source: Reprinted with permission from Bradshaw et al (2008)

al, 2009). Finally, semi-degraded and replanted forests can be critical in supplying local human communities with a source of forest products (timber and non-timber), providing a buffer around intact 'core' areas – the concept embodied in UNESCO's Biosphere Reserves. Well-managed buffer areas can be vital in protecting the long-term integrity of remaining areas of undisturbed forest, as even where they exist, protected areas are often threatened by encroachment and illegal extraction in regions that are undergoing widespread deforestation (e.g. Wittemyer et al, 2008).

It is instructive to set the scene for thinking about monitoring forest biodiversity by first taking stock of the extent of humanity's influence on forest ecosystems around the world, and the importance of these impacts for biodiversity. The rest of this chapter provides a brief synopsis of the importance of different modified forest systems for biodiversity conservation (see also Koh and Gardner, 2009). Although many factors contribute to the value of differently managed forest systems for biodiversity conservation, it is often helpful to have in mind a gradient of structural complexity as a crude proxy of conservation value that moves from intact native forest, through selectively logged forest, secondary forests, agroforestry systems, and plantations (with clear-cuts and agricultural land representing the loss of all forest structure) (Figure 1.2). It is important to bear in mind that a gradient of 'structural integrity' also exists within each of these four systems that can have important implications for biodiversity conservation, including: different levels of logging intensity in managed native forests; differently aged secondary forests; and a wide variety of agroforestry and plantation forests that combine different planting densities and species.

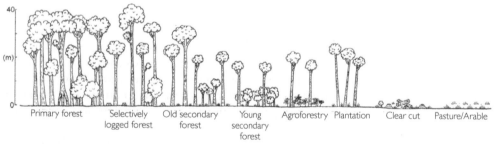

Figure 1.2 A gradient of structural integrity in human-modified forest systems

Source: Drawing courtesy of Phil Judge

Much of this book draws upon examples from natural forests managed for timber and plantations (both timber and cellulose) but, as described below, other systems, like agroforests and regenerating forests, can provide important conservation services in many areas of the world. In each case, I have sought to identify the relative importance of different forest land uses, summarize key elements of our understanding of biodiversity impacts, and highlight major areas of uncertainty in management.

BIODIVERSITY IN LOGGED FORESTS

As of 2005, approximately one third of the world's forests, a total of 1.3 billion hectares, were designated primarily for timber production (FAO, 2006). In 2006, member nations of the International Tropical Timber Organization (ITTO) exported over 13 million cubic metres of tropical non-coniferous logs worth US$2.1 billion, making a substantial contribution to the economies of these nations (ITTO, 2007). Logging activity on this massive scale has resulted in huge areas of forest being degraded following the selective removal of high-value trees, and the collateral damage associated with tree felling and extraction. Asner et al (2005) estimated that in the Brazilian Amazon between 1999 and 2002, the area of rainforest annually degraded by logging was approximately the same as that which was clear-felled for agriculture (between 12 and 19 million ha).

Although many different types of logging activity are likely to have a negative impact on the structure and composition of the forest, the extent of this impact depends on the logging intensity, including the number of trees removed per hectare, length of the rotation time, and site management practices. The density of felled trees in conventional logging programmes varies among regions and management regimes, from as few as one tree every several hectares (e.g. mahogany, *Swietenia macrophylla* in South America) to more than 15 per hectare in lowland dipterocarp forests of south-east Asia (Fimbel et al, 2001). In the last few decades, reduced impact logging (RIL) techniques have been developed that extract fewer, more carefully selected trees, and involve controlled harvesting techniques (e.g. preliminary inventories, road planning, directional felling) to greatly minimize deleterious effects (Fimbel et al, 2001; Putz et al, 2008).

Differences in how forests are managed determine the extent to which logging negatively affects wildlife, with impacts felt through changes to the structure and composition of the forest environment, including alterations in tree size structure, a shift towards early successional vegetation, changes in composition of fruiting trees, fragmentation of the canopy, soil compaction and alteration of aquatic environments. In general, broad patterns of wildlife response can be explained by differences in the intensity of logging activity, as well as the amount of recovery time elapsed before a study is conducted (Putz et al, 2001).

While there is no evidence of any species having been driven extinct by selective logging, there are abundant data showing marked population declines and local extinctions in a wide range of species groups (Fimbel et al, 2001; Mijaard and Sheil, 2008). Arboreal vertebrates appear to be particularly badly affected through loss of nesting and food resources. Both Sekercioglu (2002) and Thiollay (1997) reported losses of approximately 30 per cent of forest-dependent birds from logged areas in Uganda and Sumatra respectively. Felton et al (2003) reported depleted numbers of adult orang-utans in selectively logged peat forest in Kalimantan compared to neighbouring intact sites. Bats also appear to be especially sensitive to even low levels of logging as changes in canopy cover and understorey foliage density have knock-on effects on foraging and echolocation strategies (e.g. Peters et al, 2006). A number of studies have identified a close link between the magnitude or intensity of structural logging disturbance and the extent to which species assemblages diverge from those typical of undisturbed forests, as was found in the case of moths in North America (Summerville and Crist, 2002) and ants in Brazil (Vasconcelas et al, 2000). Nevertheless, for many taxa the impacts of selective logging are far less severe, even under conventional management regimes (i.e. those not employing RIL techniques; see Chapter 3). For example, Lewis (2001) found that logging at a density of six stems per hectare had little effect on the diversity and structure of butterfly assemblages in Belize, while Mijaard and Sheil (2008) concluded that only a few terrestrial mammal species have shown marked population declines following logging in Borneo.

BIODIVERSITY IN REGENERATING FORESTS

In most areas of the world, regrowth or secondary forests regenerate naturally on abandoned agricultural land if human disturbance declines. Following centuries of human disturbance, the total area of regenerating forest has increased rapidly. Indeed, for parts of the world that have suffered widespread historical deforestation, secondary forests comprise the majority of remaining forest area (e.g. east coast of the US, much of Western Europe, and areas of high human population density like Singapore). In the tropics in the 1990s, secondary regrowth reclaimed one-sixth of all primary forests that were clear cut during the same decade (Wright, 2005).

Dunn (2004) analysed data from 39 tropical data sets of biodiversity responses to fragmentation and concluded that species richness of some faunal assemblages can recover to levels similar to mature forest within 20–40 years, but that recovery of species composition is likely to take substantially longer. The recovery of biodiversity in secondary forests varies strongly between different species groups depending on their life histories, with species responses generally falling into three categories (Bowen et al, 2007): (i) species that decline in abundance or are absent from regrowth due to specialist habitat requirements; (ii) old-growth forest species that benefit from altered conditions in regenerating forest and increase in abundance or distribution; and (iii) open-area species that invade regenerating areas to exploit newly available resources. These conclusions are mirrored by the results of the comprehensive Jari study in north-east Brazil which found that 41 per cent of old-growth vertebrate and invertebrate species were lacking from secondary forests of 12–18 years of age, and that species responses varied strongly among and within taxonomic groups (Barlow et al, 2007a).

The general lack of data and the context-dependent nature of existing studies on biodiversity recovery in secondary forests severely limit our ability to make general predictions about the potential for species conservation in tropical secondary forests (Chazdon et al, in press). However, we can conclude that secondary forests are likely to be more diverse the more closely they reflect the structural, functional and compositional properties of mature forest, and are set within a favourable landscape context (Chazdon, 2003; Bowen et al, 2007). In particular, the conservation of old-growth species in secondary forests will be maximized in areas where: extensive tracts of old-growth forest remain within the wider region; older secondary forests have persisted; post-conversion land use was of limited duration and low intensity; post-abandonment anthropogenic disturbance is relatively low; seed dispersing fauna are protected; and old-growth forests are close to abandoned sites (Chazdon et al, in press).

The conservation value of a secondary forest should increase over time as old-growth species accumulate during forest recovery, but older secondary forests are poorly studied and long-term datasets are lacking. Existing chronosequence studies of regenerating forests demonstrate that biotic recovery occurs over considerably longer time scales than structural recovery, and that re-establishment of certain species and functional group composition can take centuries or millennia (DeWalt et al, 2003; Liebsch et al, 2008). Despite this uncertainty, regeneration represents the only remaining conservation option for many regions of the world that have suffered severe historical deforestation. An estimated 350 million ha of the tropics are classified as degraded due to poor management (Maginnis and Jackson, 2005). While the natural recovery of this land is not inevitable, there is encouraging evidence that judicious approaches to reforestation can greatly facilitate the regeneration process and enhance the prospects of biodiversity in modified landscapes (Chazdon, 2008).

BIODIVERSITY IN AGROFORESTRY SYSTEMS

Agroforestry is a summary term for practices that involve the integration of trees and other woody perennials into arable and pastoral farming systems through the conservation of existing trees, their active planting and tending, or the tolerance of natural regeneration in fallow areas (Schroth et al, 2004). Its main purpose is to diversify production for increased social, economic and environmental benefits, and has attracted increasing attention of scientists working at the interface between integrated natural resource management and biodiversity conservation, especially in tropical countries (Schroth et al, 2004; Scherr and McNeely, 2007). Farmers in many traditional agricultural systems have maintained or actively included trees as parts of the landscape for thousands of years to provide benefits such as shade, shelter, animal and human food (McNeely, 2004). Agroforestry is also taking on an increasingly important role in sub-tropical and temperate countries, and during the last 15 years has generated considerable interest in Australia, where large areas of native forest where rapidly felled during the 19th and 20th centuries (Box 1.3).

Agroforestry can benefit biodiversity conservation in three ways: (i) the provision of suitable habitat for forest species in areas that have suffered significant historical deforestation; (ii) the provision of a landscape matrix that permits the movement of species among forest remnants; (iii) and the provision of livelihoods for local people, which may in turn relieve pressure on remaining areas of primary forest. Because of their high levels of floristic diversity and complex vegetation, agroforests represent a mid-point in forest structural integrity between monoculture plantations and primary forest (Schroth and Harvey, 2007). In areas of the tropics

---

Box 1.3 Agroforestry in Australia

The origins of agroforestry in Australia are quite different from those in tropical less-developed countries. While many tropical agroforestry systems have ancient origins and have played a vital role in the provision of key food and construction resources, the concept of integrating trees and shrubs into farming systems in Australia came from a relatively recent tradition of 'land-care enthusiasts' during the last century. Nevertheless, the concept of agroforestry in Australia has developed rapidly during recent years, facilitated by the establishment of the Joint Venture Agroforestry Program in 1993, with funding from a range of government and industry groups. The objective of this programme was to support the development of a profitable agroforestry industry that also delivered tangible environmental benefits. Progress has been achieved through the support of a number of innovative projects, such as the Australian Master TreeGrower Program which provides an education and outreach programme for landholders who are interested in growing and managing trees on their farms.

An example of a successful agroforestry programme is the Otway Agroforestry Network (www.oan.org.au), which was set up and ran by local farmers and has supported tree-growing programmes on production farms through a variety of incentives, including to generate marketable timber and non-timber tree products, shelter stock, control erosion and dryland salinity, attract native birds and enhance property values. The use of tree-planting to limit dryland salinity in this case study serves to highlight how the benefits of agroforestry can often be quite specific to a given place.

---

that have lost the majority of old-growth forest, the dominant near-forest vegetation is frequently comprised of some form of agroforestry, highlighting the enormous importance of these systems for conservation in some regions, including shade-coffee in Central America and the Western Ghats, shade-cacao in the Atlantic Forest of Brazil, jungle rubber in the Sumatran lowlands, and home-gardens in countries across the world.

The majority of studies that have examined the biodiveristy value of agroforestry systems have found that although some species are invariably lost following conversion of native habitat, a large proportion of the original fauna and flora is maintained when compared to more intensified agricultural land uses. In reviewing the results of 36 studies, Bhagwat et al (2008) found that patterns of similarity in species composition between agroforest plots and areas of native forest ranged from 25 per cent (herbaceous plants) to 65 per cent (mammals). Although existing studies have not revealed any clear pattern regarding which groups of species are unlikely to be conserved within agroforestry systems, it appears that rare and range-restricted species are often those that suffer the greatest declines following forest conversion, while those that increase in abundance are often open-habitat and generalist taxa (Scales and Marsden, 2008). However, even species that are usually only found in areas of native vegetation can use agroforests to move between forest remnants, as is the case for two species of sloth in Costa Rica that frequently use shade-cacao plantations as a source of food and resting sites (Vaughan et al, 2007).

Differences in the amount of biodiversity that is retained in different agro-forestry systems can often be explained by differences in the intensity of past and present management regimes (Bhagwat et al, 2008). For example, the effect of management intensification on biodiversity is clearly demonstrated by the marked loss of forest species following the simplification of shade-coffee plantations and a decrease in the density and diversity of shade trees (reviewed by Philpott et al, 2008; Table 1.1). In addition, the ability of agroforestry systems to maintain a significant proportion of the regional biota depends on the maintenance of sufficient areas of natural habitat, both to support highly sensitive species (Schroth and Harvey, 2007) and to provide source populations (Anand et al, 2008).

BIODIVERSITY IN TREE PLANTATIONS

As for agroforestry systems, tree plantations have the potential to make an important contribution to biodiversity conservation for two key reasons: (i) they more closely reflect the structural complexity of native forest than many more intensive agricultural land uses; and (ii) they occupy a large area of once-forested land in many parts of the world. The total area of the plantation forest estate in 2005 was about 109 million ha, and is continuing to increase by approximately 2.5 million ha per year (FAO, 2006). In the tropics alone, the total coverage of plantation

Table 1.1 Description of main coffee management systems

Type of coffee production	Tree composition	Tree richness	Canopy cover (%)	Coffee density	Shade strata	Canopy management techniques
Rustic coffee	Native forest canopy	25	> 90	Low-medium	3	Minimal canopy intervention
Traditional polyculture coffee	Some forest trees and some planted timber and fruit trees	10–20	60–90	Low-medium	3	No or little pruning of the shade canopy
Commercial polyculture coffee	Mostly planted canopy trees (timber and fruit trees) and N-fixing legumes, few very abundant genera	5–10	30–60	Medium-high	2	Regular pruning of canopy, removal of epiphytes
Shade monoculture coffee	Canopy dominated by one species or genus of tree (i.e., *Inga* spp.)	1–5	< 30	High	1	Regular pruning of canopy, removal of epiphytes
Sun coffee	Rare isolated trees or without tree canopy	0	0	High	0	Na

Source: From Philpott et al (2008)

forestry increased from approximately 17 million ha in 1980 to 70 million ha in 2000 (FAO, 2006). As demands for timber and wood fibre continue to increase around the world, it is highly likely that these upward trends will persist or even accelerate.

Many tree plantations have been traditionally labelled as 'green deserts', and are presumed to be hostile to native species and largely devoid of wildlife (Kanowski et al, 2005). However, closer inspection of available data indicates that while it is certainly true that some intensively managed plantation monocultures offer very little value to biodiversity, other plantation systems can provide valuable species habitat, even for some threatened and endangered taxa (Hartley, 2002; Carnus et al, 2006). This apparent contradiction is explained in part by marked differences in the levels of biodiversity that can be supported by different types of plantation. For example, there is a stark contrast in the conservation value of industrial monocultures of exotic species that often have little or no intrinsic value for native forest species, compared with complex multi-species plantations that encompass remnants of native vegetation and are managed as a mosaic of differently aged stands (Hartley, 2002; Lindenmayer and Hobbs, 2004; Kanowski et al, 2005). However, a second reason why many plantations are incorrectly presumed to be biological deserts is that human perceptions of habitat quality are often distinct from how native species themselves perceive the landscape (Lindenmayer et al, 2003b).

Although few comprehensive and robust field studies have been conducted to examine the conservation value of plantations, those that exist suggest that, under certain conditions, the numbers of species inhabiting these areas may be greater than expected. For example, a very thorough study in north-east Brazilian Amazonia found that *eucalyptus* plantations contained nearly half of the regional forest fauna, although it is very unlikely that all of these taxa could maintain viable populations if the neighbouring primary forest was cleared (Barlow et al, 2007a).

The value of a given plantation forest for conservation is partly determined by how it is managed. For example, at the stand level, many studies have found that faunal diversity in tree plantations is strongly influenced by the maintenance of structural attributes, such as snags and dead wood, and the tolerance of succession by native plant species in the understorey (Hartley, 2002). More floristically and structurally complex plantations provide more resources for many forest species (e.g. fruit feeding butterflies, Barlow et al, 2008). At the landscape scale, spatial heterogeneity in stand management and age has been shown to be a key factor in determining the overall level of diversity within a given plantation forest (Lindenmayer and Hobbs, 2004; Lindenmayer et al, 2006).

AN ECOSYSTEM APPROACH TO FOREST CONSERVATION

It is evident from the above summary that, due to their widespread extent and relative structural complexity, modified forests that are either currently, or have once been managed for some form of human use represent an important and valuable asset for biodiversity conservation. They provide valuable habitat in their own right, help connect primary forest remnants and buffer reserves, and protect areas against direct exploitation for timber and non-timber forest resources. Indeed for many parts of the world, managed, degraded or replanted forest areas (even of exotic species) are all that remains to safeguard the future of regional biodiversity and they should not be overlooked by conservationists and land managers (Daily, 2001; Lindenmayer and Franklin, 2002; Bawa et al, 2004; Gardner et al, 2009; Figure 1.3). However, the value of a human-modified habitat depends not only on its position along a gradient of land-use intensification (Figure 1.2) but also upon its wider landscape context (Norris, 2008; Fig. 1.4), and in particular the amount of native forest that remains in the adjacent area (Gardner et al, 2009).

In summary, conservation scientists are now in broad agreement that we need to move beyond a near exclusive focus on protected areas, and embrace opportunities for biodiversity conservation that emerge from an integrated approach to managing entire landscapes and regions, commonly known as the ecosystem approach (Box 1.4). The ecosystem approach is embodied by a research perspective that incorporates human activities as integral components of ecosystems, and places a strong emphasis on understanding the coupled social-ecological dynamics of modified lands (Palmer et al, 2004; Sayer and Maginnis, 2005a).

Figure 1.3 A cacao-dominated landscape, known locally as 'cabruca', in the north-east of Brazil

Source: Photograph courtesy of Renata Pardini

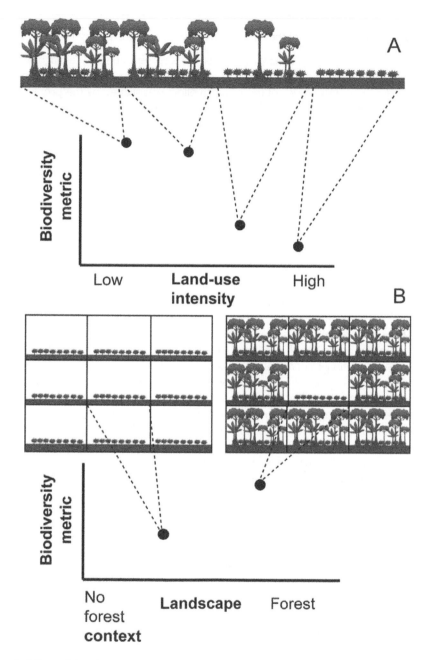

Figure 1.4 Describing the relationship between biodiversity and increasing land-use intensification (A), and the wider landscape context (B).

Source: Redrawn with permission from Norris (2008)

Note: The term 'biodiversity metric' is used to describe any statistic that might be used to quantify biodiversity value, such as the number of native species. The land-use schematic in A) illustrates an agroforestry production system (e.g. coffee) that ranges in intensity from (low intensity) traditional shade systems with relatively extensive tree cover to (high intensity) full-sun systems devoid of shade. The land-use schematic in (B) illustrates a full-sun production system in contrasting landscape contexts – one in a landscape consisting of other full-sun farms (no forest), and the other in a landscape consisting of traditional shade farms with extensive tree cover (forest)

Box 1.4 The origins of the ecosystem approach to forest conservation

Since conservation science began as a formal research discipline some 30 years ago, the focus of attention has been on the design of reserves and protected areas as the pre-eminent tool for biodiversity conservation. Much of this research effort was shaped by the 'problem-isolation paradigm' (Maginnis et al, 2004). This paradigm defines an allocation approach to the management of forest ecosystems, and attempts to satisfy conservation and development requirements through the adoption of independent and isolated strategies: timber production is achieved through clear-felling or intensive silvicultural practices, while biodiversity conservation has been largely limited to the establishment and protection of reserve networks.

An alternative paradigm, enshrined in the 1996 Ecosystem Principles of the Convention on Biological Diversity (CBD) (although the general concept has much earlier origins; Sayer and Maginnis, 2005a, is founded on the belief that an integrated ecosystem approach to the protection, management and restoration of the total forest area will produce greater net benefits for biodiversity conservation than is possible by pursuing biodiversity conservation and timber production separately (Franklin, 1993; Grumbine, 1994; Noss, 1996; Parrish et al, 2003; Aldrich et al, 2004; Sayer and Maginnis 2005a). Arguments in favour of the ecosystem approach reflect the conviction that conservation cannot be considered a 'set-aside' issue (Lindenmayer and Franklin, 2002), and that the future of many of the world's forest species depends upon the effectiveness of strategies for protecting biodiversity in areas outside reserves. Central to this approach is the recognition that in order to be effective, the conservation of biodiversity must encompass a continuum of management approaches, from the establishment of reserves of primary forest to a host of off-reserve conservation measures, including the maintenance of landscape connectivity, buffer zones around reserves, landscape heterogeneity, stand complexity, sensitive management of natural disturbance regimes and protection of aquatic ecosystems (Pimentel et al, 1992; Franklin, 1993; Fimbel et al, 2001; Lindenmayer and Franklin, 2002; Perfecto and Vandermeer, 2008; Gardner et al, 2009; Figure 1.5). The particular suite of management strategies that may be adopted varies widely from place to place as 'ecosystem approaches' are understood differently, and are driven by different environmental and social trade-offs in different parts of the world (Sayer and Maginnis, 2005a).

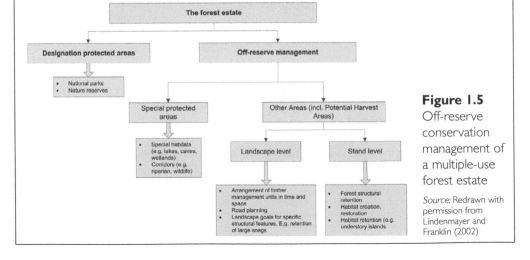

Figure 1.5
Off-reserve conservation management of a multiple-use forest estate

Source: Redrawn with permission from Lindenmayer and Franklin (2002)

CHAPTER 2

The Origins and Development of Ecologically Responsible Forest Management

Synopsis

- Managing forests for biodiversity conservation is part of a more general goal to maintain the diverse range of economic, social and environmental benefits that humans derive from forest ecosystems. This goal is commonly referred to as sustainable forest management (SFM) and has its foundations in the 1992 Rio Declaration on Environmental Development. Global commitment to SFM was reaffirmed by a resolution of the UN General Assembly in 2007 on the Sustainable Management of All Forest Types.
- SFM is a subjective and evolving concept that should not be interpreted as a practical guide for management, but rather as a mechanism for creating a sense of community, connection and purpose, and a guiding vision for social and political discourse.
- Ecologically responsible forest management draws upon the vision provided by SFM, and is based on implementing minimum practice standards to reduce biodiversity loss, while continuously striving to improve management performance with respect to long-term conservation goals.
- Criteria and Indicators (C&I) are the foremost mechanism for defining responsible forest management and reporting on progress. Third party certification has been one of the most important policy developments in efforts to translate global-level C&I into auditable management standards that can be implemented within individual forest landscapes.
- Management standards can be systems based, requiring the adoption of certain management practices in order to achieve compliance, or performance based, requiring the maintenance of certain forest attributes above minimum levels (irrespective of the method used). In practice, most management authorities use a combination of both approaches

A primary purpose of monitoring forest biodiversity, and the central thesis of this book, is to identify ways in which forest systems can be managed to support conservation goals. Managing forests to promote biodiversity conservation is part of a more general goal to ensure the long-term maintenance of the diverse range of benefits, including economic, social and environmental, that humans derive from forest ecosystems. This goal has traditionally fallen under the banner of SFM. The political process underpinning the evolution of SFM has been instrumental in developing the value basis upon which many forest management goals are currently set, as well as providing a template for the development of management standards for on-the-ground implementation. Biodiversity conservation has emerged as a universally recognized forest value that should be preserved through management.

The purpose of this chapter is to review the origins and ongoing development of SFM, including the meaning of important and related terms and concepts, and the role of key intergovernmental and voluntary institutions that have been involved in the process. First I outline the varying definitions of SFM and the related concept of ecologically sustainable forestry. Then I discuss the problems of the term 'sustainable' and how SFM is more appropriately viewed as a guiding management vision for more responsible management approaches rather than a measurable standard. Finally, I discuss the origins and importance of C&I as a basis for standards in forest management, and the emergence of forest certification as a voluntary mechanism for ensuring compliance with minimum performance standards.

THE ORIGINS OF SUSTAINABLE FOREST MANAGEMENT (SFM)

The idea of sustainable development in its broadest sense was first formalized by the Brundtland Commission of 1987, in the report entitled *Our Common Future*, and was defined as '*development which meets the needs of current generations without compromising the ability of future generations to meet their own needs*'. This in turn formed the basis of SFM, which was first visibly conceived in the Forest Principles of the 1992 Rio Declaration on Environmental Development. They recognized that in order to meet the social, economic and ecological needs of present and future generations, the management of forests had to progress beyond an exclusive focus on maintaining sustainable timber yields (www.un.org/documents/ga/conf151/aconf 15126-3annex3.htm).

Drawing upon the basis provided by the Rio Forest Principles, a large number of definitions of SFM have subsequently emerged from a variety of international organizations and processes that are concerned with the responsible management of native forests around the world (Box 2.1). While sometimes vague, all of these definitions encompass the need to protect and maintain the environmental values of forests. Some explicitly recognize the importance of biodiversity. As is also clear from the Rio Forest Principles, there are two elements to environmental sustainability: (i) land use (referring to the local management practices that are implemented in a given area forest stand); and (ii) land management (referring to the management of multiple stands across an entire landscape). These dual elements provide recognition of the fact that not all management goals can be achieved in the same locations (thereby meaning that attention to whole landscape planning is necessary), and that achieving some goals requires careful on-the-ground management at the level of individual forest stands (Poore, 2003; Zarin et al, 2007; Chapter 15).

Box 2.1 Internationally recognized definitions of SFM

Different definitions of SFM have been provided by different international organizations and processes concerned with the responsible management of forests. All of these definitions find their origin in the 1992 Rio Forests Principles, but attain varying levels of detail depending on their political origins. For example, the highest, UN-level definition lacks the more detailed listing of valued elements contained within the cross-European definition (Kunzmann, 2008). Four of the most prominent definitions of SFM are reproduced below (for a more complete list see: www.fao.org/docrep/005/y4171e/Y4171E49.htm):

International Tropical Timber Council (1992) www.itto.int

Sustainable forest management is the process of managing permanent forest land to achieve one or more clearly specified objectives of management with regard to the production of a continuous flow of desired forest products and services without undue reduction of its inherent values and future productivity and without undue desirable effects on the physical and social environment.

This definition is elaborated with the associated explanatory text:

Forest-related activities should not damage the forest to the extent that its capacity to deliver products and services — such as timber, water and biodiversity conservation — is significantly reduced. Forest management should also aim to balance the needs of different forest users so that its benefits and costs are shared equitably.

Ministerial Conference on the Protection of Forests in Europe (MCPFE, Helsinki 1993) www.foresteurope.org/

'Sustainable management' means the stewardship and use of forests and forest lands in a way, and at a rate, that maintains their biodiversity, productivity, regeneration capacity, vitality and their potential to fulfil, now and in the future, relevant ecological, economic and social functions, at local, national, and global levels, and that does not cause damage to other ecosystems.

Food and Agriculture Organization (FAO) (2007) www.fao.org/forestry

The 2007 State of the World's Forests 2007 report of the FAO does not provide a definition of SFM. Instead regional assessments are based on seven thematic elements that together can identify *progress towards* SFM, namely:

- *extent of forest resources;*
- *biological diversity;*
- *forest health and vitality;*
- *productive functions of forest resources;*
- *protective functions of forest resources;*
- *socio-economic functions; and*
- *legal, policy and institutional framework.*

> ## UN – Non-legally binding instrument on all types of forests (2007)
> www.un.org/esa/forests
>
> *Sustainable forest management, as a dynamic and evolving concept, aims to maintain and enhance the economic, social and environmental values of all types of forests, for the benefit of present and future generations.*

SFM is a dynamic concept and since its early roots subsequent definitions have increasingly reflected a growing awareness of the biodiversity crisis by encompassing increasingly detailed issues relating to environmental conservation (Kneeshaw et al, 2000; Brown et al, 2001; Sheil et al, 2004; FAO, 2006). For example, working in Australia, Lindenmayer et al (2006) define the emerging concept of *ecologically* SFM for native forests as:

'...*perpetuating ecosystem integrity while continuing to provide wood and non-wood values; where ecosystem integrity means the maintenance of forest structure, species composition, and the rate of ecological processes and functions within the bounds of normal disturbance regimes.*'

The more explicitly conservation-orientated concept of ecologically sustainable forestry as defined by Lindenmayer et al (2006) shares its roots with much earlier thinking and research on the idea of ecosystem management (Grumbine, 1994). Indeed, it is arguable that ecosystem management provided the conceptual basis for both SFM and the ecosystem approach (Schlaepfer et al, 2004). Ecosystem management has its roots in the early ecological thinking in the US during the 1930s and 1940s and is defined by the central goal of maintaining ecological integrity, which is comprised of five specific goals: (i) maintaining viable populations; (ii) ecosystem representation; (iii) maintaining ecological processes (i.e. natural disturbance regimes); (iv) protecting evolutionary potential of species and ecosystems; and (v) accommodating human use in light of the above (Grumbine, 1994).

Two key ideas arise from all the definitions of SFM and serve to guide the debate around more responsible approaches to managing forest ecosystems. First that SFM is a balancing act between the need to simultaneously maximize ecological, economic and social values, and second, that economic forestry activities must operate within the natural constraints of the ecological system (McDonald and Lane, 2004). It is also clear that the dual concepts of the ecosystem approach (Box 1.4) and SFM are closely intertwined (Wilkie et al, 2003; Schlaepfer et al, 2004). Official recognition of the convergence of these two originally distinct ideas was achieved following the Seventh Conference of the Parties to the Convention on Biological Diversity in 2004, which linked the national forest and biodiversity action programmes of member states through recognition of SFM as a means of applying an integrated ecosystem approach to forest conservation and management (Wilkie et al, 2003; CBD, 2004; McAfee et al, 2006).

SUSTAINABLE FOREST MANAGEMENT (SFM) AS A GUIDING VISION VERSUS A MEASURABLE STANDARD

One of the significant benefits to have emerged from the discourse around SFM is that the dialogue process itself initiated a strong debate on the practicalities of managing human-use forests with regard to long-term biodiversity conservation goals. Many commentators have raised concerns that the 'catch-all' nature of SFM (or sustainable development in general) makes it an inherently fuzzy concept that is subject to internal contradiction and is extremely difficult to define and implement in practical terms. However, a fundamental element of the most recent definition of SFM, as provided by the UN Instrument on Sustainable Management of all Forest Types, is that it is a **dynamic and evolving concept**. In practice, what this means is that we do not yet have a convincing working definition of what ecologically sustainable forestry actually means in terms of managing forest landscapes for multiple values (Lindenmayer and Franklin, 2003; Wintle and Lindenmayer, 2008). This makes demonstration of 'sustainability' (e.g. as set against a management or certification standard) impossible from a practical perspective.

Nevertheless, this practical limitation does not mean that the concept is redundant. There is now broad acceptance within the scientific and forestry community that the concept of SFM has its greatest utility as a powerful and guiding vision rather than a quantifiable standard of management performance (e.g. Kneeshaw et al, 2000; Yamasaki et al, 2002; Lindenmayer and Franklin, 2003; Sayer and Maginnis, 2005b). McCool and Stankey, 2001 (cited in Raison et al 2001) propose that sustainability embraces what can be described as a 'guiding fiction', and while its precepts cannot be proven or measured they serve to create a sense of community, connection and purpose that together help trigger necessary political and social discourse. In particular, debate about sustainable forest use requires society to think about the future and its relationship with nature, and to explicitly consider the consequences and implications of the alternative choices which confront it.

This more realistic and pragmatic approach is the basis for the concepts of *continuous improvement* and *adaptation*: management processes should at first be based on minimum practice standards that are then subject to adaptation and revision following the collection of new information on the relative impacts and effectiveness of alternative strategies (Lindenmayer and Franklin, 2003; Wintle and Lindenmayer, 2008). Existing knowledge is always imperfect and natural systems are inevitably going to change so there is no purpose or value in holding some *a priori* definition of sustainability as sacred (Sayer and Campbell, 2004). Recognizing the need for forest management to be adaptive highlights the fundamental importance of monitoring. From a conservation perspective if we are serious about understanding the state of the world's forests and the extent to which existing management strategies are inadequate in ensuring the long-term maintenance of ecological integrity, then the monitoring process represents one of the most fundamental challenges facing the future of forest management.

These arguments form the philosophical basis adopted in this book, as laid out in the third section (Chapters 8–15) in an operational framework for monitoring forest biodiversity. They also help define the concept of **ecologically responsible forest management** as a management approach that draws upon the vision provided by the SFM discourse, implements minimum practice standards where available and appropriate, while continuously striving to improve management performance with respect to long-term conservation goals.

Responsible management of non-native forests

As discussed in Chapter 1, it is important to maximize opportunities for biodiversity conservation across the entire forest estate and not just within areas of native forest (whether managed for production or conservation) (see also Koh and Gardner, 2009). For many areas of the world, the most extensive areas of forest-like habitat that may remain within a given region are comprised of isolated fragments, plantations and areas of permanent agriculture and agroforestry. In recognition of the conservation potential of these modified forest systems, sustainability concepts and approaches have extended beyond the scope of native forests and applied to a number of different forms of responsible management for plantation (both timber and non-timber) and permanent agroforestry systems (e.g. cocoa and shade-coffee) (Poulson et al, 2001; FSC, 2002; Elliot, 2003; Table 2.1). Clearly the term SFM has even less practical application to planted forests than it does to native forests in light of the environmental losses that are inevitably incurred following initial clearing of the land, and certification of these systems can only ever form one part of a landscape-wide strategy for biodiversity conservation.

C&I IN FOREST MANAGEMENT

Following establishment of sustainable use as a policy goal, the most significant and innovative tool for promoting SFM in both native and plantation forests has been the development of principles, criteria and indicators (PC&I) (or just criteria and indicators, C&I) (CBD, 1996; Prabhu et al, 1999; Poulson et al, 2001, Raison et al, 2001; CICI, 2003; UNFF, 2004; Holvoet and Muys, 2004; Finegan, 2005; Wijewardana, 2008). After more than a decade of international forestry dialogue, there is now broad acceptance of the role of C&I as the foremost mechanism for defining and promoting SFM (Sheil et al, 2004), as well as a key instrument in guiding monitoring and evaluation programmes and reporting progress, including for biodiversity (Buck and Rametsteiner, 2003). Stork et al (1997) and Prabhu et al (1999) provide the following definitions of PC&I that broadly reflect those adopted throughout the forest management literature:

- **Principles**: *Fundamental truths or laws [that are] the basis of reasoning or action.*
- **Criteria**: *Intermediate points to which the information provided by indicators can be integrated and where interpretable assessment crystallizes.*
- **Indicators**: *Any variables or components of a forest ecosystem or management system that are used to infer the status of a particular criterion.*

Table 2.1 Responsible management systems for non-native forests

Modified forest system	Work towards sustainable management practices and relevance for biodiversity conservation	Further information and relevant authorities
Timber plantations	Plantation management is a critical issue for the world's forests. Whether viewed as a sustainable fibre supply, a carbon sink, or a means for ameliorating deforestation pressure in remaining native forests, plantations are controversial and often divisive. The implications of plantation forests for biodiversity are poorly understood and vary markedly depending on plantation type and landscape context (Barlow et al., 2007; Brockerhoff et al., 2008; and see Chapter 1) yet it is clear that sensitive management practices at landscape- and stand-level scales can make a significant contribution to their value for conservation. The Forest Stewardship Council (FSC) has led work in strengthening the standard for responsible management of plantations and to date more than 7 million ha of plantations and 30 million ha of mixed forest have been certified (see Box 2.5 for relevant Principles and Criteria (P&C)). Following criticism from conservation organisations and a public consultation process in 2008, the FSC has recently revised its management standards for plantation forests (see Box 2.4)	FSC (www.fsc.org) and in particular see the recent FSC Plantations Review (http://www.fsc.org/plantations review.html)
Shade-coffee (and other food-crop plantations)	More than 25 million people in the tropics depend on coffee, a crop that is the economic backbone of many countries and the world's second most traded commodity after oil. Coffee is farmed on about 12 million ha worldwide, and most of the farms are in areas regarded as globally important priorities for conservation (e.g. much of Central America, the Atlantic Forest of Brazil and the Western Ghats of India). Around the world, coffee is traditionally produced under a diverse and dense canopy of shade trees as a complex agroforestry system. It has been well demonstrated that the floristic and structural diversity of these shade plantations can make a significant contribution to forest biodiversity retention (Perfecto et al., 2005; Philpott et al., 2008). The Rainforest Alliance has partnered with the Sustainable Agriculture Network (SAN) to provide a certification mechanism for responsibly managed coffee farms, and help develop a path towards sustainable use. A revised Sustainable Agriculture Standard, including clear requirements for 'ecosystem conservation' and 'wildlife protection' was released in February of 2008 by SAN to encompass certification for all food-crop plantations, including coffee but also cocoa, bananas and citrus	Sustainable Agriculture Network and the Rainforest Alliance (www.rainforest-alliance.org). Mostly working in Latin America

Table 2.1 Responsible management systems for non-native forests (continued)

Modified forest system	Work towards sustainable management practices and relevance for biodiversity conservation	Further information and relevant authorities
Oil Palm	The African oil palm, *Elaeis guineensis*, is one of the world's most rapidly expanding crops, and has the highest yields and largest market share of all oil crops. While cultivation has historically focused in Malaysia and Indonesia, oil palm is increasingly grown across the lowlands of other countries in Southeast Asia, Latin America and Central Africa. The few studies available show that oil palm is a poor substitute habitat for the majority of tropical forest species, particularly those of conservation concern (Fitzherbert et al, 2008). In response to consumer concerns about deforestation, the Roundtable on Sustainable Palm Oil was formed from an industry-NGO collaboration in 2003. Under this scheme, members commit to environmental and social standards for responsible palm oil production, including an assurance that no forests of High Conservation Value will be cleared for plantations. A guidance document containing P&C for sustainable oil palm production was published for field testing in 2006 and the first certified oil palm products reached the markets in late 2008	Round table on Sustainable Palm Oil (www.rspo.org)

C&I for forest management have been developed at a variety of levels, including international, regional or sub-national, and forest management unit (FMU) levels.

International criteria and indicator processes

At the international scale, C&I provide a policy vehicle for generating consensus and commitment towards more responsible management practices within individual nations. At this level, indicators serve to identify the system attributes that need careful management to ensure that values are maintained, for example for biological diversity in the case of the European (MCPFE) C&I process (Table 2.2).

At the time of writing, around 150 countries are participating in one or more of nine regional or international C&I processes for SFM that are used to monitor national policy and improve informed decision-making (UNFF, 2004, reviewed in Holvoet and Muys, 2004; Dudley et al, 2005; McAfee et al, 2006; Wijewardana, 2008). In order of their establishment these agreements include the ITTO (1992), Helsinki Process of the MCPFE (1993), the African Timber Organization (1993), the Montreal Process (1995), the Amazon Cooperation Treaty (Tarapoto Agreement, 1995), Dry Zone Africa (1995), Near East Process (1996), Lepaterique Process for Central America (1999) and the Bhopal–India Process (1999). These international

Table 2.2 Criterion 4 and associated indicators relating to biological diversity of the Ministerial Conference on the Protection of Forests in Europe (MCPFE)

Indicator		Description
4.1	Tree species composition	Area of forest and other wooded land, classified by number of tree species occurring and by forest type
4.2	Regeneration	Area of regeneration within even-aged stands and uneven-aged stands, classified by regeneration type
4.3	Naturalness	Area of forest and other wooded land, classified by 'undisturbed by man', by 'semi-natural' or by 'plantations', each by forest type
4.4	Introduced tree species	Area of forest and other wooded land dominated by introduced tree species
4.5	Deadwood	Volume of standing deadwood and of lying deadwood on forest and other wooded land classified by forest type
4.6	Genetic resources	Area managed for conservation and utilisation of forest tree genetic resources (in situ and ex situ gene conservation) and area managed for seed production
4.7	Landscape pattern	Landscape-level spatial pattern of forest cover
4.8	Threatened forest species	Number of threatened forest species, classified according to IUCN Red List categories in relation to total number of forest species
4.9	Protected forests	Area of forest and other wooded land protected to conserve biodiversity, landscapes and specific natural elements, according to MCPFE protection categories

processes have, over the past decade, refined C&I to the point that there is now substantial conceptual agreement regarding the meaning and overall goal of SFM at a policy level (CICI, 2003; McDonald and Lane, 2004; Stem et al, 2005). Nevertheless, around 65 countries, some of which host large areas of native forest (e.g. many central Asian countries) still do not belong to any C&I process, (Wijewardana, 2008). Moreover, it is important to recognize that not all of these nine processes represent comparable levels of commitment towards SFM, and in fact only three of them (MCPFE, Montreal Process and ITTO) have made substantial progress on reporting how forests are actually being managed within member nations. Only a small fraction of the global forest estate can be considered as being under a responsible system of management, with the situation being particularly poor in the tropics (ITTO, 2006; FAO, 2007; Box 2.2).

Set against the enormous challenge of promoting SFM at the global scale is the promise offered by the recently negotiated non-legally binding instrument on all types of forests that was formerly adopted by the General Assembly of the UNs at the end of 2007 (Box 2.3). The fact that many countries have yet to seriously engage in SFM means that this international agreement provides a significant window of opportunity for improving forest management practices, including the preservation of biodiversity, across the world in the coming decade.

Box 2.2 Status of tropical forest management

The *Status of tropical forest management 2005* report (ITTO, 2006) evaluates forest status in producer member countries of the ITTO. A 1988 survey by ITTO found that less than 1 million ha of tropical forest was being managed in accordance with good forestry practices. The new ITTO report considers changes in the subsequent 17 years in the 33 ITTO member countries that produce tropical timber (Table 2.3).

The ITTO encourages countries to adopt responsible forest management in areas of permanent forest estate, through which the inherent values of the forest are maintained (or at least not unduly reduced), while revenues are earned, people are employed and communities are sustained by the production of timber and other forest products and services. ITTO has also developed C&I for monitoring and assessment.

Depending on the reader's perspective the ITTO report identified both significant progress towards good forest practice in the tropics but also major challenges that lie ahead. More forests have been committed to permanent forest estate for production or protection, and more are being managed under what is considered as responsible use. Although this is encouraging, the proportion of natural, production forest under responsible management is still very low (less than 5 per cent of the total area) and is distributed unevenly across the tropics and within countries.

Table 2.3 Assessment of SFM in ITTO member nations

Region	Production							Protection			All	
	Natural				Planted							
	Total area	With management plans	Certified	Sustainably managed	Total area	With management plans	Certified	Total area	With management plans	Sustainably managed	Total area	Sustainably managed
Africa	70,461	10,016	1 480	4 303	825	488	0	39,271	1 216	1 728	110,557	6 031
Asia and the Pacific	97,377	55,060	4 914	14,397	38,349	11,465	184	70,979	8 247	5 147	206,705	19,544
Latin America and the Caribbean	184,727	31,174	4 150	6 468	5 604	2 371	1 589	351,249	8 374	4 343	541,580	10,811
Total	353,565	96,250	10,544	25,168	44,778	14,315	11,773	461,499	17,837	11,218	858,842	36,386

At present, the natural permanent forest estate in Africa, Asia and the Pacific, and Latin America and the Caribbean is estimated to cover 110, 168 and 536 million ha respectively, a total of 814 million ha in the 33 ITTO member producer countries. Of the permanent forest estate in Latin America and the Caribbean, nearly half (271 million hectares) is within Brazil. One major difficulty in interpreting these figures is that estimates of total forest area also vary according to source. At the high end of the range of estimates, Africa has 274 million hectares of forest (40 per cent of which is in the permanent forest estate); at the low end, 234 million ha (47 per cent of which is in permanent estate). In Asia and the Pacific, the figures are 316 million ha (65 per cent) and 283 million ha (73 per cent), respectively; in Latin America and the Caribbean, have 931 million ha (58 per cent) and 766 million hectares (71 per cent) respectively.

Source: Text and data based on FAO (2007)

Box 2.3 Non-legally binding instrument on sustainable management of all types of forests

Building upon earlier international negotiations in forest management, the UNs Forum on Forests recently succeeded in achieving consensus among member nations and adopted the 'non-legally binding instrument on all types of forests' (April 2007, New York). This was then subsequently adopted by the General Assembly of the UNs on 17 December 2007. The instrument is the last step of many in the attempt to create a coherent international legal framework regarding forests and their uses. However, as a Resolution of the General Assembly and as set out expressly in its title, it is non-legally binding, a soft law document (Kunzmann, 2008). The instrument represents nearly 15 years of negotiations within the international community to conclude a universally accepted and comprehensive document governing forest management.

At the heart of the instrument are four Global Objectives on Forests. These reaffirm the member states' following shared global objectives on forests and commitment to work globally, regionally and nationally to achieve progress towards their achievement by 2015:

Global objective 1 Reverse the loss of forest cover worldwide through SFM, including protection, restoration, afforestation and reforestation, and increase efforts to prevent forest degradation.

Global objective 2 Enhance forest-based economic, social and environmental benefits, including by improving the livelihoods of forest-dependent people.

Global objective 3 Increase significantly the area of protected forests worldwide and other areas of sustainably managed forests, as well as the proportion of forest products from sustainably managed forests.

Global objective 4 Reverse the decline in official development assistance for SFM and mobilize significantly increased, new and additional financial resources for the implementation of SFM.

Resistance from individual countries (particularly the US and Brazil) resulted in a document that lacked as much clarity or 'teeth' as other members would have wished (see Kunzmann, 2008) and much of the language simply repeats commitments that have already been captured in the 1992 Rio Convention, the sixth Session of the UNFF or other international organizations. However, the value of this instrument lies in the advantage that it ties together the most important rules and standards of forest policy in one document and that it aims to realize more responsible approaches to forest management instead of limiting itself to a mere repetition of global objectives on forests (Kunzmann, 2008).

Source: The full official UN document can be found at:
www.fordaq.com/www/news/2007/UN_Instrument%20on%20all%20types%20of%20forests.pdf

C&I at the level of forest management units: Management standards and certification

To move the concept of SFM beyond political consensus, it is vital that implementation is achieved at the scale of individual forest landscapes (Raison et al, 2001). While the principles of forest management as defined by international C&I processes apply effectively to national and global scales, the suitability of the concepts erode when scaling to the operational level of individual landscapes (Guynn et al, 2004). Progress towards sustainability can only be assessed when related to a valued product or condition (Poore, 2003), and the challenge in implementing C&I at the local scale lies in identifying a minimum standard of management procedures that will be sufficient in maintaining the forest values identified by individual indicators (e.g. Table 2.2 in the case of European forests). The description of an indicator on its own provides little guidance to forest managers unless it is associated with a minimum standard. However, once clearly defined, C&I provide the core element of a management standard, which can then be employed by auditors to evaluate management performance.

The central challenge in implementing responsible forest management has been one of scale: how to adapt C&I developed for national-level use to the scale of forest management units where decisions are made by forest managers and can be adapted over time (Ozinga, 2001; Raison et al, 2001). One of the major policy developments relevant to the implementation of management initiatives at the local scale, which has occurred in parallel with the development of international C&I processes, is the emergence of third-party forest certification (Spellerberg and Sawyer, 1996; Hickey, 2004; Nussbaum and Simula, 2005; and see Cashore et al (2006) for a recent review of forest certification.) The origins of forest certification lie in attempts to improve forest management in tropical developing countries and promote access to environmentally sensitive Western markets (Rickenbach and Overdevest, 2006). They function through a voluntary submission to third-party inspection against agreed management standards (Nussbaum et al, 2002; Nussbaum and Simula, 2005). International C&I processes and certification schemes differ fundamentally in concept and structure but operate in a complementary top–down (policy frameworks) versus bottom–up (market pressure) fashion (Freer-Smith and Carnus, 2008). Monitoring of progress towards more sustainable use is important at both levels – national-level indicators contribute towards the development and regular updating of policy instruments, while trends in indicators at the local level help adjust forest management prescriptions over time to meet national goals.

Although there are at least seven major certification schemes worldwide (Nussbaum and Simula, 2005), the FSC is currently the only one that has a rapidly increasing and truly global coverage as well as multilateral support from the conservation community (Gullison, 2003; Rametsteiner and Simula, 2003; Dickinson et al, 2004; Leslie, 2004). Nevertheless, recent scrutiny by international conservation NGOs has cast doubt over the integrity of the FSC-certification process with regards to biodiversity conservation and other sustainability criteria, and a complete review of FSC P&C is currently ongoing (Box 2.4). This controversy adds to the challenge of promoting ecologically responsible forest management while also highlighting the need to develop monitoring programmes in the context of real-world constraints.

Box 2.4 Evaluation of FSC performance in maintaining integral standards of responsible forest management

Since its conception in 1993, FSC has created the leading model for credible certification of responsible forest stewardship worldwide. It was uniquely founded on the collaboration and shared commitment of representatives from a wide range of environmental, social and economic organizations to create, maintain and further evolve a transparent and credible certification system for responsibly managed forests. Leading international conservation NGOs formed part of the founding membership, including Greenpeace, WWF and Friends of the Earth.

The reality of rapid growth, combined with the complexities linked to responsible forest management and product labelling, has led to FSC facing an increasing number of challenges. In particular there have been a growing number of accusations of inappropriate certification of particular forest managers, sparking dissent among some of its original founders, including Friends of the Earth in the UK which released a statement in September 2008 stating that it 'is deeply concerned by the number of FSC certifications that are now sparking controversy and threatening the credibility of the scheme. We cannot support a scheme that fails to guarantee high environmental and social standards. As a result we can no longer recommend the FSC standard' (see www.fsc-watch.org/archives/2008/09/22/Friends_of_ the_Earth).

In their recent report entitled *Holding the line with FSC*, which serves to evaluate growing criticism over the integrity of management standards, Greenpeace (2008) highlighted the particular need for more guidance materials and training in relation to auditing, managing and planning for environmental impacts (Criterion 6, see Box 2.5).

In response to external criticism, FSC is currently undertaking a comprehensive review of many issues, and of the more than 80 recommendations presented to FSC by Greenpeace in mid-2007, FSC completely agrees with over half, with a few fully implemented already and most partially completed. While an explicit programme of work is underway to help, there still remains an urgent need for further and detailed attention to the assessment of environmental and biodiversity impacts in FSC-certified forests. A number of external agencies are working to assist in this endeavour, including individual scientists and research organizations such as ProForest (www.proforest.net/). It is hoped that this book will contribute towards realizing the changes needed in the design and implementation of biodiversity monitoring programmes.

Management standards are intended to synthesize what is known about best practice and their development is made on the basis of the best available knowledge and a multi-stakeholder decision-making process (Nussbaum et al, 2002). This process has been most notably conducted by forest certification authorities concerned with both native and plantation forests (e.g. FSC, 2002; Box 2.5), as well as independent forest research organizations (most notably the Centre for International Forest Research (CIFOR) (Prabhu et al, 1996; Stork et al, 1997; CIFOR, 1999; Prabhu et al, 1999; Box 2.6).

Box 2.5 P&C of the FSC that relate to biodiversity conservation and monitoring

Principle 6: Environmental impact

Forest management shall conserve biological diversity and its associated values, water resources, soils, and unique and fragile ecosystems and landscapes, and, by so doing, maintain the ecological functions and the integrity of the forest.

- 6.1 Assessment of environmental impacts shall be completed – appropriate to the scale, intensity of forest management and the uniqueness of the affected resources – and adequately integrated into management systems. Assessments shall include landscape-level considerations as well as the impacts of on-site processing facilities. Environmental impacts shall be assessed prior to commencement of site-disturbing operations.

- 6.2 Safeguards shall exist which protect rare, threatened and endangered species and their habitats (e.g. nesting and feeding areas). Conservation zones and protection areas shall be established, appropriate to the scale and intensity of forest management and the uniqueness of the affected resources. Inappropriate hunting, fishing, trapping and collecting shall be controlled.

- 6.3 Ecological functions and values shall be maintained intact, enhanced or restored, including: forest regeneration and succession; genetic, species and ecosystem diversity; natural cycles that affect the productivity of the forest ecosystem.

- 6.4 Representative samples of existing ecosystems within the landscape shall be protected in their natural state and recorded on maps, appropriate to the scale and intensity of operations and the uniqueness of the affected resources.

Principle 8: Monitoring and assessment

Monitoring shall be conducted – appropriate to the scale and intensity of forest management – to assess the condition of the forest, yields of forest products, chain of custody, management activities and their social and environmental impacts.

- 8.1 The frequency and intensity of monitoring should be determined by the scale and intensity of forest management operations, as well as the relative complexity and fragility of the affected environment. Monitoring procedures should be consistent and replicable over time to allow comparison of results and assessment of change.

- 8.2 Forest management should include the research and data collection needed to monitor, at a minimum, the following indicators: regeneration and condition of the forest; composition and observed changes in the flora and fauna; environmental and social impacts of harvesting and other operations; costs, productivity and efficiency of forest management.

Source: Text from FSC (2002)

Box 2.6 Principles, criteria and indicators (PC&I)) of the Center for International Forestry Research (CIFOR) that relate to biodiversity conservation

Principle 2: Maintenance of ecosystem integrity

C 2.1 The processes that maintain biodiversity in managed forests are conserved.

- I 2.1.1 Landscape pattern is maintained.

- I 2.1.2 Change in diversity of habitat as a result of human interventions should be maintained within critical limits.

- I 2.1.3 Community guild structures do not show significant changes in the representation of especially sensitive guilds, pollinator and disperser guilds.

- I 2.1.4 The richness/diversity of selected groups show no significant change.

- I 2.1.5 Population sizes and demographic structures of selected species do not show significant change, and demographically and ecologically critical life-cycle stages continue to be presented.

- I 2.1.6 The status of decomposition and nutrient cycling shows no significant change.

- I 2.1.7 There is no significant change in the quality and quantity of water from the catchment.

- I 2.1.8 Enrichment planting, if carried out, should be based on indigenous locally adapted species.

C 2.2 Ecosystem function is maintained.

- I 2.2.1 No chemical contamination to food chains and ecosystem.

- I 2.2.2 Ecologically sensitive areas, especially buffer zones along watercourses, are protected.

- I 2.2.3 Representative areas, especially sites of ecological importance, are protected and appropriately managed.

- I 2.2.4 Rare or endangered species are protected.

- I 2.2.5 Erosion and other forms of soil degradation are minimized.

C 2.3 Conservation of the processes that maintain genetic variation.

- I 2.3.1 Levels of genetic diversity are maintained within critical limits.

- I 2.3.2 There is no directional change in genotypic frequencies.

- I 2.3.3 There are no significant changes in gene flow/migration.

- I 2.3.4 There are no significant changes in the mating system.

Variability in individual forest management schemes is reflected in the two basic approaches to standard development that are used by certification authorities: systems-based standards versus performance-based standards, and their association with different types of indicators (Nussbaum et al, 2002; Leslie, 2004; Guynn et al, 2004; Nussbaum and Simula, 2005). Systems standards specify only the management practices necessary to achieve compliance against agreed guidelines (e.g. RIL techniques (Putz et al, 2008)), and are designed to reflect the type and status of an organization's current management strategy. In contrast, performance standards are designed to reflect a level of actual management performance that must be achieved through the maintenance of certain valued attributes. While they are invariably considered to be more rigorous (Cashore et al, 2006), performance standards often fail to offer specific guidance on the method by which such targets should be achieved. However, the credibility of systems standards has been criticized because organizations that operate under the same management strategies but in different situations may exhibit differing levels of performance (Ozinga, 2001). As a result, the International Standards Authority (ISO 14000) only permits the use of product labelling in association with performance standards. Although performance standards are vital towards achieving ecological success in forest management, uncertainty in the scientific basis of ecologically responsible management means that the setting of reliable performance-based indicators continues to represent a significant challenge to forest managers around the world.

Although the majority of certification schemes are considered to be largely performance based (Nussbaum and Simula, 2005), in reality each approach provides complementary benefits, and most certification schemes employ them in combination under a philosophy of continuous improvement (Leslie, 2004; Cashore et al, 2006). In the case of the FSC, generic *guiding* P&C (FSC, 2002; Box 2.6) are used to provide the starting point for developing regional or national Forest Stewardship Standards that are used as guidelines for regional management systems (FSC, 2004). To tailor the suitability of these generic P&C, specific indicators are selected that are appropriate to the regional management context, with each indicator being required to specify the minimum thresholds or levels of performance that can be measured during an evaluation (FSC, 2004). To accommodate the fact that national and regional standards do not exist for many areas of the world, certification bodies (including the FSC) are often required to develop proxy or interim standards that are based upon C&I developed elsewhere (Nussbaum and Simula, 2005; and see Chapter 7).

CHAPTER 3

The need for forest biodiversity monitoring

Synopsis

- We generally have a poor understanding of the prospects for biodiversity in human-modified landscapes, or the adequacy of existing management standards to meet conservation goals.
- Decisions about minimum practice standards for ecologically responsible forest management often rely more upon expert opinion and subjective judgement than a logical appraisal of empirical evidence through monitoring and evaluation.
- The purpose of biodiversity monitoring is to provide a linchpin between ultimate management goals and the ongoing management process, the guiding hand by which conservation objectives can be translated into improved on-the-ground management standards.
- For monitoring to make a meaningful contribution towards improving management three criteria need to be satisfied: (i) it must be purposeful with respect to clear goals and objectives; (ii) it must be effective with respect to a program's ability to deliver on stated objectives; (iii) it must be realistic with respect to its ability to operate within the context of real-world constraints.

The purpose of forest biodiversity monitoring programmes is to respond to an information need and help develop more ecologically responsible management strategies that enhance opportunities for conservation. This chapter focuses on the underlying justification for monitoring as the driver of the entire adaptive management (AM) process, and the mechanism by which scientific uncertainty in the conditions for ecologically responsible management can be reduced.

SCIENTIFIC UNCERTAINTY AND BIODIVERSITY CONSERVATION IN HUMAN-MODIFIED FOREST ECOSYSTEMS

Decades of research have equipped us with substantial knowledge regarding general principles of 'best-practice' for forest management (e.g. Hunter, 1990, 1999; Lindenmayer and Franklin, 2002; Fischer et al, 2006; Lindenmayer et al, 2006; Meijaard et al, 2006; Meijaard and Sheil, 2008; Putz et al, 2008). These efforts have focused on the importance of maintaining native forest cover, maximizing landscape hetero-geneity, maintaining connectivity, and maximizing the structural integrity of forest stands. They largely represent an attempt to mimic, through management inter-ventions, the natural structure of forest landscapes. However, the work that underpins such recommendations is piecemeal and for large areas of the world our knowledge of the prospects for biodiversity conservation in modified forest land-scapes remains rudimentary. This scarcity of scientific information is well demonstrated by the fact that the majority of what we know about the biodiversity implications of different types of forest disturbance and use originates from a rela-tively small number of research sites (Gardner et al, 2009; Box 3.1).

Box 3.1 Geographic biases in research effort on tropical forest
biodiversity conservation

It is impossible to know everything about everywhere. Ecologists therefore consistently depend upon a process of knowledge transfer between places and across time to provide the theoretical and empirical context necessary to interpret biodiversity patterns and inform management strategies. Understanding the validity of this transfer process is a research prior-ity in its own right. In comparison to other parts of the world, the biodiversity of tropical forests remains particularly poorly understood. Severe geographical biases in the distribution of research effort across the tropical forest biome clearly highlight the problems facing conser-vation scientists and practitioners working in poorly studied landscapes and regions (Fig. 3.1). Because of such biases, there is often a strong dependency on the research findings from a small number of very well studied localities. Famous examples include the Biological Dynam-ics of Forest Fragments Project in Brazil, Las Cruces and Los Tuxtlas research stations in Costa Rica, and Danum Valley in Sabah, Malaysian Borneo. While it is true that some research find-ings will have broad implications, there is a danger that false conclusions can be propagated from a limited set of context-specific studies or localities, leading to inappropriate knowledge transfers and paradigms that could hamper the development of effective forest management strategies in less well studied areas of the world. Specific examples of the inappropriate gener-alization of research findings are commonplace, including the lumping of human-modified forest systems into single ecological categories (e.g. 'secondary forests', 'plantations' and 'agro-forestry') despite high levels of internal heterogeneity in biophysical properties and management regimes (Gardner et al, 2007a).

Figure 3.1 Research intensity in modified tropical forest landscapes

Source: Figure redrawn with permission from Gardner et al. (2009)

Despite being founded on solid general principles, a general lack of data means that the effectiveness of many management practices for biodiversity conservation is unclear (Lindenmayer et al, 2006). For example, while it may be clear that the inclusion of no-take reserves within a production landscape is beneficial for biodiversity (and therefore essential in promoting ecologically sustainable forestry), what is far less clear is the relationship between the size, shape and spatial configuration of reserve areas within the wider landscape matrix and the conservation services they provide. Another area of high uncertainty in forest management concerns the ecological impacts of selective logging (Box 3.2).

For some modified forest systems, very few biodiversity data are available. A good example of this is tropical secondary forests that cover a significant part of deforested land in many countries (FAO, 2006). Understanding the potential importance of tropical secondary forests for conservation has attracted much research attention from ecologists and conservation biologists, as well as considerable controversy. Debate was recently sparked by Wright and Muller-Landau (2006) who proposed that the regeneration of secondary forests in degraded tropical landscapes is likely to avert the widely anticipated mass extinction of native forest species. However, other researchers have highlighted serious inadequacies in the quantity and quality of species data that underpin this claim, casting doubt on the potential for secondary forest to serve as a 'safety net' for tropical biodiversity (Gardner et al, 2007a; Laurance, 2007). The only way to resolve this burning question and evaluate the future of tropical forest species in regenerating lands is to monitor the response of those species to different forms of forest recovery, including the ability of management approaches to enhance the reforestation process.

While the lack of scientific understanding surrounding biodiversity in managed forests is particularly evident in the tropics, it is also true that we are a long way from having sufficient knowledge to ensure the persistence of biodiversity in comparatively well studied temperate systems. For example, in reviewing the SFM plan for the North Island of New Zealand, Gillman (2008) concluded that the plans for assessing, monitoring and maintaining natural values, including biodiversity, are inadequate. Similarly, Hickey and Innes (2008) in evaluating sustainable forest management in British Columbia, Canada, concluded that a number of key challenges remain in the development of an effective indicator framework that is capable of demonstrating progress towards a sustainable management system for the region. In a recent review of boreal forest management for biodiversity conservation, Bradshaw et al (in press) also identified more research as being key to the development of effective conservation strategies, including the retention of structural complexity following logging and the management of fire and large-scale disease outbreaks.

Identifying ways in which different types of forest use and management can help safeguard the future of forest biodiversity and inform more sustainable land-use strategies represents a major research priority for the 21st century. This challenge is particularly complicated by the fact that modified forest systems lack any natural ecological analogue – and their ecological dynamics and composition may therefore be poorly reflected by studies of undisturbed forests (Hobbs et al, 2006; Gardner et al, 2009; Box 3.3).

Box 3.2 Biodiversity research and the management of selectively logged forests

Research on how to ensure long-term timber yields in managed forests is significantly more advanced than research focused on biodiversity conservation (Grieser Johns, 2001), and as a result, existing forest management guidelines fall significantly short of prescribing the necessary conditions for effective biodiversity conservation (Grieser Johns, 2001; Zarin et al, 2007; Putz et al, 2008).

   While a growing number of studies have compared patterns of diversity inside and outside logged areas of forest, relatively few have succeeded in evaluating the contribution of particular management practices for biodiversity conservation. For example, the practice of RIL is now widely accepted as a basis for ecologically responsible forestry (Putz et al, 2008), and a number of comparative studies have demonstrated that RIL forests maintain notably more forest species than areas logged using conventional techniques (Lewis, 2001; Meijaard and Sheil, 2008; and see Chapter 1). However,

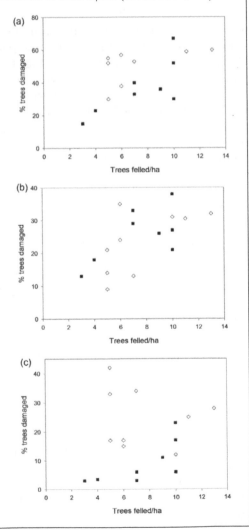

while it is clear that different RIL practices (e.g. felling, timber extraction) result in far less structural damage to the forest than conventional logging (Sist et al, 2003; Fig. 3.2), very few studies have succeeded in linking specific management strategies (e.g. as related to logging intensities, minimum cutting diameters, timber extraction techniques, etc.) to changes in actual biodiversity. The majority of guidelines and performance standards (existing or recommended) for legal or voluntary forest management at all levels have yet to be evaluated against biodiversity data.

   Regulations, standards and management proposals are occasionally supported by empirical data that show thresholds of impact on forest structure (but not actual biodiversity, e.g. Sist et al, 2003). However, in the majority of cases they are either based on the opinion of experienced foresters and ecologists, or are essentially political decisions (e.g. the case of the Brazilian Forest Code).

Figure 3.2 The relationship between felling intensity and tree damage between RIL and conventional logging techniques

Source: Redrawn with permission from Sist et al. (2003)
Note: (a) between felling intensity and total trees damaged in RIL (filled squares) and in CNV (open diamonds), (b) between felling intensity and proportion of trees damaged by felling RIL and CNV, (c) between felling intensity and proportion of trees damaged by skidding.

Box 3.3 Biodiversity conservation in novel forest ecosystems

The juxtaposition of many interconnected structural, compositional and functional changes to forest ecosystems has led to the recognition that human-modified landscapes host increasingly novel species assemblages, and patterns of species interactions that are unlikely to have evolutionary analogues (Hobbs et al, 2006). There is also growing evidence to suggest that the rate of many ecological processes may be both magnified and accelerated in modified forest landscapes, with unpredictable implications for the maintenance of biodiversity (Laurance, 2002). Novel systems may foster new patterns of species loss as extinction is most likely to occur when new threats or combinations of threats emerge that are outside the evolutionary experience of species, or threats occur at a rate that outpaces adaptation (Brook et al, 2008).

Although the definition of what constitutes a 'novel ecosystem' remains somewhat arbitrary, their emergence follows the selective loss and gain of key taxa, the creation of dispersal barriers, or changes in system productivity that fundamentally alter the relative abundance structure of resident biota (Hobbs et al, 2006). Two compelling examples are the creation of 'new forests' in Puerto Rico that are comprised of species assemblage structures that have not previously been recorded from the island (Lugo and Helmer, 2004), and the fundamental alteration of the 3D structure of native Hawaiian rainforests following the establishment of alien plant species (Asner et al, 2008). Understanding the structure and function of novel ecosystems is of fundamental importance in evaluating patterns of biodiversity change, and the effectiveness of conservation strategies within modified forest landscapes (Chazdon, 2008).

The lack of a strong science basis for responsible forest management means that many choices regarding performance standards and the structure and function of monitoring programmes are made on the basis of expert, yet subjective opinion, rather than a logical appraisal of empirical evidence within a standardized framework (Kneeshaw et al, 2000). This, in turn, has led to considerable variation in the choice of indicators and definition of management standards for biodiversity conservation and other forest values (Holvoet and Muys, 2004; McDonald and Lane, 2004), disagreement as to what should be considered responsible management by certification authorities (Nussbaum et al, 2002), and a widespread failure in many countries to develop standards and implement effective forest management initiatives, especially in the tropics (McGinley and Finegan, 2003; McDonald and Lane, 2004; UNFF, 2004).

While some broad-brush management guidelines can be reliably applied to most of the world (e.g. wider forest corridors in plantation landscapes and longer rotation cycles in logged forest will generally benefit local wildlife), differences in biophysical and anthropogenic landscape context and the historical legacy of human disturbance will always mean that no single blueprint will ever exist. Moreover, there are important differences between regions in the dominant forms of forest use and associated cultural and socio-economic factors that determine the potential success of conservation actions. As concluded by Sayer and Maginnis

(2005a), the future of more sustainable approaches to forest management (including considerations for biodiversity conservation) lies in recognizing a plurality of context-dependent management approaches that are effective in different real-world situations.

Adopting an adaptive approach to managing risk and reducing uncertainty through monitoring is vital to improving our understanding of the relative costs and benefits of alternative management options, and ensuring that ecological criteria are adequately represented in management standards (Wintle and Lindenmayer, 2008). In response to this need, the design of monitoring programmes has emerged as an area of high-priority for forestry research (Buck et al, 2003; CICI, 2003; UNFF, 2004, 2006; Wijewardana, 2008).

THE PURPOSE OF BIODIVERSITY MONITORING AS A GUIDE TO MANAGEMENT

The goal of ecologically responsible forest management is to minimize, wherever possible, the impact of human activity on forest biodiversity and associated ecological processes. This is achieved through the development of management strategies that employ and revise minimum practice standards in an attempt to guarantee progress towards long-term conservation goals (see Chapter 2). The central motivation behind this book is the identification of ways in which biodiversity monitoring can be effective in facilitating and guiding this process.

To properly envision the purpose of monitoring, it is first necessary to position the monitoring process appropriately within the wider context of the full management system. Forests are managed primarily for some kind of economic production (timber, plantation harvest, non-timber forest products, etc.). However, incorporating biodiversity conservation as an additional management goal entails the same generic procedure as managing for any other valued attribute. It requires:

- the establishment of a clear set of goals, objectives, indicators and targets;
- the employment of existing knowledge to design and implement management standards appropriate for achieving progress towards set goals;
- the assessment of management-related impacts to determine whether agreed minimum levels of management performance have been attained;
- the evaluation of management standards, as well as any additional unplanned human impacts or threatening processes, to determine the extent to which existing standards are sufficient in achieving progress towards long-term conservation goals and where the greatest opportunities for improvement lie; and
- adaptive management i.e. the adjustment of management practices on the basis of recommendations from monitoring to enhance long-term prospects for biodiversity conservation.

It is immediately clear from this simplified description of the management process that when done properly monitoring is not an isolated exercise. *In a very real sense monitoring and management represent two sides of the same coin; neither exercise makes proper sense unless it is accompanied by the other.* In essence, the purpose of biodiversity monitoring is to provide a linchpin between ultimate management goals and the ongoing management process, the guiding hand by which conservation objectives can be translated into improved on-the-ground management standards. To achieve this broad purpose monitoring serves two specific and inter-related functions. First, it is an assessment tool that is used to assess the status or condition of biodiversity in a managed system (wherever conservation management strategies may exist). In so doing it can provide an assessment of management performance and compliance against predetermined standards. Without reliable information on the status and trends of a forest, it is impossible to expect that managers can ensure the conservation of viable populations of native species and the maintenance of key ecological processes (Noss, 1999). Second, it is an evaluation tool that is used to compare the effectiveness of alternative management actions, whether existing or potential – thereby providing a means to both validate the adequacy of existing management standards and identify directions for future progress towards more sustainable systems of use (see Chapter 4 for a typology of monitoring approaches). In both cases, the makeup of a meaningful monitoring programme can be qualified in terms of three criteria (Fig. 3.3; and see Stem et al, 2005):

1   **Purposeful** A monitoring programme needs to be founded on clear goals and objectives that serve to justify the investment of limited resources and provide an incentive for acting upon any management recommendations that may emerge from the monitoring process.

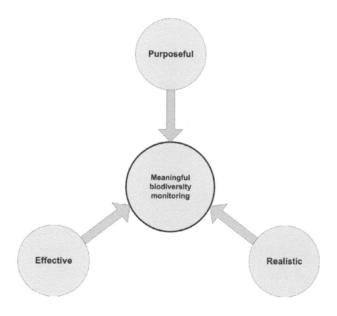

Figure 3.3 The three components of meaningful monitoring

2   **Effective** The design and implementation of a monitoring programme needs to be such that it can deliver on the stated objectives. In the case of monitoring forest biodiversity, being able to report simply on the status and trends of the elements that we value for conservation is neither a sufficient nor cost-effective activity. In addition, we should strive to identify clear and traceable links between the elements we value, and the underlying human activities and management practices that drive observed changes and can be adjusted in response to stakeholder demands.

3   **Realistic** To be successful, a monitoring programme needs to be viable in the context of real-world constraints. In particular, this requires that monitoring programmes are effective in generating a useful return on the investments in time and money that are made, are able to function with available technical and logistical support, and are embedded within sufficiently reliable institutions to be sustainable in the long term.

The individual ingredients of a meaningful monitoring programme that meets these three basic criteria include not only scientific but also practical and institutional considerations. A key factor for success lies in overcoming the organizational insularity that currently exists between researchers, managers and bureaucrats (Field et al, 2007). This insularity can be broken down by developing monitoring programmes that are founded in sound scientific principles, yet are also cognizant of the real-world practical and social constraints that ultimately determine the longevity and effectiveness of the programme.

CHAPTER 4

A Typology of Approaches and Indicators for Monitoring Forest Biodiversity

Synopsis

- There are three main types of monitoring, each serving a complementary function in the overall management system: (i) *implementation monitoring* to assess compliance with minimum practice standards; (ii) *effectiveness monitoring* to assess the status of the managed system and report on whether minimum performance standards for conservation have been achieved; and (iii) *validation monitoring* to improve our understanding of the cause–effect relationships that link human impacts with changes in biodiversity, and determine the extent to which existing management practices are adequate for meeting long-term conservation goals.

- Indicators provide the practical tool by which changes in management practices and ecological responses can be measured and evaluated against agreed standards and long-term goals.

- Management practice indicators serve a prescriptive function and define minimum practice standards. Performance indicators serve an evaluative function and are used to assess ecological responses to management and help define performance-based standards and conservation goals.

- Performance indicators can be both indirect in the form of measures of forest structure and function and direct in the form of biological indicators and/or individual target species.

- Biological indicators are used to estimate the impacts of management on parameters that cannot easily be measured directly, including the distribution and abundance of non-sampled species and overall ecosystem condition as measured by deviation from some more desirable system state. Target species complement biological indicators and include endangered species, and species of particular ecological, economic or cultural importance.

Inconsistency in language and a lack of clear definitions is a major obstacle to communication and understanding in science. The field of biodiversity monitoring presents no exception, and confusion around the terminology used to describe different approaches to monitoring, and the purpose of different types of indicators, has been a major contributing factor to the slow pace of development in monitoring across many parts of the world (Sheil et al, 2004). This chapter attempts to help rectify this situation by presenting a clear typology that distinguishes both between different monitoring approaches and different types of indicator and target species that are used to quantify and evaluate observed changes.

MONITORING APPROACHES

There are three broad types of monitoring approach that are relevant to managed forests and natural resource systems in general (Noss and Cooperrider, 1994; Lindenmayer and Franklin, 2002), and different approaches serve different and complementary functions in the overall management process (Table 4.1). In general, the approach taken depends on whether those responsible for managing the forest manager are concerned with:

- **Implementation monitoring** Knowing whether recommended management guidelines and practices are being adhered to (i.e. as relevant to system-based management standards). Implementation monitoring does not involve the actual monitoring of biodiversity, but rather just records whether management is in line with agreed practice.
- **Effectiveness monitoring** Knowing the condition (status and trends) of a measured management outcome (i.e. as relevant to performance-based management standards). Effectiveness monitoring asks whether a particular conservation management target has been achieved in a given area but does not question the reason for any success or failure. Implementation and effectiveness monitoring perform complementary functions within a comprehensive management audit to assess compliance against agreed standards.
- **Validation monitoring** Whether they wish to validate the extent to which particular management interventions are having the desired effect (i.e. management-orientated research). Validation monitoring goes a critical step beyond effectiveness monitoring and asks whether a particular conservation goal has been achieved *for the right reasons.*

Implementation monitoring

Implementation monitoring is the most basic type of monitoring, and has as its central function the assessment and demonstration of forest management compliance against agreed minimum standards. It is based on the assessment and reporting of individual management plans, interventions and actions that together form the management standard, including landscape (e.g. roads, connectivity) and stand level (e.g. cutting schedules, chemical use, seasonal logging restrictions) management activities. An obvious criticism of implementation monitoring is that it does not determine whether a management plan is successful in achieving biodiversity conservation goals. It is entirely possible, either because of scientific

Table 4.1 Approaches to monitoring the performance of forest management systems for biodiversity conservation

Level of commitment to biodiversity monitoring and evaluation	Description	Synonyms	Types of indicators used (see Table 4.2 for full typology)
Recognition and acceptance of agreed P&C for responsible forest management	Does not involve monitoring Often expressed through commitments to non-legally binding conventions or multilateral agreements. For example, the non-legally binding instrument on all types of forest (see Box 2.3)		None
Implementation monitoring	A systems-based approach concerned with assessing the compliance of management activities against agreed standards	Compliance monitoring, Operational monitoring, Process monitoring, Process evaluation	Management practice indicators
Effectiveness monitoring (performance assessment)	Concerned with assessing the status of a management target as compared to performance standards. Effectiveness monitoring asks whether a particular conservation management target has been achieved in a given area but does not question the reason for success or failure. Can be either species or non-species based but there are strong arguments to support the use of structural indicators for effectiveness monitoring as they are easy to link directly to management. Lacks any formal framework for learning, and provides a largely reactive approach to management	Impact monitoring, Impact evaluation, Status and trends monitoring	Indirect or direct performance indicators (and/or target species)
Surveillance monitoring	Direct monitoring or "surveillance" of biodiversity trends over time. Can be useful in capturing unpredictable changes, for understanding background levels of variability in indicator values, and for evaluating non-spatial human impacts (e.g. climate change). However, because it is not explicitly concerned with evaluating differences in the management system, its function is entirely reactive and it is only able to make a minimal contribution to learning	Status and trends monitoring, Effects-orientated monitoring,	Direct performance indicators (and/or target species)

Table 4.1 Approaches to monitoring the performance of forest management systems for biodiversity conservation (continued)

Level of commitment to biodiversity monitoring and evaluation	Description	Synonyms	Types of indicators used (see Table 4.2 for full typology)
Validation monitoring	Incorporates both a systems- and performance-based approach to assessing changes in forest through the simultaneous monitoring of both management and ecological variables. Actively links management interventions to changes in biodiversity and ecological integrity (as measured by a hierarchy of coupled structure and biological indicators). Testable hypotheses and carefully designed sampling regimes are employed to help establish causal relationships between management interventions and ecological variables, thereby assisting in evaluating the relative effectiveness of alternative management strategies. Incorporates a formal plan for learning, which enables results to be effectively integrated into revised management plans. Management-orientated validation monitoring represents a central component of AM.	Targeted monitoring, Strategic monitoring, Adaptive management (AM), Formative evaluation	Management practice and indirect and direct performance indicators (and/or target species)

Note: This framework of monitoring approaches is hierarchically nested, such that more advanced approaches tend to encompass the assessment of indicators used by more superficial programmes. As more resources become available, new activities and elements may be added to the basic protocol
Source: Compiled with reference to Noss and Cooperrider (1994), Lindenmayer and Franklin (2002), Stem et al (2005), Nichols and Williams (2006), Stadt et al (2006) and Margoluis et al (2009)

uncertainty surrounding the ecological implications of management or because of the influence of unpredictable and external factors (e.g. natural disturbances or human impacts elsewhere in the same watershed), to successfully implement all the agreed management strategies yet fail to conserve biodiversity. Thus, implementation monitoring on its own is inadequate for reporting on management performance, and needs to be complemented by the direct measurement of valued attributes themselves (Lindenmayer et al, 2002a).

Effectiveness monitoring

Effectiveness monitoring assesses the status and trends of ecological attributes within a management system, and is essential for evaluating compliance with performance-

based management standards (Nussbaum et al, 2002; Table 4.1). Effectiveness monitoring relies upon indirect or direct measurements of management performance
which provide the basis for assessing changes in ecological condition. Somewhat
unfortunately, the term 'effectiveness' can be misleading as it gives the impression
that any observed changes are the direct *effect* of a particular management intervention. However, an observed change may be the result of a number of different
processes, not all of which are necessarily linked to management. For example, it is
possible for the conservation value of a site to either degrade or improve for reasons
that are entirely independent of the management process – for example, due to
natural disturbances, such as fire and hurricanes, or the influence of natural multi-
annual cycles in the distribution and abundance of species populations. Unless the
purpose of monitoring is simply to provide a mechanism for assessing compliance,
effectiveness monitoring (as it is commonly defined) is rarely sufficient on its own,
especially if managers are under legal obligation to demonstrate the consequences
of their actions, make predictions of management performance in other areas or
identify opportunities for future improvements.

Validation monitoring

Validation monitoring links changes in management with changes in valued attributes, and is the only approach able to measure whether specific management
actions produce the desired effect. Such an approach has great appeal to
conservation practitioners who are not satisfied with achieving the right results for
the wrong reasons (Stem et al, 2005).

Validation monitoring provides the opportunity to determine which management impacts are detrimental towards the conservation and restoration of
biodiversity, while also identifying ways in which management strategies can be
further developed and adjusted to mitigate against such impacts in the future
(Lindenmayer et al, 2006). Although validation approaches can vary (Stem et al,
2005; Table 4.1), they are generally concerned with employing testable hypotheses
and carefully designed sampling regimes to help establish causal relationships
between management interventions and ecological variables, and in so doing
comprise fundamental components of AM (Holling, 1978; and see Chapters 8–15).
Understanding how to achieve a meaningful validation monitoring programme in
the context of meaningful objectives and real-world constraints and practicalities is
a central purpose of this book.

Surveillance monitoring

A fourth form of monitoring that is not linked to the assessment or evaluation of the
management system in any specific way but simply constitutes a status report of
general trends in biodiversity over time at a particular site (Table 4.1). This type of
monitoring, termed surveillance or status monitoring, can be useful in acting as a
warning device of unpredictable changes in biodiversity, for understanding background levels for variability in control sites, and for evaluating non-spatial human

impacts (e.g. climate change). Surveillance monitoring can also contribute towards 'state of the environment' reports (e.g. for external interested parties, whether government or civil society). This book is concerned with identifying ways in which monitoring can meaningfully engage with management, either as a device for measuring performance or to evaluate alternative hypotheses about how management practices could be improved. Because surveillance monitoring is disengaged from the management process, it is of limited use either for measuring performance or for identifying the underlying causes of any change in biodiversity (unless the drivers are very simple and conspicuous), and it is given little further attention here (see Chapter 5 for more details).

MONITORING INDICATORS

In discussing environmental indicators in general, Hammond (1995, p1) defines an indicator as:

> '...*something that provides a clue to a matter of larger significance or makes perceptible a trend or phenomenon that is not immediately detectable. [. . .] Thus an indicator's significance extends beyond what is actually measured to a larger phenomena of interest.*'

Indicators provide the *practical tool* by which changes in management practices, system attributes and ecological processes can be measured, and minimum performance standards or thresholds evaluated. Consequently, the choice of indicator is fundamental in defining the approach to both monitoring and management (Lindenmayer and Burgmann, 2005). It is important to clearly distinguish the concept of an indicator from that of an attribute: attributes are all encompassing and relate to any quantifiable element of concern, whereas indicators only represent the subset of attributes or attribute features that are used as surrogates of other valued attributes.

To assist in relieving the significant confusion that surrounds the indicator concept, Table 4.2 provides a typology of indicator types that helps reveal problems of synonymity and identifies the role of different indicators within a hierarchical pressure-state-response framework. In general terms, it is useful to consider two main types of indicator: those whose function is prescriptive, and those whose function is evaluative (Kneeshaw et al, 2000; Rempel et al, 2004) – although these concepts can be confounded depending on the perspective of the observer (see Table 4.2). Management policy and process indicators are both prescriptive in that they are used to measure or verify the existence or implementation of certain policies and management strategies. Management-practice indicators are termed as such because they are used to measure the implementation of management practices. In a similar sense, they are also referred to as 'pressure' indicators (e.g. Hagan and Whitman, 2006). In contrast, performance indicators (Table 4.2) are evaluative in the literal sense that they are used to evaluate changes in management performance. I distinguish the concepts of *indirect* versus *direct* performance

Table 4.2 A typology of indicators for monitoring forest management systems

Indicator	Synonyms	Description	Examples	Key advantages	Key disadvantages
Policy indicator	*Planning indicator*	To assess the existence of an adequate institutional framework to support planning, research and development, and reporting. Can be relevant to both stand- and landscape-management issues	• Existence of clearly defined management objectives • Existence of a comprehensive landscape and stand management plan • Availability of trained individuals for monitoring purposes	• Fundamentally important for the reliable implementation of best practice guidelines and AM • Easy and cheap to assess • Fully comparable across forest landholdings	• Provides no assessment of actual management practices or performance
Management-practice indicator	*Management-process indicator, Pressure indicator, Control indicator*	To assess the extent of intervention of an explicit management *practice* that forms part of an overall management strategy. It is *prescriptive* in that it defines the type and level of *treatment* or *control* applied by a management intervention. The use of appropriate process indicators reflects a proactive approach to promoting responsible forest management. Process indicators are used as the basis of systems-based standards for forest management	• Logging intensity • Stand structural retention practices • Length of rotation cycle • Reserve allocation and design	• Easy to define • Available for immediate implementation based on knowledge of 'best-practice' management • Provides the most suitable proxy for the assessment and early detection of changes to ecological resources • Cheap to monitor • Readily transferable among forest landholdings • Requires no *a priori* ecological knowledge	• Only represents the application of a minimum or best practice standard and doesn't measure management performance or evaluate the effectiveness of alternative strategies
Performance indicator	*Condition indicator, Health indicator, Response indicator, Environmental indicator, Monitoring indicator, Outcome-based indicator*	To assess the *performance* of forest management towards meeting long-term conservation goals. In general, they are considered to be *evaluative* in that they assess the *response* of a system attribute to a particular management practice. Performance indicators can provide both indirect (i.e. measures of forest structure and function) and direct (i.e. measures of biodiversity) assessments of the target criteria and depending on the perspective of the stakeholders they can be considered as end-points in themselves (e.g. target species). The use of appropriate performance indicators in the evaluative sense reflects a reactive approach to promoting responsible forest management. Performance indicators are used as the basis of performance-based standards for forest management. Individual target species can sometimes provide a complementary indicator function (e.g. keystone species) or may just be worthy of monitoring in their own right (e.g. endangered species and pest species)			

Table 4.2 A typology of indicators for monitoring forest management systems (continued)

Indicator	Synonyms	Description	Examples	Key advantages	Key disadvantages
● Indirect performance (e.g. structural or disturbance) indicator	*State indicator, eco-physical indicator*	Indirect assessment of ecological condition through the measurement of structural (landscape and habitat) and functional forest attributes. Reflects a coarse-filter approach in assessing progress towards sustainability based upon implicit hypotheses of the relationship between the indicators and valued conservation attributes. Can be either evaluative or prescriptive depending on the context (see text)	● Habitat structural complexity ● Area of native forest cover ● Habitat fragmentation ● Availability of dead wood ● Disturbance regime (e.g. fire and grazing) ● Soil contaminant loads	● Transparently linked to management actions ● Directly evaluates stand and landscape-scale heterogeneity and structural diversity ● Relatively cheap and simple to monitor	● Assumes an adequate understanding of the relationship between structural/spatial habitat attributes and the persistence of forest biodiversity
● Direct performance (biological) indicator	*Bioindicators response indicator, taxon-based indicator, predictor taxa*	Direct measurement of attributes of interest as defined by a particular conservation goal. Varyingly operate as surrogates of other attributes of the target criterion (e.g. other species, or a measure of ecological integrity). Reflects a fine-filter approach in assessing progress towards responsible management. Direct performance indicators can be thought of as purely evaluative	● Environmental indicators ● Cross-taxon response indicators ● Ecological disturbance indicators ● Target species	● Provides a direct measurement of change in actual conservation values	● Not always clear how they are linked to management activities ● Often captures only a subset of a much broader conservation goal

Source: Compiled with reference to Stork et al (1997), McGeoch (1998), Caro and O'Doherty (1999), Kneeshaw et al (2000), Lindenmayer et al (2000) and Rempel et al (2004)

indicators, and recognize that both are commonly used to measure management performance for forest certification (e.g. FSC, 2002; and see Newton and Kapos, 2002). Because of the difficulties in reliably linking management impacts to changes in the distribution and abundance of species, direct species-based performance indicators are often more useful as evaluators of performance standards rather than direct measures of compliance (see Chapter 8).

Management practice indicators

Management indicators directly measure management practices, including stand- and landscape-level interventions. Together they detail which aspects of a forest landscape are managed (i.e. as distinguished from un-managed activities and external threats). The assessment of management practice indicators forms the basis of an implementation monitoring programme that is used to evaluate compliance against an agreed management standard

Management performance indicators

When the consequences of management are uncertain (a condition that is almost always true) performance-based indicators provide a more transparent and comparable assessment of management accountability than is possible with process-based indicators.

Forest managers very rarely have the capacity or expertise necessary to directly manage populations and species. Instead, they manipulate structural and functional aspects of the ecosystem, which in turn have a variety of consequences for biodiversity (Lindenmayer and Franklin, 2002; Lindenmayer and Fischer, 2006; Gardner et al, 2009). To accommodate this fact, managers of forests, like other ecosystems, address the problem of biodiversity conservation through a combination of coarse and fine-filter management approaches (Hunter 1990).

Coarse-filter management is focused on creating or maintaining the ecosystem structures and ecological processes required for the persistence of a wide range of species. By contrast, species-specific or fine-filter management addresses the direct resource or habitat needs for particular target species that are not addressed through coarse-filter approaches. The extent to which coarse-filter approaches ensure adequate resource provision for a large number of species is poorly understood for much of the world, and requires testing through validation monitoring (Table 4.1). Coarse- and fine-filter management approaches can be evaluated through monitoring programmes using a combination of what I term here indirect and direct performance indicators.

Indirect performance indicators provide the foundation for a performance-based standard and are intended to represent the proximate drivers of any observed changes in biodiversity (Table 4.2; and see Chapter 11). In the majority of cases, indirect performance indicators are made up of stand- and landscape-level indicators of forest structure (*sensu* Lindenmayer et al, 2000), which provide a readily accessible and quantifiable indication of ecologically relevant forest management impacts. Some indicators of ecological processes, such as major disturbance regimes (e.g. fire, flood, pest-outbreaks) and changes in soil or water contaminant loads, can also represent valuable indirect indicators of performance.

The key requirement of any indirect performance indicator, whether structural or process based, is that any changes in biodiversity can be clearly linked to management impacts through a logical chain of cause and effect. Chapter 6 discusses some important considerations in the use of structural indicators as proxies of biodiversity, while Chapter 8 presents arguments in support of their role as the primary mechanism for reliably assessing management performance. Chapter 11 identifies a number of approaches that are commonly used to select and measure structural indicators in the field.

Direct performance indicators differ from indirect performance indicators because they directly measure changes in the valued attributes rather than their perceived ecological requirements (e.g. native species rather than habitat availability). For the purposes of this book, direct performance indicators can be considered as synonymous with biological indicators and target species (see Table 4.2). Chapter 12 identifies a number of approaches that are commonly used to select and measure biological indicators and target species in the field. A comprehensive ecological monitoring programme may also encompass indicators of ecological processes that play important roles in the maintenance of biodiversity (e.g. leaf litter decomposition (Ghazoul and Hellier, 2000); or soil properties (Curran et al, 2005)), yet whose links to management activities are poorly understood.

Biological indicators

Chosen carefully, biological indicators can make an invaluable contribution to monitoring because they are the only method of synthesizing the overwhelming complexity of ecological systems, and are therefore the most effective tool for linking conservation science to policy (UNEP, 2000). They also provide the highest level of information quality because biological indicators are valued attributes in their own right. Financial and logistical constraints mean that it is impossible to measure all elements of biodiversity (Lawton et al, 1998; Gardner et al, 2008a), and biological indicators that can operate as surrogates for changes in ecosystem health or condition or the distribution and status of other species provide a practical solution to an otherwise intractable problem (Margules et al, 2002, Niemi and McDonald, 2004).

As surrogates of change in the condition and/or diversity of forest ecosystems, biological indicators can be used in a regulatory sense to provide an early warning signal of impending environmental change, as well as in a diagnostic sense, as an aid to interpreting the ecological consequences of alternative management strategies (Dale and Beyeler, 2001; Lawton and Gaston, 2001; Niemi and McDonald, 2004). The two fundamental requirements for all biological indicators are: (i) that they reflect something that cannot be measured directly, while also providing more information than that which relates only to the indicators themselves; and (ii) their measurement in the forest is logistically and financially feasible. Beyond this, the concept of an indicator species or species group can adopt a myriad of different meanings (Caro and O'Doherty, 1999; Lindenmayer et al, 2000). Although the semantics of biological indicators has a highly confused history in the ecological literature (Caro and O'Doherty, 1999; Caro, 2010), I follow McGeoch (1998, 2007) in recognizing three broad and overlapping categories of biological indicator that each correspond to conceptually different applications, namely: environmental indicators, biodiversity indicators and ecological indicators (see also Landres et al, 1988; Lawton and Gaston, 2001; Box 4.1).

Environmental indicators

Environmental indicators are species, or groups of species, that provide a predictable and quantifiable measure of an environmental state or impact on some abiotic or physical parameter of interest that may be difficult or expensive to measure directly. They have most commonly been applied to indicate levels of pollutants and toxins in water, but also other measures such as soil fertility (McGeoch, 1998, 2007). Related terms include bioassays, accumulator species and biomarkers.

Biodiversity indicators

Biodiversity indicators operate, as the name suggests, as surrogates of biodiversity, and are described as species or groups of species whose distribution or level of diversity reflects some measure of diversity of other taxa (i.e. their distribution is highly congruent with the distribution of other, unrelated species) (Noss, 1990; McGeoch, 1998, 2007). Despite often having weak theoretical and empirical support (Lindenmayer et al, 2000, 2002a; and see Chapter 6), this concept has dominated much of the discussion surrounding biological indicators. The term biodiversity indicator is commonly employed in reference to large-scale conservation planning assessments where understanding spatial patterns of species congruency is central to developing an effective network of protected areas. However, the concept also encompasses what I term here as 'cross-taxon disturbance response indicators' (and see Caro, in press), which are applicable to monitoring systems and relate to these individual species or species groups that are used to capture the impacts of disturbance on other species or species groups (e.g. Barlow et al, 2007a; Gardner et al, 2008a).

Ecological indicators

Ecological indicators are species that demonstrate the effect of environmental change and degradation on biota or biotic systems (Kremen, 1992; McGeoch, 1998, 2007; Bani et al, 2006; Howe et al, 2007). As some researchers have employed a more general usage for ecological indicators (e.g. Noss, 1999; Niemi and McDonald, 2004), here I use the more specific term of 'ecological disturbance indicators' with respect to their application for evaluating the ecological consequences of human disturbance in modified systems (see also Caro, in press).

Ecological indicators may or may not capture the specific responses of other species to disturbance, but their primary utility is in providing a species-based gauge of the otherwise difficult-to-quantify holistic concepts of ecological condition and integrity, where measurements are made as some form of deviation from a reference or minimally disturbed state (see discussion in Chapter 9). The implication of a loss of integrity as signalled by a decline in ecological disturbance indicators is derived from what we know about what such species do. Indicator species groups that are both sensitive to environmental change, and are known to perform important ecological functions, make excellent ecological disturbance indicators as they provide the most reliable inferences about the ecological and functional implications of disturbance.

Box 4.1 Status of research on bioindicators

McGeoch (2007) recently published a review of research conducted between 1998 and 2005 on all the major categories of bioindicators. The Science Citation Index on Web of Science was searched for entries for the period 1998 to 2005 using the following keywords: bioindicat*, indicator species, surrogate and biodiversity, ecological indicat*, environmental indicat* and biodiversity indicat*. After deletion of inappropriate references, the total remaining was 2311, each of which was then categorized according to the above definitions. Environmental bioindicators (accounting for 48 per cent of the total number of studies) were found to be generally at a more advanced stage of development than either the ecological indicators (38 per cent) or biodiversity indicators (8 per cent). The large number of environmental indicators is due in particular to freshwater monitoring schemes involving macroinvertebrates. From a taxonomic perspective, the literature is dominated by studies on plants (particularly lichens as pollution indicators) and invertebrates (including insects), which constitute over 65 per cent of all publications. Within the Arthropoda, the hexapods encompass the vast majority of studies, with the Coleoptera and Hymenoptera most frequently represented. The Coleoptera, especially ground, tiger and dung beetles, are particularly well recognized as ecological bioindicators and have also been extensively tested in biodiversity assessments and monitoring.

Because a primary goal of biodiversity conservation is to improve our understanding of the consequences of anthropogenic-induced stressors on native biota, ecological disturbance indicators as defined here represent arguably the most critical objective in the field of biological indicators (McGeoch, 1998, 2007; Pearce and Venier, 2006; and see Chapters 6 and 12). An important way in which ecological disturbance indicators are quite distinct from environmental indicators is that they assess the effects or ecological consequences of environmental change (the indicator *itself* is of intrinsic interest), whereas environmental indicators are used more simply as a gauge of change in a particular abiotic environmental parameter (McGeoch, 1998). Because ecological disturbance indicators indicate functional changes to an ecological system (as measured by a deviation from a reference condition), they reveal insights into the consequences of forest management that cannot be gained from direct measurement.

Focal species

In addition to the indicator concepts as described above, there are additional types of indicators that operate as partial surrogates for biodiversity yet have been defined to have a more specific usage and fall under the general category of 'focal' species (e.g. Lambeck 1997; Caro and O'Doherty, 1999; Box 4.2). Focal species have been developed largely in response to criticisms of the biodiversity indicator concept as based on patterns of species congruency. They are characterized as species that have particular ecological requirements, the protection of which can help ensure the conservation of other species, encompassing the concepts of umbrella species, keystone species and resource or process-limited species (Mills et al, 1993; Lambeck, 1997; Noss, 1999; Box 4.2). Under a framework developed by Lambeck (1997) (see also Noss, 1999; Box 4.2), focal species are used to identify specific threats, and the species most sensitive to each threat are then used to define the minimum acceptable level at which that threat can occur. Consequently, they encompass elements of both the biodiversity and ecological indicator species concept, yet as noted by Niemi and McDonald (2004), focal species tend to differ from ecological disturbance indicator species because they do not necessarily serve to measure ecological condition, nor do they convey a clear stress–response relationship.

Box 4.2 Focal species

Lambeck (1997) and Noss (1999) have proposed a number of categories of 'focal' or target species that can act as complementary targets for biodiversity monitoring programmes. The original recommendation from Lambeck (1997) is that a suite of focal species should be chosen, each of which can be used to define different attributes that must be present in a landscape if it is to retain its native biota. Within each of the species categories, the species that is most sensitive is presumed to act as umbrella species for the others in the same category. Noss extended Lambeck's (1997) original list to include keystone species (Mills et al, 1993) as those species that demonstrate exceptionally strong interactions with other taxa or ecosystem processes (see also Chapter 12).

- **Area-limited species** Species that require the largest patch sizes to maintain viable populations. These species typically have large home ranges and/or low population densities, such as many mammalian carnivores.

- **Dispersal-limited species** Species that are limited in their ability to move from patch to patch, or that face a high mortality risk in trying to do so. These species require patches in close proximity to one another, movement corridors, or crossings across barriers, such as roads. Flightless insects limited to forest interiors, lungless salamanders, small forest mammals, and large mammals subject to roadkill or illegal hunting are among the forest species in this category.

- **Resource-limited species** Species requiring specific resources that are often or at least sometimes in critically short supply. These resources may include large snags, nectar sources, fruits, and so on. The number of individuals the region can support is determined by the carrying capacity at the time when the critical resource is most limited (Lambeck, 1997). Hummingbirds, frugivorous birds, and cavity-nesting birds and mammals are in this category.

- **Process-limited species** Species sensitive to the level, rate, spatial characteristics or timing of some ecological process, such as flooding, fire, wind transport of sediments, grazing, competition with exotics, or predation. Plant species that require fire for germination or to escape competition are among the many possible species in this category.

- **Keystone species** Ecologically pivotal species whose impact on a community or ecosystem is large, and disproportionately large for their abundance. Examples in forests include cavity-excavating birds and herbivorous insects subject to outbreaks (see also Chapter 12).

Source: Text based on Noss (1999)

Target species of particular conservation and management concern

There are a number of individual species which, for a variety of reasons, may deserve to be monitored for their *intrinsic interest*, and not because they necessarily indicate patterns of any other species or ecological processes, or condition. For the purposes of this book, I term all such species 'target species'. Many legal and voluntary forest management guidelines give particular emphasis to the role of target species of conservation concern (i.e. endemic, threatened and endangered species) as a focus for biodiversity monitoring and evaluation programmes. Identification of threatened and endangered species may be made on the basis of regional, national and global listings, each of which involves a distinct set of selection criteria. However, the globally accepted standard for classifying extinction risk and identifying threatened species is the IUCN Red List (Mace et al, 2008). In addition, other species that are deserving of particular management attention include invasive and pest species that may have significant impacts on local biodiversity, as well as species that are of particular economic or cultural importance for local people, for example non-timber forest products, such as palms, nuts and game meat (Godoy and Bawa, 1993) or 'flagship species' that are used to motivate conservation action (Caro and O'Doherty, 1999).

PART II

Challenges Facing Forest Biodiversity Monitoring

CHAPTER 5

Challenges to Monitoring: Problems of Purpose

Synopsis

- Uncertainty in the goals and objectives of forest conservation undermines management effectiveness and creates uncertainty in the purpose and design of biodiversity monitoring programmes.
- Many biodiversity monitoring programmes are essentially surveillance exercises that fail to deliver useful guidance for management because they are viewed as an end in themselves, and are not purposefully designed in the context of the wider management process.
- Monitoring done badly can be worse than no monitoring at all as it carries the risk of generating unreliable knowledge, squandering limited resources and eroding credibility in the practical value and relevance of the entire monitoring process among both managers and scientists.
- Limited resources mean that most monitoring programmes represent a trade-off between the quality of sample data and the need to ensure the long-term feasibility of the programmeme.
- Inconsistency in terminology has presented a major challenge to efforts to communicate the purpose and objectives of biodiversity monitoring to both participants and end users.

purpose, _n_. Oxford English Dictionary: _'to serve one's purpose: to be of use or service in effecting one's object; to be capable of bringing about a desired result'_
Biodiversity monitoring programmes need to be founded on a clear sense of purpose if they are to be successful in delivering information that can guide improvements in management and justify the investment of limited resources. This chapter builds upon the context provided by Chapters 3 and 4 to highlight some of the challenges involved in developing a monitoring programme that is fit for purpose. First, I highlight how uncertainty in the underlying conservation goals and objectives of forest management can undermine efforts to develop a meaningful monitoring programme. Second, I discuss how a common failure of biodiversity monitoring programmes to make a meaningful contribution to improving forest management has generated a growing crisis of credibility in the value and purpose

of the entire process. I illustrate how many monitoring programmes essentially operate as surveillance exercises that fail to link observed changes in biodiversity to underlying drivers and hence are limited in how they can act as a guide for responsible management. I also discuss the need for monitoring not only to generate useful information for management, but also to generate information in a cost-efficient manner. Although cost is rarely afforded upfront consideration in monitoring, achieving goals and objectives within a limited budget is central to the purpose of any management endeavour, and particularly so for biodiversity monitoring, which is often accused of being a resource-intensive exercise. I close the chapter by emphasizing the importance of adopting clear terminology as a basis for developing a successful monitoring programme.

THE CHALLENGE OF SETTING CONSERVATION GOALS AND OBJECTIVES AS A BASIS FOR MANAGEMENT AND MONITORING

The establishment of clear goals and objectives is central to the success of any conservation project. Conservation goals are visionary conceptual statements that reflect societal values and management intent. Goals also provide the basis for setting specific and quantifiable objectives that are necessary for defining management standards and designing evaluation frameworks for monitoring. However, conservation science and practice is hampered by a widespread failure to clearly define conservation goals (Salafsky et al, 2002). Furthermore, even where goals are stated, quantitative objectives are typically set without consistency or scientific rigour (Tear et al, 2005). Uncertainty regarding goals and objectives undermines management effectiveness and creates uncertainty in the purpose and design of biodiversity monitoring programmes, including the choice of indicators used to assess and report on management performance, and the identification of management areas that are priorities for evaluation (Gibbs et al, 1999; Yoccoz et al, 2001; Failing and Gregory, 2003).

Criteria and indicators (C&I), which are the universal mechanism by which forest management standards are set, often fail to define clear goals and objectives for biodiversity conservation (Wintle and Lindenmayer, 2008). Despite the rapid expansion of the forest certification industry in many countries (Gullison, 2003; Nussbaum and Simula, 2005), for much of the world the main focus of research attention has been limited to national and regional-scale C&I processes. For most countries, C&I for forest management are far from being a finely developed management tool (Poulson et al, 2001; Rametsteiner and Simula, 2003). This is especially true in the case of ecological criteria that are invariably given less rigorous attention in the development of management standards than economic or social criteria (Bennett, 2000). One consequence of this is that the selection of many C&I is often based on the values, opinion and experience of an evaluator or auditor rather than empirical evidence (Pokorny and Adams, 2003; Guynn et al, 2004; Dudley et al, 2005). The common lack of an objective and transparent method for selecting C&I

has led to a large number of often poorly defined approaches being used in practice, resulting in a loss of confidence in C&I by many forest managers (Ghazoul, 2001), and a high level of variability among different forest certification schemes (Nussbaum et al, 2002; Pokorny and Adams, 2003; Hickey et al, 2005; Box 5.1).

Box 5.1 Disparity in (C&I) sets for SFM

The diversity of C&I sets is a source of considerable confusion and uncertainty in how monitoring programmes should be designed and implemented to measure progress towards management goals. Pokorny and Adams (2003) evaluated the diversity of five C&I sets for application to forest management programmes in the Brazilian Amazon (sourced from: CIFOR generic template; CIFOR Adaptive Collaborative Management (ACM) for the state of Pará; FSC; ITTO; and the Tarapoto Process of the Amazon Cooperation Treaty).

This study clearly demonstrated that although all the C&I sets were designed to guide and promote SFM, and addressed the ecological, social and economic dimensions of sustainability, they exhibited very different thematic foci. For example, the CIFOR set included more C&I relating to the direct evaluation of environmental impacts than all other sets put together. By comparison, the FSC set focused more strongly on the evaluation of forest operations (i.e. a systems-based management standard). The divergence in content and purpose between the C&I sets was further highlighted by comparing patterns of coincidence in indicator types used for assessment. For example, comparing the FSC and Tarapoto sets, only 27 per cent of indicator types evaluated were common to both, while FSC, ITTO and CIFOR had much higher levels of similarity, and complete concordance in up to 80 per cent of the topics. In addition, to demonstrating a high variability in thematic focus among C&I sets, the study also showed a general lack of evidence in the theoretical validity of different indicators and their practical applicability for field use. To increase objectivity and transparency, C&I need to be developed to more closely reflect logistical reality in the field and offer greater consideration of differences in management and geographical context (Pokorny and Adams, 2003).

Although these diverse C&I sets were developed with somewhat different purposes in mind, they all claim to fall under the banner of SFM (see Chapter 2). Consequently it is possible to imagine the development of very different management standards depending on the choice of the foundational C&I.

Beyond the selection of an appropriate set of C&I, an additional challenge in defining the requirements for responsible forest management is the verification of changes in indicator levels such that it is possible to determine whether a minimum level of performance (i.e. a desired indicator condition) has been achieved. The literature on forest management and monitoring systems uses terms such as 'verifier', 'benchmark', 'threshold' and 'minimum performance standard' to refer to this additional level of specificity (Prabhu et al, 1999). Verifiers add meaning, precision and often site-specificity to an indicator, providing a quantitative target or objective that managers need to achieve in order to demonstrate compliance with a given performance standard.

Despite the importance of verifiers, the science of natural resource management generally performs poorly when charged with the question of 'how much is enough' (Tear et al, 2005). Wintle and Lindenmayer (2008) have argued that the majority of existing forest management and certification schemes are deficient in their capacity to demonstrate progress towards ecologically sustainable forest management exactly because they lack measurable management objectives for biodiversity (see also Pokorny and Adams, 2003; Hickey and Innes, 2008; Villard and Jonsson, 2009). The lack of measurable objectives hinders the transparent evaluation of management performance through monitoring, and renders managers largely unaccountable for their performance. In addition, it permits leniency in the interpretation of satisfactory indicator levels, providing the potential for a rapid divergence in performance standards among different regions and management authorities.

A GROWING CRISIS OF CREDIBILITY IN THE VALUE AND PURPOSE OF MONITORING

On the surface the majority of monitoring programmes present themselves as providing information on management performance by reporting on the status and trends of biodiversity against some agreed standard (i.e. effectiveness monitoring) and/or helping to guide improvements in management that will enhance biodiversity prospects in the future (i.e. validation monitoring) (see Chapter 4 for discussion of monitoring approaches). However, there is very little evidence of where this basic approach has been successfully implemented and monitoring is fully integrated as part of a responsible forest management system (Lindenmayer, 1999; Lindenmayer and Franklin, 2002; Rempel et al, 2004; Lindenmayer and Likens, 2009). The quality of biodiversity monitoring is frequently poor, even in the case of developed countries that have well-advanced ecology and forest research programmes. For example, Lindenmayer (2003) summarized the situation for state forests in Australia by saying that '*the record on monitoring is appalling for all government agencies responsible for forest management*'. Outside of North America, Europe and Australia the situation is even worse, and the majority of biodiversity assessments in managed forest systems are often limited to rapid impact assessments.

Many monitoring programmes quickly become disengaged from their stated purpose, and the implementation of a biodiversity monitoring programme as a demonstration of compliance with regulatory or a certification standard takes precedence over the actual process of evaluating management performance. One common problem is the lack of a defensible framework for selecting monitoring indicators that are appropriate for evaluating management, such that the wrong variables are often measured in the wrong situations, and with poor precision and reliability (Noss and Cooperrider, 1994; Vos et al, 2000; Rempel et al, 2004; Nichols and Williams, 2006; Wintle and Lindenmayer, 2008; Lindenmayer and Likens, 2009).

These shortcomings mean that many biodiversity monitoring programmes are based on a surveillance type philosophy that is reactive rather than proactive and not encouraging of a capacity for adaptation and learning in management. While surveillance monitoring can help in acting as an early warning device of undesirable change, and provide a valuable mechanism for raising awareness about environmental degradation (Chapter 4), it can often do more harm than good as a tool for improving management (Box 5.2).

Box 5.2 The dangers of a surveillance-style approach to biodiversity monitoring

Many biodiversity monitoring programmes have failed to deliver guidance for improved management practices because they are considered to represent an end in themselves, and have not been purposefully designed in the context of the wider management process. As a consequence many monitoring programmes have evolved into an omnibus process of routine surveillance that collects unfocused data on a multitude of problems with the expectation that they will be useful in revealing past mistakes (Nichols and Williams, 2006). Such programmes typically rely upon a piecemeal collection of indicators that are disconnected from the management process, and in the event that they collect data that is meaningful to management, this information is often retrospective and therefore of limited use in developing cost-effective strategies for biodiversity conservation.

The results of surveillance monitoring can do more harm than good as the use of inappropriate indicators that have been selected out of context from the management process has the potential to yield unreliable information, where a set of false conclusions are mistaken for knowledge (Wilhere, 2002). The development of erroneous connections between cause and effect has been termed 'superstitious learning' (Fazey et al, 2005), and it can lead to damaging complacency in management and research. A particular danger here is the chain-reaction way in which research ideas often develop, and the fact that once foundations of unreliable or erroneous knowledge are established, future lines of enquiry may be based on these potentially untenable premises (Romesburg, 1981). This leads to the development of highly subjective and misleading paradigms regarding the effects of certain types of management strategy or intervention (i.e. the opposite of an evidence-based approach to conservation (Sutherland et al, 2004b). Moreover, the reactive nature of management under a paradigm of surveillance monitoring can mean that precaution is held as an excuse for inactivity in the face of uncertainty, and pre-emptive action is prevented by the lack of strong links between target biodiversity attributes and management practices (see Sutherland, 2006).

A steady accumulation of poorly conceived biodiversity monitoring programmes that are disconnected from the underlying management system and have limited, if any, value for enhancing forest conservation opportunities carries the danger of generating a crisis of credibility in value and relevance of the entire monitoring process (Sheil, 2002; Field et al, 2007; Lovett et al, 2007; Lindenmayer and Likens,

2009). This situation can generate a number of worrying consequences including the following:

- A squandering of resources (both money and time devoted by trained personnel) that could be valuably directed towards tackling other problems in conservation and landscape management.
- An erosion of trust in the contribution of science to management by funding bodies and political agencies. Without the agencies that fund biodiversity monitoring work witnessing an appreciable return-on-investment, the likelihood of future resources being dedicated to support long-term monitoring activities will become increasingly jeopardized. Institutional impediments to establishing and maintaining long-term funding and logistical support for monitoring remain a major challenge, and monitoring programmes are often the first budget items to be downsized when savings need to be made (Wintle and Lindenmayer, 2008).
- A loss of interest in leading conservation scientists to become involved in monitoring activities, which many perceive as mundane, regulatory processes that are not associated with cutting-edge science. This is concerning as the scientific community makes a vital contribution to biodiversity monitoring, not only through the provision of technical expertise but also frequently through matched funding and in-kind support (e.g. research students and laboratory assistants).
- A failure to capitalize on a narrow window of opportunity for biodiversity conservation in managed forest landscapes before ongoing intensification drives further and potentially irrecoverable species losses.

However, it should be recognized that designing a meaningful monitoring programme can be far from a trivial exercise. It is arguable that biodiversity monitoring is deserving of particular attention in comparison to other sustainability criteria (e.g. other environmental values such as water and carbon but also many economic and social criteria) exactly because it suffers from particular technical and institutional challenges, including:

- the inherent complexity of biodiversity, encompassing not only compositional aspects of populations, species and assemblages but also structural and functional diversity;
- the complex patterns of cause and effect that relate human activities, management strategies and biodiversity and are often associated with indirect interaction effects and feedback mechanisms;
- the need to implement biodiversity monitoring over long timescales in order to capture delayed or lag effects and species interactions that play out over intergenerational timescales;
- the need to implement biodiversity monitoring over large spatial scales to take account of source-sink and colonization processes that determine patterns of species distribution and abundance across entire landscapes;

- the high operational cost driven by the diverse responses of different elements of biodiversity to changes in management and the need to monitor multiple components; and
- a common lack of sufficient political and economic incentives to support meaningful biodiversity monitoring programmes due to a failure of regulatory or market mechanisms to recognize the value of biodiversity conservation outside of reserves.

The perpetuation of the kind of endemic problems outlined above has led many scientists to frequently view biodiversity monitoring as an unrewarding activity that involves little science (reviewed in Yoccoz et al, 2001; Nichols and Williams, 2006), and the spending of limited conservation funds on ecological monitoring programmes as a misappropriation of resources that takes away from more practical conservation initiatives (Sheil, 2001; Cleary, 2006). Despite the tremendous untapped potential of monitoring and evaluation programmes to contribute towards improved forest biodiversity conservation (Rempel et al, 2004), such disillusionment is understandable as monitoring that is not linked to management is a zero-sum game; limited resources are expended without any tangible benefit in return.

Viewing cost efficiency as central to the overall purpose of biodiversity monitoring

The limited financial resources available to conservation research (James et al, 1999; Balmford and Whitten, 2003) demand that money invested in biodiversity surveys is spent as effectively as possible. Forest biodiversity monitoring programmes hold no exception to this predicament, as the cost and logistical requirements necessary to support comprehensive monitoring are non-trivial for the majority of organizations or conservation projects (Margules and Austin, 1991; Manley et al, 2004; Gardner et al, 2008a; Fig. 5.1). However, a driving motivation for writing this book is the widespread recognition that many existing biodiversity monitoring programmes, whether in forests or elsewhere, represent poor value for money. Although some excellent programmes certainly do exist, there is a general feeling that much more could be achieved with the substantial amount of human and financial resources that have been dedicated to biodiversity monitoring (Lindenmayer and Burgman, 2005; Lindenmayer and Likens, 2009).

Limitations on resources mean that many, if not all, monitoring programmes concerned with sampling forest biodiversity are forced to trade-off the desire for increased precision and spatial coverage, with the need to ensure the feasibility and sustainability of the programme itself (Margules and Austin, 1991; Danielsen et al, 2003; Brashares and Sam, 2005; Field et al, 2005; Gardner et al, 2008a; and see Chapter 16). There is no sense in designing a monitoring programme that fails to provide reliable and representative data for the amount of resources that are available, even if the initial goals are theoretically relevant for management.

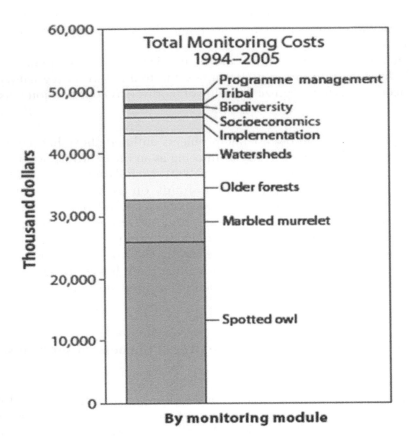

Figure 5.1 Total cost of monitoring programmes in the North West Forest Plan 1994–2005

Source: Redrawn from Rapp (2008)

Indicator choice is central to the challenge of designing cost-effective monitoring programmes. However, indicators currently used in conservation and management are often subjectively chosen in the absence of clear objectives, rely heavily on anecdotal evidence, 'expert' opinion and ease of sampling (Kneeshaw et al, 2000), and hardly ever attempt to evaluate the monetary cost of sampling different taxa using a standardized approach that accounts for differences in survey effort across different species groups (Gardner et al, 2008a). Nevertheless, different biological indicator groups can incur significantly different monitoring costs as determined by differences in field and laboratory equipment, survey time and requirements for expert identification (see Chapter 12).

The lack of a coherent cost–benefit framework for selecting cost-effective indicators (Hilty and Merenlender, 2000; Gardner et al, 2008a) coupled with the difficulties of defining efficient sampling schemes (Field et al, 2005), can either result in overly intensive programmes that cannot be sustained long enough to address questions fundamental to effective management (Danielsen et al, 2003), or in overly simplistic programmes that collect superficial data, and waste time and resources (Yoccoz et al, 2001; Margules et al, 2002). This dilemma has led to much confusion and frustration among managers and scientists alike (Hagan and Whitman, 2006), and has generated a major barrier to the implementation of effective monitoring programmes in forest management systems worldwide (UNFF, 2004; Dudley et al, 2005; Hagan and Whitman, 2006). Indeed the combination of high costs together with the observation that biological indicators often provide only delayed and ineffectual data on the consequences of management or conservation interventions, has already led some to consider that collecting species data is impractical and unsustainable (Ghazoul, 2001; Parrish et al, 2003). The failure of scientists to resolve the conflict between scientific ideals and practical reality in monitoring programmes is most evident in developing countries, where limited resources and sporadic international funding condemn many data collection efforts to failure (e.g. Danielsen et al, 2003), as well as providing an unwelcome additional barrier to achieving forest certification (Sheil et al, 2004).

THE IMPORTANCE OF DEFINITIONS AND TERMINOLOGY TO PROVIDE CLARITY OF PURPOSE

A major part of the challenge of defining the purpose and objectives of biodiversity monitoring lies in overcoming the terminology and semantics, which can readily confound attempts to explain project goals to both internal participants and outside observers. Inconsistency in language and a lack of clear definitions is a major obstacle to communication and understanding among scientists and managers (Salafsky et al, 2002; Salafsky and Salzer, 2005; Stem et al, 2005; Salafsky et al, 2008), as well as being of central importance to the production of defendable science (Murphy and Noon, 1991, Grumbine, 1994). It is difficult to imagine another problem in applied conservation science more stricken by confused semantics than SFM. Indeed, the often vague and meaningless definitions of objectives, indicators and standards have been a major barrier to the development of more responsible approaches to forest management in many parts of the world (Bennett, 2000; Ghazoul, 2001; Sheil et al, 2004).

Perhaps the most obvious example of where confusion about definitions has hindered progress in promoting ecologically responsible forest management is in the meaning of biodiversity itself (Gaston, 2009). While the difficulties in applying the concept of biodiversity to real-world management problems have been long recognized (Noss, 1990), the definition of biodiversity and biodiversity value often varies significantly among different stakeholders involved in forest management (see Chapter 10). As noted by Sheil (2001), the lack of any clear universal agreement on the meaning of biodiversity has meant that non-specialists often seek out experts for guidance. Experts are then liable to promote what they choose depending upon their own personal experience and understanding (e.g. with reference to certain taxonomic groups), and consequently generic standards do not exist. For example, it is common to see biodiversity misinterpreted as synonymous with simply species diversity or richness – a usage that trivializes the broader meaning of the term (as a hierarchical multi-scale concept that includes structural, compositional and functional elements (Noss, 1990; Fig. 5.2)), and promotes misconceptions of the nature of conservation problems (Angermeier and Karr, 1994). However, due to its inherent complexity and holistic nature, biodiversity is better considered as a 'metaconcept' similar to 'ecosystem health', and therefore by definition cannot be isolated as a specific forest attribute or management objective that should be maintained and restored (Failing and Gregory, 2003). It is arguable that biodiversity can only be considered a practically useful concept when it is clearly defined and translated into a subset of specific and meaningful indicators that can be measured and used to inform management decisions (Failing and Gregory, 2003; see Chapter 12). Nevertheless, the proliferation of terms that have been used to describe different biological indicators, together with their often ambiguous nature (see above and Table 4.2), have done little to ensure that the process of indicator selection is clear and meaningful for forest managers (Newton and Kapos, 2002; Rempel et al, 2004). Harmonizing and clarifying this terminology has been identified as one of the most important contributions that science can make towards promoting SFM (Buck and Rametsteiner, 2003), and a major focus of this book (e.g. see Chapters 4 and 8).

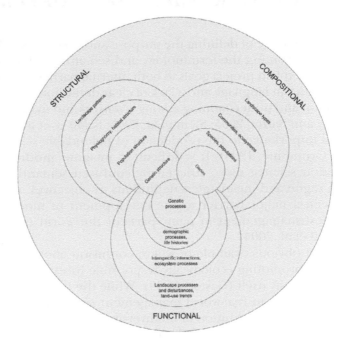

Figure 5.2 The hierarchical definition of biodiversity

Source: Redrawn from Noss (1990)

CHAPTER 6

Challenges to Monitoring: Problems of Design

Synopsis

- When designing a biodiversity monitoring programme equal attention should be paid to the relevance of expected sample data for management and the choice of indicators as to questions of sampling design and analysis.
- Measuring changes across different levels of a cause–effect chain, from changes in ultimate stressors (e.g. management practices), to changes in proximate stressors (e.g. forest structure and function) and changes in biodiversity, is accompanied by an increase in cost and technical difficulty as well as increased uncertainty in the interpretation of sample data.
- Structural indicators support a coarse-filter approach to forest management and are relatively cheap and quick to measure but offer no guarantee that conservation targets have been achieved. There are relatively few empirical tests of the importance of specific structural indicators for biodiversity.
- Species data represent the highest standard for evaluating progress towards conservation goals, yet many types of species data and biological indicators do not provide reliable sources information for management. A lack of threat information and persistent sampling difficulties mean that endangered species are rarely a viable option for monitoring. There is little empirical evidence to support the concept of a biodiversity indicator that is capable of capturing changes in other species. There is similarly little scientific support for the focal species as a practically useful concept for monitoring. Ecological disturbance indicators are useful for monitoring because their dynamics can be linked to management impacts and underlying changes in ecological condition, yet they are rarely chosen on the basis of standardised selection criteria.
- Baseline measurements of indicator variability at reference sites are essential for interpreting management impacts on biodiversity but are fraught with both practical and technical difficulties.

Once clear goals and objectives have been decided, a biodiversity monitoring programme needs to be carefully designed to ensure that it serves its intended purpose. A well-intentioned monitoring programme depends upon three design factors in order to be successful: (i) the selection of appropriate and effective indicators to report on management performance and progress; (ii) the application of a data collection strategy that provides reliable and representative sample data; and (iii) the appropriate interpretation of sample data and changes to indicator values such that a fair and comparable evaluation of management performance is possible.

Research attention has traditionally been focused on questions concerning sampling design and data collection. While sampling issues are obviously important (see Chapter 14), more fundamental and arguably less well understood concerns relate to the type of indicators that are chosen in the first place, and the way in which any observed changes in sample data should be interpreted. This is partly due to the fact that many of the problems involved in selecting appropriate indicators and interpreting changes in their values are not purely technical, but involve subjective judgements regarding the underlying value and purpose of monitoring. This chapter analyses some of the major obstacles that have challenged past attempts to select indicators for monitoring, as well as develop effective approaches to interpreting changes in biodiversity.

In order to select performance indicators that are both ecologically meaningful and practically feasible, we first need to understand the relative advantages and limitations of different alternatives within a causal framework of management impacts and biodiversity responses. The first section of this chapter identifies some of the important issues associated with the role played by both structural and biological indicators in biodiversity monitoring programmes. Particular attention is given to the problem of deciding between different species-based measurements (biological indicators *sensu lato*, but also individual species that may provide a target for management) that can provide the basis for monitoring. In the second half of the chapter, I outline some of the major challenges that face the setting of conservation objectives for responsible forestry and the interpretation of any changes to indicator values that is needed to assess management performance against such objectives. Much of this discussion focuses on the difficulties involved in selecting and quantifying reference sites that can provide a benchmark for evaluating any observed changes in biodiversity.

THE CHALLENGE OF SELECTING APPROPRIATE INDICATORS FOR BIODIVERSITY MONITORING

Despite the undisputable importance of indicators in biodiversity monitoring, there is much controversy and frustration among forest managers and stakeholders concerning their utility (Lindenmayer et al, 2000, 2002a; Failing and Gregory, 2003; Hagan and Whitman, 2006; and see Chapter 5). Indeed, the challenge of selecting indicators can often seem overwhelming and commonly results in 'indicator fatigue' among practitioners and scientists alike (Hagan and Whitman, 2006). This confusion has often resulted in the selection of indicators that are not only inconsistent among neighbouring sites (Rametsteiner and Simula, 2003), but also inappropriate for the task in hand (Lindenmayer et al, 2000; Guynn et al, 2004).

All indicators have their drawbacks yet some are better suited for certain tasks than others, and consequently the indicator selection process is one of the most crucial components of a biodiversity monitoring programme (see Chapter 5). Sampling at different levels of a causal framework, from ultimate stressors (e.g. management practices), to proximate stressors (e.g. changes to structural and functional attributes, including both habitat and landscape elements, as well as changes to major disturbance regimes, and like fire) changes in valued ecological attributes (e.g. species, populations, ecological processes), is accompanied by an increase in both cost and technical difficulty and in uncertainty when interpreting the effects of management. The more disengaged the system attributes that form the focus of a monitoring programme become from the actual management process, the more difficult it becomes to trace patterns of cause and effect.

Challenges in selecting structural indicators

Because of their ease of measurement and clear linkages to human impacts, indicators of key habitat and landscape structures have generated considerable interest, both as practical surrogates for biodiversity and as a key to understanding how to maintain biodiversity and ecological integrity in forest ecosystems (Parrish et al, 2003; McElhinny et al, 2005; and see Chapters 9 and 11). The underlying rationale is that landscapes containing forest stands with a diversity of structural attributes are considered more likely to include a variety of resources and species that utilize these resources, with the effect that there is often a positive association between elements of biodiversity and measurements and indices of structural complexity (e.g. Huston, 1994; Mac Nally et al, 2001; Tews et al, 2004; Box 6.1).

This use of habitat and landscape attributes to identify priorities for conservation represents a coarse-filter approach to management (Noss and Cooperrider, 1994; see Chapter 4), and reflects the well-established conviction that we will fail in achieving our goal of preserving biological diversity, let alone ensuring sustainability, if we focus primarily on species-specific management strategies (Franklin, 1993; Noss, 1996). Coarse-filter approaches to habitat assessment are attractive because of their cost-efficiency, but also the belief that they can be applied over large areas (including sites that are inhabited by different species) and therefore provide some kind of 'common currency' of conservation value that can be compared across sites (e.g. through forest certification schemes). This understanding has inspired the development of novel policy instruments for conservation, such as biodiversity offsetting and mitigation banking (Parkes et al, 2003; Box 6.2).

It is often assumed that monitoring forest structural indicators obviates the need to monitor elements of biodiversity directly. Indeed, many (if not the majority) of the indicators that are used in practice to evaluate forest management initiatives do not include any information on actual species and populations (Dudley et al, 2005). However, the extent to which structural indicators adequately reflect changes in the underlying attribute of concern is not always clear, and even where it is, the presence of suitable habitat is no guarantee that species of interest are actually present (Noss, 1990; Carignan and Villard, 2002; Hanksi, 2005; Box 6.2). Without direct measurements of biodiversity in the form of species and populations, an isolated focus on structural indicators carries the risk of inadvertently conserving 'empty and silent

Box 6.1 The importance of keystone structures in determining patterns of faunal diversity

The idea that habitat heterogeneity begets species diversity lies at the foundation of modern ecology. The basic hypothesis is that structurally complex habitats may provide more niches and diverse ways of exploiting environmental resources, thereby increasing local biodiversity. For much of the terrestrial world, plants play a fundamental role in structuring the physical environment, and therefore have a significant influence on the distribution and abundance of animals. However, despite considerable research showing a positive relationship between structural environmental complexity and faunal diversity in a variety of ecosystems and at a variety of scales, we still have a poor understanding of the relationship between different forms of environmental heterogeneity and biodiversity.

Tews et al (2004) recently reviewed the scientific literature on the habitat heterogeneity–animal species diversity relationship and evaluated uncertainties and biases in its empirical support. They collated 85 publications for the period 1960–2000, identifying each study by the type of heterogeneity that was described (Figure 6.1), the animal species group and ecosystem studied, and the spatial scale of the study. The analyses showed that the majority of studies found a positive correlation between habitat heterogeneity/diversity and faunal diversity, although much of the support for this relationship came from studies of vertebrates and human-modified systems. The meta study also showed that effects of heterogeneity vary depending on both the focal species group and the particular structural attribute that was measured.

An important conclusion of the review by Tews et al (2004) is that the ecological effects of habitat heterogeneity are scale dependent and can vary considerably among species groups depending on whether structural attributes are perceived by organisms as patch-scale variability in habitat condition or whole landscape fragmentation. This distinction highlights the importance of understanding the extent to which anthropocentric interpretations of landscape structure bias ecological studies of species-resource requirements (e.g. Manning et al, 2004; Fischer and Lindenmayer, 2006).

In many of the studies reviewed by Tews et al (2004), different species groups were seen to be closely linked to 'keystone structures' that have a particular influence in determining species presence (e.g. temporary flooded areas or large emergent trees). Detecting crucial keystone structures of the vegetation has profound implications for nature conservation and biodiversity management (see Chapter 12).

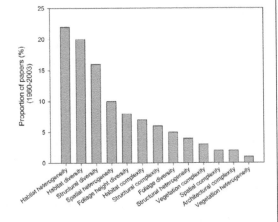

Figure 6.1 Research on the relationship between habitat heterogeneity and species diversity

Source: From Tews et al (2004)

Source: Text based on Tews et al (2004)

Box 6.2 Biodiversity offsets and the habitat hectares method

Assessments of the 'quality', 'condition' or 'status' of stands of native vegetation or habitat frequently represent an essential component of conservation assessments and management plans. However, even when soundly based upon ecological principles, these assessments are usually highly subjective and involve implicit value judgements. One more objective approach that has been recently proposed is the habitat ha (HH) method that is currently employed in Victoria, Australia, for biodiversity offset design as well as elsewhere in the world, (see www.forest-trends.org/biodiversityoffsetprogramme/). The HH method (Parkes et al, 2003) is based on explicit comparisons between existing vegetation features and those of 'benchmarks' sites, representing the average characteristics of mature stands of native vegetation of the same community type in a 'natural' or 'undisturbed' condition. Components of the index incorporate vegetation physiognomy and critical aspects of viability (e.g. degree of regeneration, impact of weeds) and spatial considerations (e.g. area, distribution and connectivity of remnant vegetation in the broader landscape). The HH method represents a coarse-filter approach that does not seek to provide a definitive statement on conservation status or habitat suitability for individual species.

   Biodiversity is a devilishly complex concept and by definition is not amenable to collapsing into a single index. As such the HH approach will benefit from further research and ongoing refinement by scientists and practitioners (McCarthy et al, 2004). Particular difficulties lie in identifying components of habitat quality that are vital in ensuring the maintenance of biodiversity, and which cannot be substituted for other elements. As in the case of all conservation applications of structural indicators of biodiversity, the HH method represents good practice based on current scientific understanding of the relationship between habitat structure and ecological condition, including the persistence of native populations and species. However, to ensure that conservation goals are met, structural indicators need to be subject to a continuous process of validation using direct tests against the distribution and abundance of species.

Source: Text based on Parkes et al (2003) and McCarthy et al (2004)

forests' (*sensu* Redford, 1992) – where the forest may appear relatively intact from satellite images or aerial photographs but is devoid of much of its original wildlife.

   This dilemma highlights the fact that 'means' and 'ends' are often confounded in the development of management strategies for biodiversity conservation (Noss and Cooperrider, 1994; Failing and Gregory, 2003). The preservation of selected structural attributes through particular management interventions, or the maintenance of structural indicators above some minimum target or threshold, does not guarantee the protection of ecological integrity or the conservation of biodiversity. Rather, it only informs about the *likelihood* that certain biological attributes (e.g. native species) will be present (Tews et al, 2004; Dudley et al, 2005).

   No definitive suite of structural indicators at stand and landscape scales exists for either temperate or tropical forests (McElhinny et al, 2005; Similä et al, 2006);

different researchers have emphasized different subsets of attributes, and relatively few studies provide quantitative evidence demonstrating the importance of any particular attribute for patterns of biodiversity (Noss, 1999; Lindenmayer et al, 2002a; McAlpine and Eyre, 2002; Hanski, 2005).

Ultimately, coarse-filter approaches to conservation that depend on habitat and landscape attributes and structures require further empirical testing (Lindenmayer et al, 2000; Groves et al, 2002) and a direct reference to the needs of resident species (Lambeck,1997, 2002). In order to ensure progress towards ecologically responsible forestry, management for the conservation of biodiversity should ideally be based on building from what we already know about species, rather than starting with untested anthropocentric assumptions about the way species perceive environmental variation (see Brooks et al, 2004a, 2004b; Fischer and Lindenmayer, 2006). Despite these concerns, managers today still require a working and defensible means by which performance can be assessed, and structural indicators provide a pragmatic and ready solution to this problem. Working to continuously improve our understanding of the relationship between the structure of forest ecosystems (and hence the validity of structural indicators) and biodiversity represents a central task for biodiversity monitoring.

Challenges in selecting biological indicators

Biodiversity conservation will ultimately succeed or fail at the level of populations and species (Balmford et al, 2003). Consequently, analyses based directly on species and populations represent the highest possible standard for assessing management performance and ecological objectives (Noss, 1990; Kremen, 1992; McGeoch, 1998; Lindenmayer et al, 2002a). Without explicit data on the diversity and distribution of species, our ability to understand the effectiveness of different management strategies (Dudley et al, 2005), or to optimize the location of high-value conservation areas (Jennings et al, 2003; Brooks et al, 2004a, 2004b), will continue to remain primitive. However, because we cannot study all species, it is necessary to focus on evaluating the consequences of management for a select number. This includes target species that represent a focus of particular conservation concern (e.g. rare, threatened species), and/or species that are used as indicators for wider patterns of ecological integrity or biodiversity (see Chapter 4 for a typology of indicators).

Resource and logistical constraints mean that it is invariably preferable to focus on achieving a reliable understanding of a relatively small number of species than to gather superficial and potentially biased information on a great many (Lindenmayer, 1999; McGeoch, 2007). Nevertheless, it is important to recognize that irrespective of the number of taxa it is possible to include as part of a monitoring programme, they will always only ever represent a tiny fraction of the total biota in any forest on Earth. Thus, while the use of a large number of species from a wide range of taxonomic and functional groups can help cover a greater diversity of species' responses to disturbance, significant assumptions are still needed in drawing general conclusions about the overall state of the system (see also Chapter 12).

As discussed in Chapter 4, there is often considerable philosophical resistance to the concept of a biological indicator by conservationists. For example, Voss and Borchardt (1992, quoted in Turnhout et al, 2007) argued that indicators:

> '... *give the impression that we know all about biological processes and their interaction. However, that is not (and probably will never be) the case and a definition of these values just shows how arrogant the human attitude is towards nature.*'

While it is easy to be sympathetic to the position that we will never know enough to properly understand and manage nature, it is essential that we adopt a transparent and pragmatic approach to evaluating the impacts of human activities on biodiversity if we are to make progress in conservation. Achieving such an approach requires a focus on elements of biodiversity for which we are able to collect representative information that can be reliably interpreted in the context of management.

While there has been a huge amount of debate and conjecture on the application of different types of species-based indicators in conservation, there has been comparatively little empirical research that can provide evidence in support of different approaches for a given application (McGeoch, 2007). The lack of a defensible evidence basis for selecting biological indicators in so many biodiversity monitoring programmes has meant that many indicators are used inappropriately (Caro and O'Doherty, 1999; Storch and Bissonette, 2003). This has resulted in the creation of disconnected monitoring cycles, which are only peripherally linked to management activities (Kneeshaw et al, 2000, Rempel et al, 2004; and see Chapter 5). Failing and Gregory (2003) go further and suggest that there are often so many mistakes made in the selection of biological indicators that their application can sometimes be counter-productive, while Simberloff (1998) concludes that monitoring programmes are likely to fail altogether if inappropriate indicator taxa are selected.

Here, I briefly review some of the salient problems that are associated with the use of different approaches to the selection of biological indicators, namely: threatened species (although they should not be strictly considered as indicators they are included here because of their common usage); biodiversity surrogate schemes (biodiversity disturbance indicators, umbrella species and focal species and guilds); and ecological disturbance indicators (see Chapter 4 for a full typology of different types of biological indicator).

Threatened and endangered species

Public pressure has persistently focused attention on the protection of selected taxa of conservation concern. As a consequence, many endangered species have attained indicator status within management performance standards by default, due either to moral or legal obligations for their conservation (Pearson, 1994). A focus on rare, threatened and endangered species is therefore a very common feature of biodiversity monitoring programmes. This is most clearly evidenced by their appearance in many sets of C&I, and forest certification standards (Holvoet and Muys, 2004; Box 6.3).

However, the last two decades of research have witnessed a growing consensus that a piecemeal approach to conservation that is focused primarily on specific elements of threatened diversity, in contrast to a more comprehensive effort to protect the integrity of the underlying ecological systems is both inadequate and wasteful (Franklin, 1993; Angermeier and Karr, 1994; Noss, 1996; Ghazoul, 2001). Moreover, the fact that we have very little understanding of the resource and habitat requirements of the vast majority of threatened forest species casts serious doubt on their suitability as a basis for monitoring management effectiveness (Box 6.3).

Box 6.3 Requirements for the protection of rare and endangered species within a regional certification standard

Criterion 6.2 of the FSC global standard on environmental impact requires that 'Safeguards shall exist which protect rare, threatened and endangered species and their habitats (e.g., nesting and feeding areas)'. This criterion forms the basis for articulating specific management guidelines and requirements for protecting rare and endangered species within individual regional and national forest management standards. One example is the FSC National Boreal Standard for Canada (FSC Canada, 2004) – a country which was been the focus of considerable research attention in forest ecology and conservation. In the Canadian standard, Criterion 6.2 is represented by a set of seven indicators and 23 associated verifiers (Table 6.1).

Table 6.1 Criterion 6.2 on environmental impact in the Canadian National Boreal Forest Standard

Criteria	Description	Verifier
6.2.1	A list of the species at risk (as identified by federal, provincial, and regional legislation/lists) known or believed to exist within the forest is presented in the plan or associated documents and is updated annually. Where a current regional list does not exist, the applicant consults with appropriate sources of information, experts, or knowledgeable individuals to generate such a list.	Annually updated lists of species of concern as presented in the plan or associated documents.
6.2.2	Habitats of species at risk (as identified by federal, provincial, and regional legislation/lists) known or believed to exist within the forest are identified by field surveys or other means and delineated on maps.	Maps showing habitat of species at risk. Documentation of the means by which maps were developed. Records of consultations with indigenous peoples, local trappers and others with knowledge of local wildlife.
6.2.3	The applicant identifies whether and how landscape-level management is accommodating the habitat needs for regional species at risk identified through indicator 6.2.1.	Results of analyses related to adequacy of landscape-level management in addressing habitat needs of regional species at risk.

Table 6.1 Criterion 6.2 on environmental impact in the Canadian National Boreal Forest Standard (continued)

Criteria	Description	Verifier
6.2.4	Plans exist, or are under development, to protect the habitat and populations of species at risk in the forest. These plans cover those species on provincial and federal lists identified through indicator 6.2.1, and those species on regional lists identified indicator 6.2.3 for which landscape-level management does not adequately address habitat requirements. The plans are authored by qualified individuals with expert input. The plans include the establishment and use of conservation zones where appropriate. The applicant is involved in plan implementation, or respects and cooperates with the implementation of the plans	Protection plans for species and habitat or a development schedule for plans Records of activities undertaken under the plans
6.2.5	Where plans identified through Indicator 6.2.4 do not exist or are incomplete or inadequate, a precautionary approach is used in management of the habitats of the relevant species at risk	Review of precautionary measures Comparison of approaches and levels of activity in neighbouring, similar forests Results of habitat modelling for relevant species, where it has been undertaken
6.2.6	The applicant provides training to all relevant forest workers on the identification of species at risk, and on appropriate measures to take when a species at risk, or sign of a species at risk (e.g., a nest), is identified during field operations	Training materials related to species at risk Training records Interviews with employees and contractors
6.2.7	The applicant cooperates fully with resource management agencies in the efforts to control illegal hunting, trapping and fishing of all species, and in accordance with the land-use planning decisions and strategies in the forest management plan	Evidence of cooperation Field inspection examining for evidence of control measures (e.g., road closures, signage, patrols by conservation officers) Interviews with conservation officers to determine the extent of effort to control Comparison of sections in the management plan on assistance to enforcement agencies in plans with actual activities undertaken by applicant

Source: From FSC Canada (2004)

What is clear from this standard is that while the criterion requires managers to 'protect rare, threatened and endangered species and their habitats', many of the indicators used to demonstrate management performance towards this objective are highly superficial and largely desk-based descriptive exercises (e.g. the requirement to demonstrate the existence of species and habitat lists, maps and conservation plans). Threatened species lists are not designed for reporting on the state of the environment and are likely to perform poorly for this purpose (Possingham et al, 2002). Where evidence of actual management effectiveness is required, the verifiers used are vague – for example, Criterion 6.2.3: *Results of analyses related to adequacy of landscape-level management in addressing habitat needs of regional species at risk.* The lack of specific and quantifiable biodiversity management objectives hinders the transparent evaluation of management, and makes it difficult to hold managers accountable for any perceived performance failure (Wintle and Lindenmayer, 2008).

Where knowledge of the habitat requirements of threatened species does not exist, the management standard requires the development of appropriate research into species and habitat conservation plans (Criterion 6.2.4). While such plans remain under development, a precautionary approach is mandated, where guidance depends upon work conducted elsewhere and on modelling (Criterion 6.2.5).

The main arguments against the use of threatened species as the primary means for evaluating forest management include the following:

- In the majority of countries (especially within the tropics) a focus on endangered species is entirely unfeasible as data on levels of rarity and conservation status are unavailable for the majority of species.
- Where information identifying threatened species is available, the often cryptic or rare nature of many species of conservation concern means that monitoring programmes are particularly costly and ineffective (Hilty and Merenlender, 2000; Finegan, 2005).
- A focus on endangered species is unlikely to provide data that can reveal wider patterns of environmental degradation, as endangered species are often threatened exactly because they suffer from species-specific threats (e.g. disease, overhunting) (Hilty and Merenlender, 2000), and specific resource requirements mean they are also likely to have distinct and idiosyncratic distribution patterns (e.g. Grenyer et al, 2006).
- Species that are classified as threatened are often done so on the basis of proxy threat criteria, such as rarity and endemism, that may not actually equate to genuine levels of species threat. In fact, sometimes species that are locally common are the most vulnerable to major human drivers of biodiversity loss (R. Pardini, *personal communication*).
- Areas characterized by high species diversity for multiple taxa are not necessarily associated with the occurrence of rare or endangered species (Possingham et al, 2002; Pearman and Weber, 2007; Grundel and Pavlovic, 2008), and similarly a focus on such species in conservation planning has been shown to be insufficient in ensuring the protection of overall species diversity (Bonn et al, 2002). In addition to differences in ecological requirements, the lack of

congruency with overall patterns of diversity displayed by many endangered species is partly due to the fact that the sample data from the study of such species, which are often difficult to detect or which infrequently inhabit suitable sites, are likely to be dominated by false-negatives (Thomson et al, 2005). This means that it is very easy to erroneously conclude that a species does not occupy a given site (thereby resulting in a failure to identify that area as worthy of management attention).

- Taxonomic instability and the problem of taxonomic inflation (Isaac et al, 2004; Mace, 2004) confounds our ability to monitor patterns of habitat condition or management performance using species-specific measures of conservation status. The constant state of flux that characterizes the taxonomy of many groups has led various researchers to strongly criticize the use of data on patterns of threat status as a valuable tool for monitoring and planning (Possingham et al, 2002; and reviewed in Mace, 2004).

The conservation of rare and threatened species and habitats is undoubtedly a primary goal of biodiversity conservation, and should remain as such. However, the requirements in terms of both data and the theoretical understanding that are necessary to demonstrate the adequate protection of threatened species, mean that it is rarely possible for the type of criteria outlined in Table 6.1 to be reliably employed as a transparent basis for evaluating management performance. Where data are collected on the status and trends of threatened species, this information may be important in highlighting the need for urgent and focused conservation attention following cases of population decline or local extinction. However, unless the cause of such population changes can be reliably linked to management actions, such data will be of limited used for monitoring effectiveness.

Biodiversity disturbance response indicators

Considerable attention in forestry, ecological and conservation research has been focused on the potential role of biodiversity disturbance indicators (i.e. taxa whose responses to management impacts are congruent with multiple other groups; see Chapter 4) in biodiversity monitoring programmes in general. The concept of biodiversity disturbance indicators as species or species groups that are able to provide reliable information regarding management impacts on unsampled taxa is extremely appealing and is persistently recommended as a priority criterion for indicator selection (e.g. Pereira and Cooper, 2006).

Early work tending to support the validity of species-based biodiversity disturbance indicators was valid largely because analyses were made at scales that are not relevant to the ecological interactions of individual taxa (i.e. regional and biogeographical scales) (Prendergast et al, 1993; Pawar et al, 2006; McGeoch, 2007; Pearman and Weber, 2007; Caro, 2010). Biodiversity indicators at this scale are valuable for large-scale conservation planning and priority setting but are of limited value for monitoring purposes. Subsequent empirical research of the relationship between patterns of species richness and diversity at scales relevant to management has failed to find convincing evidence of a high multi-taxa surrogacy value for any single group of species in either temperate or tropical systems (Oliver and Beattie, 1996; Prendergast and Eversham, 1997; Lawton et al, 1998; Oliver et al, 1998; Jonsson and Jonsell, 1999; Wolters et al, 2006; McGeoch, 2007; Similä et al, 2008; Table 6.2). The conclusion of this research is not that individual species groups are unable to convey any valuable information about other taxa; they can, but rather that no silver-bullet taxonomic group exists.

Focal species

The utility of the focal species concept as developed by Lambeck (1997; 2002) and Noss (1999, see Chapter 4) has received strong criticism itself, primarily because of the fact that we lack sufficient knowledge (regarding species distribution, abundance, dispersal ability and sensitivity to threatening processes) to implement such an approach for the majority of situations (Lindenmayer et al, 2002b; Lindenmayer and Fischer, 2003).

Empirical tests have also found that focal or umbrella species generally perform poorly as some species are inevitably limited by ecological factors that are not relevant to the focal taxon (Noss, 1999; Roberge and Angelstam, 2004). In theory, evaluating habitat requirements for a range of species that represent very different life history strategies allows this problem to be overcome. However, in reality the choice of focal taxa is often arbitrary (Andelman and Fagan, 2000), or based on their familiarity to individual researchers or stakeholders (Caro and O'Doherty 1999; Lindenmayer and Fischer, 2003). A lack of sufficient data, coupled with the role of such subjective factors and/or alternative agendas in the selection process, means that focal species often fail to perform effectively with respect to their initial purpose. For example, top predators often represent a popular choice of focal species as their conservation is often assumed to ensure the protection of the entire biodiversity of their supporting ecosystem. While there is strong evidence to support the role of predators as flagship species (i.e. species capable of attracting considerable stakeholder attention and conservation funding e.g. lions, jaguars and tigers), there are currently very few quantitative tests of their efficacy as surrogates of the distribution and abundance of other species (Caro et al, 2004; Sergio et al, 2008).

One adaptation of the idea that may hold promise is the 'landscape species' concept as proposed by Sanderson et al (2002) which provides a conceptual, spatially explicit methodology for thinking about how the biological requirements of focal species interact with requirements for human use in the same landscape. This approach is currently being trialled by the Wildlife Conservation Society in a number of sites around the world.

Ecological disturbance indicators

Ecological disturbance indicators are particularly important in forest biodiversity monitoring programmes because their dynamics are linked to specific management interventions and underlying changes in environmental condition (Kremen et al, 1994; McGeoch, 1998, Pearce and Venier, 2006; McGeoch, 2007). For example, understorey insectivorous birds are known to be particularly sensitive to forest

Table 6.2 Studies of multi-taxa responses to landscape change in tropical forests

Study	Taxonomic groups	Land-use types and landscape features	Multi-taxon congruency* Species richness†	Community structure ‡
Lawton et al. (1998)	Birds, butterflies, flying beetles, canopy beetles, canopy ants, leaf-litter ants, termites and soil nematodes	Old-growth, secondary forest, plantations forest, clear-cut	11%	
Schulze et al. (2004)	Trees, understorey plants, birds, butterflies, and dung beetles	Old-growth, secondary forests, agroforestry systems and annual crops	48%	
Harvey et al. (2006)	Birds, bats, butterflies and dung beetles	Riparian and secondary forest, forest fallows, live fences, and pastures	20% (birds – 38%)	
Faria et al. (2007)	Ferns, frogs, lizards, birds and bats	Old-growth, shade cacao plantations		25% (birds – 37%)
Barlow et al. (2007)	Bats, birds, dung beetles, epigeic arachnids, fruit flies, fruit-feeding butterflies, grasshoppers, large mammals, leaf-litter amphibians, lizards, moths, orchid bees, scavenger flies, small mammals, trees and lianas	Primary forest, secondary forest and *Eucalyptus* plantations	22% (lizards – 44%)	39% (butterflies – 56%)
Bassett et al. (2008)	21 focal taxa from across seven orders of arthropod: Coleoptera; Diptera; Hemiptera; Hymenoptera; Mantodea; Neuroptera; Orthoptera	Old and young secondary forest, savannah clearances, and garden cropland	3%	35%
Kessler et al. (in press)	Trees, lianas and herbs, epiphytic liverworts, birds, butterflies, lower canopy ants and beetles, dung beetles, bees and wasps and their parasitoids	Mature forests and three types of cacao agroforests	0–17%	12–18%
Uehara-Prado et al. (2009)	Fruit-feeding butterflies and five other orders/families of arthropod (Araneae, Carabidae, Scarabaeidae, Staphylinidae, and epigaic Coleoptera excluding aforementioned families)	Undisturbed old growth and "disturbed" forest	13%	66%
Pardini et al. (2009)	Ferns, trees, frugivorous butterflies, leaf-litter frogs and lizards, bats, small mammals and birds	Second-growth forests, shade cacao plantations and interiors and edges of large and small mature forest remnants	0.18%	

Notes: * Mean correlation coefficient of species richness responses
† Mean correlation coefficient of community structure responses (based on Mantel tests of similarity matrices)
‡ Number and taxon in brackets denotes the level of cross-taxon response congruency for the highest performing single taxon
Source: Redrawn with permission from Gardner et al (2009)

degradation and fragmentation in tropical forests (Barlow et al, 2002; Sekercioglu et al, 2002), and changes in these taxa can be used to gauge changes in forest condition following particular management interventions (e.g. Mason, 1996). Furthermore, insectivorous birds are very well studied in many parts of the world and are known to perform key ecological functions and processes (Sekercioglu, 2006), meaning that the ecological implications of changes to their distribution and abundance can be interpreted with a lot more confidence and utility than is possible for many other species groups.

While it is arguable that all species have inherent biological and ecological features that warrant evaluation and monitoring, the majority of studies that have recommended particular taxa as being suitable ecological indicators have been based largely on anecdotal evidence or expert opinion (McGeoch, 1998, 2007; Kneeshaw et al, 2000; Carignan and Villard, 2002). In many cases, ecological indicators are selected primarily on the basis of their ease of sampling (e.g. songbirds and 'large butterflies', Stork et al, 1997; and see Noss and Cooperrider, 1994). Reliance on the opinion (expert or otherwise) of a small group of individuals, rather than employing an objective selection process (e.g. Kremen, 1992; Dufrene and Legendre, 1997), is an unsatisfactory approach, as the assumptions and beliefs are not documented and available for questioning by others (Sutherland, 2006). Where certain taxa have been identified as suitable on the basis of defendable criteria, few studies have conducted rigorous empirical tests of their sensitivity to gradients of environmental disturbance, and almost none have simultaneously evaluated the performance of key selection criteria across multiple taxa (McGeoch, 1998; Caro and O'Doherty, 1999; Carignan and Villard, 2002). Far more common is it for individual studies to focus on emphasizing the advantages of a particular taxon in isolation, resulting in different lobby groups, each advocating the use of their own study taxon as a suitable ecological indicator of disturbance (Carignan and Villard, 2002), including: Lepidoptera (Kremen, 1992, 1994; Brown, 1997); tiger beetles (Pearson and Cassola, 1992); carabid beetles (Dufrene and Legendre, 1997; Niemelä, 2000; Rainio and Niemelä, 2003); dung beetles (Halffter and Favila, 1993; Spector, 2006); ants (Andersen and Majer, 2004); spiders (Cardoso et al, 2004); and birds (Bibby, 1999; Venier and Pearce, 2004).

Despite the likely merits of many species groups for forest monitoring and evaluation programmes, the selection of taxa that perform *optimally* in relation to key criteria can only be achieved through systematic multi-taxa field tests. Cost-effective ecological disturbance indicators (i.e. those that provide ecologically meaningful data yet are relatively cheap to sample) have seen little application in biodiversity monitoring programmes to date (Gardner et al, 2008a; Chapter 12).

Finally, it is important to note that a species group that shows a high value as an ecological disturbance indicator may perform poorly as an indicator of change in other taxa (Gardner et al, 2008a). Different species and species groups possess different niche requirements, and their response to particular management impacts or a given set of environmental conditions, may vary considerably (Lindenmayer et al, 2002b; Manning et al, 2004; Fischer and Lindenmayer, 2006), while also being strongly scale-dependent (Storch and Bissonette, 2003; Cushman and McGarigal, 2004; Bani et al, 2006; Hess et al, 2006). This conflict highlights an important obstacle that has consistently impeded the indicator selection process – trying to satisfy too many requirements simultaneously. Unfortunately, there are no easy solutions when it comes to biological indicators (Landres, 1983; Lindenmayer and Fischer, 2003; Storch and Bissonette, 2003) and attempts to address multiple monitoring objectives under the remit of a limited set of indicator species invariably leads to the collection of superficial data that is of limited utility for any purpose. A more detailed discussion of the importance of prioritizing indicator selection criteria in tackling particular objectives is given in Chapter 12.

SETTING MANAGEMENT OBJECTIVES AND INTERPRETING INDICATOR CHANGE IN BIODIVERSITY MONITORING PROGRAMMES

Once a set of appropriate indicators have been chosen, the quantification of management objectives in the form of minimum targets or standards of indicator condition is the next step in the assessment of management performance.

Conservation management objectives should be decided based on stakeholder values, the amount of available resources and limitations in scientific knowledge regarding the conditions necessary for achieving a given objective. The difficulty lies in balancing these different contributing factors, and in particular setting the lofty goal of ecological sustainability against frequently inflexible economic constraints. Even setting aside the issue of how to resolve conservation trade-offs when reviewing data from monitoring programmes, identifying minimum requirements for biodiversity persistence as a basis for management is fraught with problems. Central to this challenge is the need to understand levels of biodiversity (whether relating to populations, species or habitats) in the absence of human impacts and forest management activities i.e., the baseline or reference condition.

An ecological system is considered to have high integrity when its dominant biological and ecological characteristics (composition, structure, and functional elements and ecological processes) occur within their natural limits of variation, and can withstand or recover from the majority of natural or anthropogenic perturbations (Parrish et al, 2003; and see Chapter 9). The rational behind this statement is that species are likely to be best adapted to disturbance effects that most closely reflect the natural disturbance regime within which they evolved (Franklin, 1993). Knowledge of undisturbed or little-disturbed ecological baselines is therefore necessary for understanding the consequences of human impacts on natural systems (Arcese and Sinclair, 1997; Ghazoul and Hellier, 2000). The measurement of indicator variability in a biodiversity monitoring programme can only make sense in the context of reference points or baselines against which the ecological significance of any change or pattern can be determined, and management performance be assessed and evaluated (Kremen et al, 1993; Bibby, 1999; Di Stefano, 2001; Carignan and Villard, 2002; Niemi and McDonald, 2004).

In keeping with these arguments, many monitoring agencies (Niemi and McDonald, 2004), conservation projects (Landres et al, 1999) and forest certification authorities (Nussbaum and Simula, 2005; Higman et al, 2005) prefer or require the establishment of a quantitative benchmark for measuring and regulating change in ecosystems.

Benchmarks generally require both a measure of the average condition of the indicator in an undisturbed reference area, as well as an associated measure of variability to account for natural levels (or 'thresholds') of spatial and temporal heterogeneity (Stork et al, 1997; Ghazoul and Hellier, 2000; Yamasaki et al, 2002; Finegan, 2005). The use of critical limits of natural variability to define baselines reflects the precautionary principle (Newton and Oldfield, 2005), where natural conditions are taken as the best proxy of minimum requirements to ensure ecological sustainability. Management successes and failures are measured through a statistical analysis of trends in indicators, and judgements of progress towards 'sustainability' are made based on whether selected indicators are acceptably maintained within or close to critical-limits of (pre-disturbance) natural variation within the landscape (Stork et al, 1997; Prabhu et al, 1999; CIFOR, 1999; Ghazoul and Hellier, 2000; Di Stefano, 2001; Karsenty and Gourlet-Fleury, 2006; see Chapter 15).

Benchmarks or standards of natural variability are considered to be useful in monitoring forest management practices, because they are thought to: enable management to be identified as effective or not; improve the interpretation of regulatory and certification requirements; and promote transparency and coordination in the quality and performance of management and monitoring across different sites (Spellerberg and Sawyer, 1996). This definition of baselines is supported by a general desire among resource managers for certainty in decision-making (Ludwig et al, 1993; Richter and Redford, 1999) and the requirement for clarity in performance standards by management authorities (see Nussbaum and Simula, 2005). Despite strong arguments in support of measurable targets as the basis for evaluating management performance, a general lack of knowledge concerning adequate indicator targets is frequently identified as a cause of failure in management (Wintle and Lindenmayer, 2008; Villard and Jonsson, 2009; and see Chapter 5). The oft–stated desire to somehow 'measure sustainability' through the definition of baselines and acceptable thresholds of variability, has produced a potentially detrimental emphasis on finding a single, absolute answer to the problem of sustainable resource use (Salafsky and Margoluis, 2003; see also Chapter 2 for discussion of the concept of sustainability as applied to responsible forest management systems).

For both practical and technical reasons, this approach is plagued with difficulties and has become increasingly controversial (Nichols and Williams, 2006). For the remainder of the chapter, I discuss five such interrelated difficulties, namely: difficulties in measuring what is natural; difficulties of incorporating inevitable management impacts into performance assessments; difficulties in defining geographic boundaries for management; the dangers of perverse incentives associated with an over-reliance on undisturbed baselines; and technical difficulties in measuring indicator trends. Evaluating such difficulties is helpful in identifying ways to design more effective and realistic monitoring programmes.

Difficulties in measuring what is 'natural'

A key challenge in promoting SFM lies in defining baselines and determining what comprises a 'sustainable ecosystem state' (Prato, 2005). Defining what is 'natural' is difficult given our limited ecological understanding of the majority of species, their habitats and the extent of historical human disturbance (Hunter, 1996; Parrish et al, 2003). Baselines can be defined using a long-term regional average, conditions measured at the start of the programme, or some theoretical optimal or maximum value (Watt, 1998). Although significant progress has been made towards acquiring a long-term (i.e. paleoecological) perspective on 'what is natural' regarding biodiversity (Willis and Birks, 2006; Willis et al, 2007), our understanding is still woefully inadequate for the majority of practical applications in most parts of the world. Research on extant patterns of diversity has shown that high levels of spatial and temporal variability are the norm for many ecological variables, like population abundances and species composition (Hanksi and Gilpin, 1991; Dixon et al, 1998; Barrows et al, 2005; Hanksi, 2005; Figure 6.2).

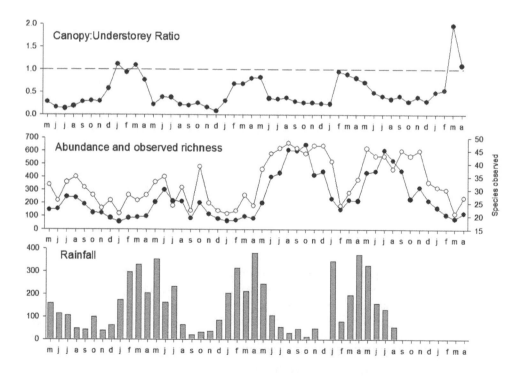

Figure 6.2 Inter-annual variability in butterfly communities at an undisturbed tropical forest site in the Brazilian Amazon

Source: W. Overall, unpublished data

Furthermore, widespread and long-term environmental degradation has led to a common problem of 'shifting baselines' in attempts to quantify conditions that predate major disturbances (Box 6.4). However, despite the importance of these issues, there are very few long-term (i.e. multi-decadal scale) studies that have monitored ecological responses to forest disturbance that can be used as a basis for comparison in individual monitoring programmes (Dunn, 2004; Gardner et al, 2007a). Consequently, it is extremely difficult to decipher the wider conservation significance of local changes in indicator levels, which means that site-level trend analyses that focus on potentially arbitrary baselines may be limited in their relevance to efforts to improve management (Dixon et al, 1998).

Difficulties of incorporating inevitable management impacts into performance assessments

Di Stefano (2001, p142) states that:

> '... *in effect assessments of sustainable forestry can be regarded as a form of environmental impact assessment, where the null hypothesis is that some human activity has no significant detrimental effect on the environment.*'

Similar statements can be found throughout the literature on natural resource management but is the expectation and requirement that management will have *no* detrimental effect realistic or even constructive? I suggest that it is neither. Even should it be possible to define critical limits of natural variability, it seems unreasonable to expect this to provide a fair 'measure' of management performance as human activities of any form are highly likely to have *some* effect on ecological systems (Robinson, 1993; Bawa and Seidler, 1998; Allen et al, 2003; Shile et al, 2004; Finegan, 2005).

Box 6.4 Shifting biodiversity baselines

One important consequence of large-scale regional deforestation and forest degradation is that ecologists are limited to measuring contemporary human impacts against a continuously shifting baseline. Few time for series data exist long-term species. A compilation of available studies that have measured species trends on multi-decadal scales in tropical forests questions the accuracy of baseline data from so-called 'pristine' forests by revealing a persistent and marked pattern of species loss over the last century in different locations across the world (Gardner et al, 2009; Table 6.3). The fact that many modified forest landscapes exhibit such dynamic baselines poses a serious challenge to ecologists and foresters who are responsible for managing biodiversity monitoring programmes and who need to readjust to continuously shifting goalposts across both space and time.

Table 6.3 Shifting baselines from long-term tropical forest study sites

Study	Region	Species group	Time period	Observed biodiversity loss at local scale	Suggested cause of biodiversity loss
Castalletta et al. (2000)	Singapore	Birds	1923–1998	31% of total original species (203), and 67% loss of original forest species (91)	Forest loss (only 5% cover remains)
Trainor (2007)	Damar Island, Indonesia	Birds	1890s–2001	17% loss of original forest species (38)	Forest loss and degradation (75% forest cover remains)
Kattan et al. (1994)	Colombian Andes	Birds	1911–1990	31% loss of original forest species (128)	Forest loss and fragmentation
Sodhi et al. (2006)	Bogor Botanical Garden, Indonesia	Birds	1931–2004	59% loss of original species (97)	Forest loss, isolation and disturbance
Robinson (1999)	Barro Colorado Island, Panama	Birds	1923–1996	36% loss of original forest and edge species (193)	Isolation effects
Sigel et al. (2006)	La Selva, Costa Rica	Birds	1960–1999	Approximately 20% species increased (open habitat generalists) and 20% species decreased (understorey specialists) in abundance	Area and isolation effects
Hanski et al. (2007)	Madagascar	Dung beetles	1953–2006	43% loss of original species (51)	Forest loss (only 10% cover remains)
Escobar et al. (2008)	La Selva, Costa Rica	Dung beetles	1969–2004	Decreased species richness and community evenness	Isolation and habitat loss in neighbouring landscape
Mortensen et al. (2008)	Guam, Pacific Ocean	Birds, lizards, bats	1960s–2003	81% loss of original forest birds (11), 50% loss of native lizards (12) and 66% loss of native bats (3)	Introduced brown treesnake (*Boiga irregularis*)
Whitfield et al. (2007)	La Selva, Costa Rica	Amphibians and lizards	1970–2005	75% decline in total densities of leaf-litter amphibians and lizards	Reduction in leaf-litter due to climate change

Source: From Gardner et al (2009)

Very few attempts have been made to define critical limits of variation in accordance with recommended performance indicators in forest management. In one example, Aguilar-Amuchastegui and Henebry (2007) used a novel approach based on variography of vegetation index data from remote sensing to identify a logging threshold of four trees per hectare as representing a change in forest structural heterogeneity (as an indicator of management performance) that is significantly outside the natural range of variability (based on 95 per cent confidence intervals). A nice addition to this study was that the authors also related this threshold to a significant change in the richness and diversity of forest-inhabiting dung beetles.

However, for many situations, it is impossible to conceive that economic activity could be sustainable while also maintaining the system within natural limits of variability. This is especially true for severely modified or converted forests (e.g. plantations and agroforestry systems, but also many heavily logged forests). Consequently, sustainability goals cannot be considered an 'all-or-nothing issue' (Sheil et al, 2004; see also Chapter 2), and the acceptance of forest exploitation must be accompanied by an acceptance of some degree of ecological change. Resolution of this problem is often found in the adoption of 'acceptable' (rather than natural) limits of indicator variability (Parrish et al, 2003; Stoddard et al, 2006; Oliver et al, 2007; and see Chapter 9). Although such recommendations are sensible enough, to a large degree they reflect a subjective or value-based judgement, concerned equally with people and institutions as it is with science (Walters and Green, 1997; Bergeron et al, 1999; Yamasaki et al, 2002; Finegan et al, 2004).

Difficulties in defining geographic boundaries for management

Variability in ecological, social and economic factors makes it very difficult to define the geographic boundaries of forest 'management zones' within which the same set of critical thresholds could be appropriately adopted (Leslie, 2004). This problem is widely acknowledged following attempts to devise international versus regional performance-based standards for forest certification schemes (see Nussbaum and Simula, 2005).

The dangers of perverse incentives associated with an over-reliance on undisturbed baselines

Many forested areas of the world are already severely degraded and lack suitable control sites that are able to provide a 'natural' baseline (Arcese and Sinclair, 1997; Parrish et al, 2003; Brook et al, 2006). Profound differences in the extent of historical disturbance in forest ecosystems can render conflicting results when evaluating progress towards sustainability in different areas of the world (Soberón et al, 2000). The setting of high performance standards as a minimum target can be counterproductive as they are only likely to attract organizations working in areas where little or no conservation improvements are necessary i.e., managed forests that are already in good condition, and are likely to stay that way (Rameststeiner and Simula,

2003). By contrast it is often the most degraded areas that can generate the greatest conservation benefits from new management initiatives. This problem is recognized by the certification community, and different threshold standards are recommended in different situations depending on the intensity of management and the background conditions (see Nussbaum et al, 2002; FSC, 2004).

Technical difficulties in measuring indicator trends

Management standards that are based on maintaining a system within natural limits of variability evaluate performance by measuring significant deviations from a benchmark. However, there are major sampling and analytical difficulties in accurately measuring what can be defined as a 'significant' change in patterns of species abundance and diversity (Dixon et al, 1998; Urquhart et al, 1998; Prato, 2005; Nichols and Williams, 2006). Traditional (and still dominant) approaches to the analysis of trend-type data are founded in the use of null hypotheses and significance testing (Prato, 2005). However, such a reductionist approach to problem solving has been varyingly described as overly conservative, as it suffers from requiring a 'disproportionate burden of [empirical] evidence' in order to reject a particular null hypothesis (e.g. that of 'no change outside critical limits', see Williams, 1997, and also Chapter 15). Strong temporal variability in patterns of abundance and diversity that are independent of any forest management intervention mean that long-term datasets may be required to detect trends (e.g. Pearce and Venier, 2005). Furthermore, significance levels are often set arbitrarily in statistics, and to be useful an *a priori* effect size of *ecological* significance (i.e. one that would result in a long-term impairment to the resilience of the system) needs to be identified. Yet, we often have no strong empirical or theoretical basis on which to decide what level of change in an indicator value is acceptable, and once again the decision is largely a subjective one that depends upon the level of rigour desired by stakeholders (Ghazoul and Hellier, 2000; Sheil et al, 2004; Finegan, 2005; Dudley et al, 2005).

The challenge of measuring management performance against the goal of ecological sustainability

The foregoing discussion clearly demonstrates that there are considerable difficulties in measuring and applying the concept of critical limits of natural variability to forest management systems, even for relatively low-impact selective logging operations. It is possible that the ambitious goal of ecologically sustainable forestry may be unfeasible by definition (Lindenmayer et al, 2006), as a focus on maintaining forest systems within 'critical limits' represents an unreasonably strong sustainability perspective that is incompatible with long-term economic goals (see Karsenty and Gourlet-Fleury, 2006; and see Chapter 2). The technical difficulties associated with measuring significant levels of deviation from natural or acceptable limits of variability can provide a smokescreen behind which both managers and auditors with limited technical expertise can hide when it comes to providing an objective assessment of performance standards for certification – a situation that can lead to 'decision-paralysis' (Walters et al, 2003; Newton and Oldfield, 2005). Consequently, such unrealistic targets will do little to encourage either monitoring of the true effectiveness of management practices (as managers will know they are committed to failure), or ways in which management can be improved in the future (as they will be largely unknown).

Despite all these difficulties in measuring changes to indicators, we are still left with a practical need to assess minimum standards of management performance. The solution lies in accepting a certain level of uncertainty and adopting an AM approach as a cornerstone of responsibility. Our knowledge and understanding of forest ecology is insufficient to justify interpreting indicators and measurements of management performance as absolute values (Spellerberg and Sawyer, 1996; Prabhu et al, 2001). We need to decide upon the desirable features and services of a given forest and manage to preserve them in the face of change. Progress in conserving forest biodiversity can be achieved through the use of carefully selected indicators and monitoring approaches that are appropriate for the distinct, yet complementary tasks of measuring current performance levels while also identifying opportunities for continuous future progress in conservation management. The third section of this book (Chapters 8–16) is focused on exploring ways in which this goal can be achieved.

CHAPTER 7

Challenges to Monitoring: Problems of Reality

Synopsis

- Cultural and institutional factors often present the greatest challenge to biodiversity monitoring because they cannot be readily solved through training or the development of new guidelines.
- Most forest management systems are based on compliance assessments against static, minimum-practice standards. AM provides a more progressive approach that recognizes uncertainty, and employs a continuous cycle of design, management and monitoring to systematically test assumptions in order to adapt and learn.
- AM is frequently misinterpreted, and formalized approaches that are based on manipulative experiments of different management treatments are often both practically and politically unfeasible. Passive approaches that overlay monitoring programmes on existing management regimes are more realistic but still require a close understanding of the social, economic and political context of the management system.
- Marked differences in the level of guidance provided by regional management standards, even within the same certification system, mean that different countries and regions are at very different stages in the development of ecologically responsible systems of forest management.
- Perhaps the most significant barrier facing biodiversity conservation in managed forests is a general reluctance among decision-makers to replace a static-command and control-style approach to management with an adaptive approach that recognizes uncertainty and the need for continuous improvement.

Biodiversity monitoring can be a costly enterprise, and often a simple lack of funding can represent one of the most difficult barriers facing the implementation of a successful programme. It is often difficult for forest managers to invest in promoting long-term sustainability perspectives when they are associated with real or perceived increases in short-term cost (Putz et al, 2000). However, money is only part of the problem. While extra funding may permit or promote improved monitoring, more money is rarely the most important factor limiting the delivery of a meaningful programme. Indeed, an often greater problem is deciding how available funds should be spent (see Chapter 5). Maximizing return on investment from limited resources requires a careful coupling of theoretical design concerns with the institutional factors that ultimately determine the success of monitoring in the real world.

It is usually the case that more can be done for the often substantial investments that are made in monitoring. Being clear about the purpose of monitoring and identifying appropriate goals and objectives is the most important step. Next it is necessary to overcome procedural and design challenges that determine whether a biodiversity monitoring programme can effectively deliver on its purpose (see Chapters 5 and 6). Nevertheless, even when provided with sufficient funding, managers may be discouraged from making the necessary investment in monitoring due to institutional weakness from regulatory or certification authorities (Ghazoul, 2001).

Real-world constraints on biodiversity monitoring relate to the factors that constrain the development of monitoring and adaptive management within a real-world institutional and regulatory operational framework (Lee, 1993; Taylor et al, 1997; Wilhere, 2002; Keough and Blahna, 2006; McAfee et al, 2006). In many ways, institutional barriers to forest conservation and management initiatives present an even greater challenge to forest managers than purely technical barriers, because they cannot be readily solved through training or the development of new guidelines.

Considerations regarding the importance of management institutions for monitoring can be identified by distinguishing between theoretical efficiency and practical effectiveness (Sutherland et al, 2004b). Efficient approaches to gathering information are of limited value unless they are also effective within their existing social, economic and political context. The neglect of such contextual factors in the development of biodiversity monitoring has frequently led to proposals for unrealistic programmes that attract criticism for diverting scarce funds away from other conservation priorities (Sheil, 2001).

In this chapter, I discuss the importance of some of the key real-world challenges that confront biodiversity monitoring: first, with respect to achieving the broad goal of AM and second with respect to the regulatory and cultural institutions that are associated with this goal.

ADAPTIVE FOREST MANAGEMENT

Management and monitoring are two sides of the same coin. AM is about recognizing uncertainty in management and our understanding of the natural world, and placing learning at the heart of the management system to ensure a process of continuous improvement. As originally conceived (Holling, 1978; Walters and Hilborn, 1978), AM can be defined as the *'systematic acquisition and application of reliable information to improve management over time'* (Wilhere, p20, 2002). As an essential ingredient of learning, monitoring is integral to the success of AM. In the same way, monitoring is limited to acting as a simple reporting tool unless it can form part of a successful AM system. In their review of AM in certified forestry, Wintle and

Lindenmayer (p1313, 2008) concluded that *scientifically defensible certification schemes should embrace true adaptive management as an overarching framework and philosophy for management and as a minimum standard for certification*. Understanding how to maximize the value of biodiversity monitoring requires knowledge of the key elements of AM and an understanding of some of the major challenges that have confronted the implementation of AM systems to date.

The purpose and promise of AM

Adaptive management, cast in the broadest terms, describes a continuous integrated cycle of design, management and monitoring to systematically test assumptions and challenge uncertainty in order to adapt and learn (Margoluis and Salafsky, 1998). In this sense, AM can be succinctly described as a systematic and purposeful approach to learning by doing (Stem et al, 2005). AM contrasts sharply with a process of surveillance monitoring and management by trial and error (see Chapter 5), and is in many ways ideally suited to resolving the uncertainty inherent in forest management systems (Ghazoul, 2001; Lindenmayer et al, 2000; Hobbs, 2003; McAfee et al, 2006). As a fundamental component of the AM process, monitoring cannot be considered as a stand-alone activity, but should be developed as an integral part of forest management standards themselves (Wilhere, 2002; Nichols and Williams, 2006).

While the idea itself is well established (Holling, 1978), the concept of AM has received an increasing amount of support from both scientists and forest managers as being central to the successful management of natural resources (Taylor et al, 1997; Lindenmayer et al, 2000). Nevertheless, the idea of AM has been repeatedly misinterpreted (Wilhere, 2002). Salafsky and Margoluis (2004) note that AM has sometimes been taken to simply infer 'good management'; if one option is tried that doesn't work then it is logical to try something else (an uncoordinated process of trial and error). In another critique, Wilhere (2002) observes that some managers have understood AM as simply representing the willingness to change, or adopt flexible management practices.

Although frequently misinterpreted, AM is most usefully considered as a continuum of management strategies, and not an 'all-or-nothing' solution. This continuum extends between passive versus active AM (Lindenmayer and Franklin, 2002), and although a more formalized active AM approach increases the likelihood that new knowledge will be produced (Walters and Holling, 1990; McCarthy and Possingham, 2007), learning can still be promoted in more passive ways (Rempel et al, 2004; Box 7.1).

Challenges in implementing AM

Despite its long history and clear theoretical benefits, there are few examples of the successful application of AM (active or passive) for either forest management (Taylor et al, 1997; Gibbs et al, 1999; Hobbs, 2003; McGinley and Finegan, 2003;

Box 7.1 Active and passive AM

Adaptive management (AM) as originally conceived is a formalized, active approach to management where management interventions are conducted in the form of deliberate (manipulative) experiments (Holling, 1978; Walters and Hilborn, 1978). Alternative interventions are viewed as true experimental treatments that are executed and monitored under a fully replicated and randomized experimental design. Consequently, active AM is highly effective at establishing cause–effect relationships between management interventions and ecological responses (e.g. changes in indicator taxa), and the ability to simultaneously evaluate multiple alternatives provides an efficient approach towards optimizing 'best practice' policy in natural resource management (Walters and Holling, 1990; Wilhere, 2002; Allan and Curtis, 2005). The major disadvantage of active AM is that it is often unrealistically complex and expensive in real ecosystems (Wilhere, 2002; Stankey et al, 2003; Sutherland, 2006).

Passive AM reflects many of the principles associated with active AM in that it assists decision-making by relating management to ecological responses through the formulation of predictive models that can be tested and refined using monitoring data (Wilhere, 2002; Prato, 2005). However, in contrast to active AM, the lack of a formal experimental design (existence of true controls, the active manipulation of treatments and random assignment of replicate samples to treatments) means that a purely passive approach to AM has the potential to confound management and environmental effects and has a limited ability to generate reliable information on causal relationships (Wilhere, 2002). Because of this lack of any formal experimental design, passive AM is traditionally associated with the testing of a single 'best' management model (Walters and Holling, 1990; Allan and Curtis, 2005). Passive AM approaches are most useful when it is impractical and expensive to design manipulative experiments, and/or high levels of natural variability modulate the effects of management in site-specific ways (Taylor et al, 1997). Forest management systems usually fit both of these conditions. The potential value of passive AM for guiding conservation decisions depends upon the extent to which monitoring programmes can be carefully overlain upon existing management regimes. The more detailed an understanding of the management plans, forest structure and disturbance history available for a given landscape, the easier it is to effectively exploit the learning opportunities presented by such natural experiments (see Chapter 10 for a more detailed discussion).

Stankey et al, 2003) or conservation in general (Wilhere, 2002; Allan and Curtis, 2005; Sutherland, 2006). The underlying philosophy of AM is incorporated into many general forest management standards, such as the FSC International Standard, which states (Criterion 8.2), '*The results of monitoring shall be incorporated into the implementation and revision of the management plan.*' However, there is a pervasive misconception that any decision to change a management action in light of an observed (usually unexpected) change in the state of the system is, by definition AM (Wintle and Lindenmayer, 2008). Individual management standards differ in their interpretation of AM and rarely contain any formal definition or detailed guidance on how this requirement should be achieved. As such, forest management systems to date have been based more on the establishment of static, minimum practice standards than on the explicit adoption of a process of continuous improvement through AM (IUCN, 2004; Finegan, 2005).

Practical experience to date has demonstrated that the challenges of implementing AM are non-trivial (Stankey et al, 2003). McAfee et al (2006) suggest four explanations that can be implicated in explaining the failure of AM approaches to generate demonstrable improvements to biodiversity conservation, namely: technical and ecological, economic, institutional and political (see also Lee, 1993; Wilhere, 2002). Technical and ecological obstacles relate to the lack of adequate ecological information and theory, a lack of expertise, limitations in the precision and suitability of analytical tools, and the lack of protocols to systematically evaluate progress towards achieving specific management objectives. To a large extent these challenges can be overcome through research and associated technical guidance of the managers of forest monitoring and AM programmes. By contrast, economic, institutional and political obstacles relate to limitations in cost and expertise, the importance of institutional stability to enable the measurement of long-term and large-scale environmental changes, the need to maintain expectations that are grounded in realism, and the importance of engaging multiple stakeholders when reviewing the results of monitoring and evaluation programmes.

These institutional challenges are far more difficult to overcome through the provision of general training and guidance. Instead, they require a detailed understanding of the social, political and economic structures in which individual forest management systems are embedded and how active engagement between forest scientists, managers and regulators can be facilitated to overcome disciplinary and stakeholder insularity that hinders progress in management (see Field et al, 2007). Barriers to the process of knowledge transfer among scientific disciplines and between science and management/policy processes are significant, and the source of considerable frustration to many well-intentioned researchers. A failure to achieve a closer integration between science and research means that much of science will continue along the path of *idiosyncratically documenting in ever more detail the deleterious consequences of forest exploitation* (Putz and Romero, 2001), without their results being coordinated towards improving management of the very same systems.

CHALLENGES TO MONITORING ASSOCIATED WITH GOVERNANCE AND REGULATORY INSTITUTIONS

It is well understood that differences in background ecological, economic and political circumstances mean that different forests are associated with different management challenges and opportunities (McDonald and Lane, 2004; Sheil et al, 2004; Hammond and Zagt, 2006). As such, management systems and associated C&I need to be adjusted to suit local conditions (Prabhu et al, 1999, McDonald and Lane, 2004). The need for a flexible policy framework is well recognized by the FSC through the use of their guiding PC) as a template or starting point for developing regional stewardship standards (FSC, 2002, 2004). Indeed, the sensitivity of this approach to variability in regional conditions has been identified as a key advantage of the FSC system (Cauley et al, 2001; Leslie, 2004). However, while this flexibility is advantageous, many countries and regions, especially in the tropics, lack the support and technical expertise necessary to identify suitable indicators, and translate generic criteria and indicators into local management standards (Ghazoul, 2001; McGinley and Finegan, 2003; UNFF, 2004; Sayer and Maginnis, 2005c). Countries that lack established FSC standard therefore frequently rely upon 'interim standards' based on generic standards adapted for local conditions through consultation with stakeholders. The Rainforest Alliances' SmartWood programme has developed interim standards for a wide range of tropical and temperate countries (Box 7.2; and see www.rainforest-alliance.org/forestry/interim_standards.cfm).

Box 7.2 Interim standards

As part of the FSC process, regional standards are field-tested, revised and approved by the regional working group, and then submitted to FSC International for final approval. If approved, the final product is an 'FSC accredited standard'. Once accredited, all FSC-approved certifiers (like SmartWood) must use the endorsed regional standard as the fundamental starting point for FSC certification in that country/region. Certifiers may choose to be more rigorous than the regional standard, but they cannot be less rigorous. In all countries or regions not covered by an FSC-accredited forest stewardship standard, SmartWood is in the process of developing locally adapted *interim standards* for use in evaluating forest management operations. The adapted standard draws upon the FSC International Standard, FSC and other documentation on regional standard development (e.g. SmartWood Generic Standard Guidance as produced by the Rainforest Alliance, www.rainforest-alliance.org/forestry) as well as local and national forest management documents, modifying generic certification indicators to take into account the national context (e.g. legal requirements, environmental, social and economic perspectives).

Although interim standards typically provide more detail than generic FSC P&C they offer considerable leniency on the part of local forest managers as to management practices (including monitoring procedures) that are adopted.

The result of inadequate local and regional guidelines is that different countries are at markedly different stages in the development and implementation of ecologically responsible forest management systems and associated biodiversity monitoring programmes (McGinley and Finegan, 2003; CICI, 2003), and these differences are often poorly understood (Holvoet and Muys, 2004). To provide an example of this discrepancy, the clearest distinction can be found in comparing ecological elements of a forest stewardship standard between a developed temperate country (Canada) and a developing tropical country (Brazil). In Canada, FSC Principle 6, Criteria 3 (concerned primarily with impacts of forest management on biodiversity) has been developed within the national standard to encompass 19 indicators and 75 'verifiers', covering 11 A4 pages of guidelines (FSC Canada, 2004). In contrast, and although working from the same generic International Standard,

Principle 6, Criteria 3 of the FSC national standard in Brazil includes only four indicators, no verifiers and amounts to a total of 85 words of guidance (FSC Brazil, 2004). Similar discrepancies can be found when comparing other ecological criteria.

It is inevitable that development of responsible forestry will be greater in some parts of the world than others, and to be successful any certification authority needs to be sufficiently flexible to account for this. However, the fact that there is such a large discrepancy is cause for considerable concern. Customers of forestry products in Canada who are used to seeing the FSC logo on products they purchase from the domestic market will believe that imported products from Brazil that carry the same logo are from forests managed and evaluated to a similar standard. It is possible that they may be, but not without substantial extra initiative on the part of an individual forestry enterprise in Brazil.

Where institutional support is weakest, significant responsibility for making decisions is naturally delegated to local managers. One consequence of this is that the decision-making process will invariably have access to only a narrow knowledge and experience base, and there may be little coordination among sites regarding key management and monitoring decisions, including the selection of indicators. This situation is well evidenced by the Canada–Brazil discrepancy; the fact that far less guidance is provided in the case of Brazil means that individual forest managers have far more leniency to interpret their responsibilities towards meeting individual criteria.

CULTURAL CHALLENGES TO MONITORING

Perhaps the most challenging barrier facing the implementation of a successful biodiversity monitoring programme – that is, one that both delivers scientifically rigorous data and is effective in influencing the management process – is that of culture. By this I mean the way in which different people who are involved in the monitoring process interact (or do not interact) with each other, and the extent to which institutional norms, personal behaviour and multiple agendas serve to define and condition the nature of these interactions. Overcoming the many challenges posed by such cultural issues is extremely challenging as they are often difficult to identify in the first place, and when they can be identified major changes in human behaviour may be required for them to be overcome.

Cultural challenges to biodiversity monitoring exist both within the scientific community and between scientists and the management and regulatory authorities that are the end users of monitoring data. A major challenge within science is that the traditional publication and grant-based award system generally favours initiatives that result in a narrow set of outcomes and offer the greatest chance of being published, while at the same time often penalizing attempts to increase the practical relevance and broaden the disciplinary scope of research (Uriarte et al, 2007). This conflict of interest is well recognized and represents a major barrier in delivering biodiversity monitoring that can make a meaningful contribution to conservation (*Nature*, 2007; Chapron and Arletazz, 2008). The competitive nature of the academic system can also act as a barrier to the sharing of data, expertise and ideas among researchers working in different places and institutions. This problem is exacerbated by a commonly found resistance within the private forestry sector to share information about the ecological impacts of management activities that may be deemed to be commercially sensitive. In addition to problems facing monitoring, the actual process of forest management also often requires close cooperation across multiple landowners in order to be successful. This is particularly true in institutionally complex landscapes that are comprised of a patchwork of stakeholders representing different state and private systems of forest ownership, whether legally established according to current national regulations or traditionally-based, all of which may have different policy goals and management strategies for conserving biodiversity (e.g. the north-west coast of the US (Spies et al, 2007; Suzuki and Olson 2008) and Queensland, Australia (McAlpine et al, 2007)).

Perhaps the deepest cultural barrier facing the development and implementation of biodiversity monitoring programmes that can make a genuine contribution to improving management is the comparatively low priority given to ecological criteria (compared to economic or social criteria) within forest management standards generally (Gullison, 2003; Hagan and Whitman, 2006). For example, forest certification authorities have been strongly criticised for failing to give sufficient accommodation to ecological components of sustainability (Bennett, 2000; Hagan and Whitman, 2006; and see Chapter 5). One example of the inadequate level of attention that is paid to ecological criteria in many certification standards is the almost exclusive focus on threatened species as a basis for monitoring work, despite the fact that for much of the world we have no idea as to the identity of these species, much less their sensitivity to management activities (see Chapter 6). This kind of problem is exacerbated when scientists contracted to advise on or implement monitoring programmes often select taxa based on their own personal expertise or the practical ease of sampling, rather than indicators that are better suited to detecting the effects of management in the most cost-effective manner (Noss and Cooperrider, 1994; Gardner et al, 2008a).

Finally, a general cultural barrier to the implementation of AM is an understandable reluctance among managers and policy makers involved in natural resource management to accept uncertainty in scientific knowledge (Lee, 1993; Richter and Redford, 1999; Allan and Curtis, 2005; McAfee et al, 2006). Through a desire to achieve economic efficiency, forest managers often appear to demonstrate a preference for command-and-control style approaches to management. However, the application of this philosophy to poorly understood systems can result in counter-intuitive and even perverse consequences if short-term practices result in an erosion of long-term resilience (Holling and Meffe, 1996). Overcoming this deep-rooted barrier, and incorporating uncertainty as an explicit component of the management process (i.e. AM) represents a major obstacle to the development of more sustainable systems of resource use.

PART III

An Operational Framework
For Monitoring Forest Biodiversity

CHAPTER 8

Clarifying Purpose: An Operational Framework for Monitoring Forest Biodiversity

Synopsis

- Implementation, effectiveness and validation monitoring perform complementary and re-enforcing roles in guiding the development of more ecologically responsible approaches to forest management.
- Selecting different indicators for monitoring based not on individually applied criteria, but on their functional relationship to each other helps to clarify their complementary and inter dependent roles, and is vital in enabling monitoring to engage with the AM process.
- Species-based indicators do not provide reliable measurements for assessing compliance against performance standards because of inherent difficulties in establishing clear links between management impacts and changes in biodiversity.
- By contrast, indicators of forest structure (and both stand and landscape scales) do provide reliable measurements of management performance because they exhibit direct and tractable responses to changes in human activity.
- Ultimately, the goal of validation monitoring is to improve our understanding of the processes that link changes in management to changes in biodiversity via intermediate changes in the structure and function of the forest. The primary role of biological indicators and target species is therefore not to act as direct indicators of performance but as *evaluators* of the performance indicators that define forest management standards. This process of validation is often missing from monitoring yet it is the only way of linking changing management practices with actual conservation goals.

For biodiversity monitoring to make a meaningful contribution to the development of more responsible approaches to forest management, it is first necessary to develop a clear definition and understanding of the purpose of monitoring (see Chapters 3 and 5). How can a monitoring programme be designed in a way that is most capable of bringing about a desired result? To what extent does a difference in underlying purpose determine the types of information and monitoring approaches that are most appropriate for a given monitoring programme? Such questions should come before any consideration of particular conservation goals or objectives because they are concerned with the fundamental way in which monitoring programmes are structured and implemented, and how they interact and engage with the wider management process. Key to successfully integrating biodiversity monitoring within the wider management process is recognition of the complementary role played by different types of monitoring approach and indicators.

In this chapter, I build on the context provided by the first two sections of the book and outline the different ways in which the process of biodiversity monitoring relates to the management of modified forest ecosystems. I clarify how implementation, effectiveness and validation monitoring programmes perform distinct yet complementary and re-enforcing roles in the development of more ecologically responsible management systems. I also analyse how different types of indicator are varyingly suited to different monitoring tasks. Finally, I draw these components together and provide a general operational framework for biodiversity monitoring that clearly illustrates the interactive and dynamic role played by different approaches and indicators.

UNDERSTANDING THE ROLE OF DIFFERENT MONITORING APPROACHES IN FOREST MANAGEMENT

In an ideal world, we would have a perfect understanding of how different management interventions impact forest biodiversity, and we could use this understanding to dictate a clear code of practice that guarantees responsible use. Assessments of management compliance against such a performance standard could be made simply by monitoring the implementation of management activities (i.e. via implementation monitoring). Such a situation is described as *linear management* (Noss and Cooperrider, 1994). However, clearly we cannot predict the effects of all management interventions, and the challenge of biodiversity monitoring is much more complex than a tick-the-box implementation exercise (see Chapter 5). Decades of research in forestry and forest ecology have generated many important management guidelines (e.g. Hunter, 1990; Lindenmayer and Franklin, 2002; and see Chapter 2), which if implemented, would result in considerable improvements to the conservation of native biodiversity in forests around the world. However, many threatening processes remain poorly understood, and in the vast majority of cases we have a poor understanding of how generic guidelines can be most effectively adapted to fit the context of an individual forest landscape, management system and biota (see Chapter 3). Which human impacts carry the greatest risk to local biodiversity and which management strategies are likely to be the most cost effective at mitigating this damage and ensuring long-term conservation success? The challenge facing those responsible for implementing meaningful biodiversity monitoring programmes is to attend to both these demands. First, to ensure that management proceeds on the basis of minimum practice standards and that the implementation of such practices is effective in delivering measurable outcomes.

And, second, that even best practice standards are constantly improved by evaluating how new and revised management strategies can contribute towards achieving long-term conservation goals. As discussed in earlier chapters, this challenge is embodied in the concept of AM and is best served by implementing an integrated multi-tiered monitoring programme (Fig. 8.1), encompassing:

- **Implementation monitoring** to assess compliance with agreed minimum practice standards by monitoring the adoption of selected management practices and impact mitigation strategies by a management authority.
- **Effectiveness monitoring** to assess changes in selected management impacts that serve as indicators of performance, and compare any observed changes against minimum standards. Implementation and effectiveness monitoring perform complementary functions of a management audit to assess compliance against agreed system- and performance-based standards respectively.
- **Validation monitoring** to evaluate the extent to which existing management strategies and associated performance indicators are effective in securing real conservation benefits, and in doing so guide a process of continuous ecological improvement in management. In essence validation monitoring describes the programme of scientific inquiry that is necessary to help identify and establish the causal relationships that link management activities, direct management impacts on the structure and function of the forest ecosystem, and ultimate changes in biodiversity and forest integrity.

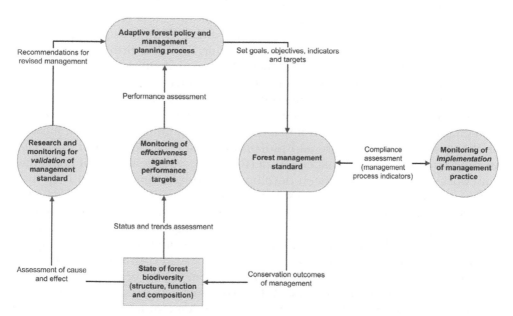

Figure 8.1 A conceptual framework of an integrated biodiversity monitoring programme for adaptive forest management

Note: To be effective in both assessing and evaluating performance, a monitoring programme should comprise three tiers: implementation monitoring of management practice compliance; effectiveness monitoring of the system state against predetermined performance indicator values; and validation monitoring to evaluate how best to achieve continued progress towards long-term conservation goals

A hybrid approach combining these three classes of monitoring activities is essential in enabling an efficient and practical assessment of management performance that is immediately available, while also affording stakeholders the ability to move beyond a static regulatory system by continually evaluating the effectiveness of alternative management strategies, and striving to achieve continuous ecological improvement.

UNDERSTANDING THE ROLE OF DIFFERENT INDICATORS IN THE MONITORING PROCESS

Indicators provide the key tool by which different elements of the monitoring and evaluation process can be logically connected to the management regime. The indicator selection process is therefore central to the definition of any biodiversity monitoring programme. However, despite its importance it is also one of the areas that has achieved the least clarity in both scientific literature and in practical applications (see Chapters 5 and 6). A major limitation of many biodiversity monitoring and research programmes lies in the highly fragmented and isolated approach taken to selecting and classifying different types of indicators, and a general failure to describe how they inter-relate within the overall management-monitoring framework (Newton and Kapos, 2002; Overton et al, 2002; Stem et al, 2005; Niemeijer and de Groot, 2008; Margoluis et al, 2009). Many biodiversity monitoring programmes combine various types of indicators (e.g. policy, process, structural and biological indicators) into uncoordinated simple lists, suggesting that they should be measured and reported together, and with little hierarchical or interactive structure (Kneeshaw et al, 2000; Rempel et al, 2004).

However, indicators only have relevance to management in conjunction with other, related indicators and in the context of a specific set of research questions (Niemi and McDonald, 2004). Developing a clear understanding of the links between particular management impacts and changes in biodiversity via measurable changes in forest structure and function provides the foundation for reliably assessing and evaluating management performance. A wide variety of conceptual frameworks and models has been developed to address this task and help make explicit the definitions and relationships among phenomena of interest and different types of indicators, with different frameworks varying in the extent to which they encompass the full integrated monitoring-management system (see Chapter 13). The most widely used and simplified approach is the pressure–state–response (PSR) framework which states that human activities exert pressures on the environment which lead to changes in the forest system (i.e. structural, functional or compositional elements of biodiversity). In turn, management authorities and stakeholders may respond to observed changes by implementing revised and/or new management strategies. Figure 8.1 represents such a simple conceptual framework that makes explicit the dynamic links between human impacts, changes in the forest system and feedbacks from monitoring, while at the same time highlighting the complementary role played by different monitoring approaches in capturing and reporting on observed changes. Within this broad framework it is possible to develop a conceptual model that further unpacks the causal relationships linking changes among different indicators of the ecological system (e.g. Stork et al, 1997; and see Chapter 13).

The level of detail that is used to define conceptual models that link human activities and changes to biodiversity is determined in part by the monitoring objectives, and in part by the availability of prior ecological knowledge regarding the mechanisms responsible for driving observed changes. However, irrespective of the number of stages that are detailed, selecting indicators based not on individually applied criteria, but on their functional relationship to each other helps to clarify their complementary and interdependent functions, and is fundamental in enabling the monitoring framework to engage with the process of AM. Different monitoring approaches require different combinations of indicators and there is an increase in their data requirements and structural complexity as you move through implementation, effectiveness and validation monitoring (Fig. 8.2).

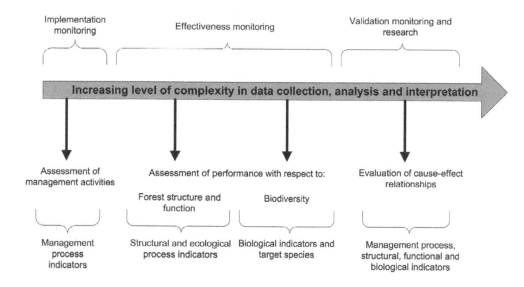

Figure 8.2 Differences in the indicator requirements and complexity of the process of data collection, analysis and interpretation for the three main approaches to monitoring a forest management system

Note: Each of the three approaches of implementation, effectiveness and validation monitoring serve distinct yet complementary functions in an integrated monitoring system. Validation monitoring is equivalent to long-term applied research. See text for a more detailed discussion on the differences between monitoring approaches
Source: Figure adapted from Stadt et al (2006)

In the initial stage of developing an integrated monitoring programme, the selection of management practice indicators forms the basis of an implementation monitoring programme that can be used to evaluate compliance against an agreed management regime (Figure 8.1). Discussion of the types and levels of practice that define any given management standard is beyond the scope of this book. However, a wealth of information and guidelines already exists to inform how management systems can be made more ecologically responsible. One example of this is in the development of RIL techniques for the management of native forests for timber (Putz et al, 2008), and similarly detailed guidance exists for other managed forest systems, including plantation forestry (Hartley, 2002; Lindenmayer and Hobbs, 2004), agroforestry systems (Schroth et al, 2004), and multiple-use forest landscapes (Lindenmayer and Franklin, 2002; McNeely and Scherr, 2003) (see Chapter 2). However, without careful monitoring it is impossible to evaluate the adequacy of existing minimum practice standards in achieving progress towards long-term conservation goals, or identify ways in which cost-effective improvements can be made. The careful selection of performance indicators is central to achieving these goals.

Choosing the right indicators for assessing and evaluating performance-based management standards

Biological indicators and target targets (Chapter 4) have received consistently strong recommendation for the purpose of monitoring management performance (e.g. Stork et al, 1997; CIFOR 1999; Hagan and Whitman, 2006). The justification for this appears logical as it is the species themselves, and the myriad of ecological interactions and process that they support, which comprise the attributes of a forest that we actually value. Without direct measurements of biodiversity in the form of species and populations, an isolated focus on structural and ecological process indicators carries the risk of inadvertently conserving 'empty and silent forests' (*sensu* Redford, 1992) i.e., forests that appear intact from a structural perspective yet lack much of the native biota (see Chapter 6).

Yet, to be credible, the basic requirement of any indicator of 'performance' is that *'changes in the indicator have to be able to directly reflect changes in the factor that is doing the performing'* (Kneeshaw et al, 2000; Failing and Gregory, 2003). Any indicators that are (or could be) used to assess whether managers are performing responsibly therefore need to be clearly linked to the management regime itself through a conceptual framework of cause and effect (Yamasaki et al, 2002; Guynn et al, 2004; Niemi and McDonald, 2004; Niemeijer and de Groot, 2008). Furthermore, in order that improvements to management can be feasibly identified and implemented, indicators of performance must also focus on forest attributes over which managers are, at least in principle, able to exercise some direct control (Kneeshaw et al, 2000). Species very rarely, if ever, adequately meet these requirements. By contrast, indicators of forest structure (and some non-structural factors, such as disturbance regimes and soil contaminant levels) generally provide a more defensible basis for the assessment of management performance. The next two sections consider the arguments behind these recommendations in more detail.

Why species-based indicators generally do not provide reliable measures for assessing compliance against performance standards

Species are limited as indicators of management performance for reasons that relate to the inherent difficulty of making clear links between management impacts and changes in biodiversity. This difficulty is comprised of three principal elements, namely: (i) poor ecological knowledge; (ii) linking changes in biodiversity to management impacts; (iii) and technical difficulties in sampling biodiversity.

Poor ecological knowledge

For most of the world, we have a very poor understanding of the ecological require-ments of most forest species or species groups, as well as their natural levels of spatial and temporal variability in abundance. Consequently, we often have a poor appre-ciation of species sensitivities to management impacts (Lindenmayer et al, 2002b; Manley et al, 2004). This uncertainty can make it very difficult to define the kind of quantifiable objectives and targets that are needed as a basis for minimum perform-ance standards (e.g. Wintle and Lindenmayer, 2008; Villard and Jonsson, 2009).

Difficulty in linking changes in biodiversity to management impacts

A fair performance-based management standard should recognize that any changes to the forest system for which a manager is not responsible should be excluded from consideration when assessing compliance. Using species as direct indicators of performance can therefore be problematic because the distribution and abundance of species can often be influenced by external factors unconnected to the manage-ment itself. For example, endangered species may be particularly threatened by stress factors that are beyond the control of managers within a given forest (e.g. the influence of disease and climate change in causing global amphibian population declines (Pounds et al, 2006)). This problem may be particularly exacerbated in forests that are located within highly degraded landscapes, as source-sink dynamics from neighbouring areas may strongly influence the viability of many local popula-tions (Hanksi, 2005). In addition, species patterns of occupancy and abundance often fluctuate independently of any changes in management for reasons that are not always discernible (e.g. Lindenmayer et al, 2008b; and see Chapter 13). Attribut-ing any such changes to the improvement of local management practices would be unjustified, and could lead to misleading recommendations or complacency in future policy.

Technical difficulties of sampling biodiversity

Effective sampling of biodiversity is a difficult and complex procedure requiring a level of technical expertise that is not available to many landowners or managers. This source of uncertainty makes it extremely difficult to hold managers to account should certain species be shown to be in decline on their landholding.

*Why structural and other non-species based indicators generally do provide
reliable measures for assessing compliance against performance standards*

In contrast to the high levels of uncertainty that often surround research findings
on the biodiversity impacts of human activities, management activities *do* have a
clear and directly tractable influence on the structure of forest habitats (e.g. availability of dead wood, maintenance of large fruiting trees) and the composition and
configuration of the managed landscape (e.g. location and design of set-aside
reserve areas, design of forest-fragmenting road networks, buffer zones and riparian corridors) (Franklin and Swanson, 2007). Moreover, it is well established that
vegetation structural complexity is closely linked to biodiversity (Woinarski, 2007;
and see Chapter 6). In addition to being sensitive to management impacts, changes
to forest structural attributes are often highly predictable (e.g. selective logging
results in a more open forest canopy and denser understorey vegetation layer).

Accordingly, structural indicators (and associated minimum target levels) of
stand- and landscape-scale forest attributes provide some of the most suitable and
appropriate measures for assessing compliance against performance standards
(Jonsson and Jonsell, 1999; Lindenmayer et al, 2000; Allen et al, 2003; Finegan et
al, 2004; Parrish et al, 2003). In addition to structural indicators, some non-
structural indicators that are (or can be) directly subject to management can also
provide valuable and complementary information on management performance.
These include disturbance processes like fire regimes, grazing by introduced
animals and changes to hydrology. They may also include direct human-induced
threats, such as changes in large mammal abundances through hunting,
introduction of chemicals in the soil and water systems and invasion by exotic
species. Regardless of the nature of the measure, the basic criteria for qualifying as
a valid indicator of management performance is that changes in the indicator can
be directly and confidently traced back to the management regime.

*Validation monitoring and the role of biological indicators
as evaluators of management standards*

A strictly indirect approach to biodiversity monitoring that is based solely on structural
and functional attributes creates a disconnection between the focus of monitoring
action and the actual attributes that determine conservation value. This can lead to
goal displacement, where indirect but easily measured indicators of performance are
mistaken for long-term goals (Hilborn, 1992). Herein lies the paradox. The fact that
we have a poor understanding of the consequences of human activity for most species
prevents the use of species-based indicators in direct evaluations of management
performance for conservation. However, this same lack of understanding and
uncertainty means that the monitoring of biological indicators is critical in validating
the relevance and suitability of structural- and process-based performance indicators
for achieving conservation goals and in making recommendations for future
improvements in management. Therefore, and in a very practical sense, the primary
role of biological indicators and target species in biodiversity monitoring should be
not as direct indicators of performance, but as evaluators of the performance
indicators that define forest management standards (Kneeshaw et al, 2000).

Possingham and Nicholson (2007) maintain the same position in concluding that the principle role of ecological research in managed systems is to determine the empirical and theoretical relationship between measurable and manageable surrogates of environmental impacts, and real outcome measures of biodiversity. This process of validation is often missing from monitoring yet it is the only way of linking changing management practices with actual conservation goals.

Ultimately, the goal of validation monitoring is to develop a robust understanding of the causal mechanisms that link selected management activities to end biodiversity responses via intermediate changes in the structure and function of the forest. Similarly, validation monitoring can shed important light on the extent to which the management of particular forest attributes may be context-dependent or limited in generating desired results. For example, Smyth et al (2002) compared patterns of abundance of hollow-nesting bird species and habitat structure in forest sites in Queensland, Australia, and demonstrated that site-specific differences in logging and fire history can have an important effect on bird species that is independent of existing structural differences (e.g. densities of hollow 'habitat trees'). Such work illustrates the need to tailor performance standards to local conditions and continually refine recommendations for achieving progress towards long-term conservation goals. Chapter 11 addresses further considerations in the selection of structural indicators of management performance, while Chapter 13 highlights the importance of considering confounding environmental variables and ecological processes when attempting to explain patterns of cause and effect in biodiversity monitoring data. Of course, there are situations where we have a very poor initial understanding of which attributes of forest structure or function are responsible for driving observed changes in biodiversity following a particular human impact. In such cases, it is difficult to *a priori* identify reliable indicators of management performance and it may only be possible to compare responses of biodiversity to changes in forest management activities themselves (e.g. logging schedules, cutting diameters, etc.) while attempting to improve our understanding of underlying relationships as monitoring proceeds.

The above arguments on the limitations of species-based indicators as measures of management performance take nothing away from the importance of setting species-based targets for conservation management (e.g. Villard and Jonsson, 2009; and see Chapters 9 and 12), or the central importance of species data in identifying high-biodiversity value areas of forest for permanent protection as part of a landscape-wide conservation plan (Margules and Sarkar, 2007; Moilanen et al, 2009). It is often the case that the conservation of individual species provides one of the central goals and motivations for large conservation investment programmes (such as the case of the northern spotted owl in the North-West Forest Plan in the western US (see Chapter 12). However, discounting practices such as population culling, supplementary feeding and captive breeding, conservationists and forest managers do not directly manage species. Instead people manage (whether for economic or conservation gains) the contributing factors that help determine, whether directly or indirectly, the potential value – subject to the influence of external threats and the legacy of historical disturbances – of an area of forest for a component of biodiversity or individual species of particular management concern (including changes to forest structure, alteration of disturbance regimes and the introduction of foreign elements such as chemicals and exotic species). Only by designing monitoring programmes with explicit recognition of the causal framework that links management actions through direct management impacts to changes in forest biodiversity can we reliably begin to assess changes in management performance and progress towards more ecologically responsible systems of forest use. Biodiversity conservation targets, whether based on overall measures of condition from multiple indicator species or populations of individual species of management concern are essential in guiding the development of any monitoring programme and measuring the ultimate success or failure of management (Chapter 9). However, it is also necessary to constantly link progress towards such targets to recommendations for on-the-ground changes in management practice.

BRINGING IT ALL TOGETHER: IMPLEMENTING AN OPERATIONAL FRAMEWORK FOR BIODIVERSITY MONITORING AS A GUIDE TO RESPONSIBLE FOREST MANAGEMENT

An important first step in the management of complex systems is to employ a relatively simple conceptual framework that can articulate our current understanding of the system and provide a mechanism by which different stakeholders are able to interpret and discuss particular issues associated with the monitoring and evaluation process (Margoluis and Salafsky, 1998; Salafsky et al, 2002; Hobbs, 2003). Figure 8.3 provides such a framework and expands on the basic structure of a coupled management and monitoring system as illustrated in Figure 8.1 to include a simple cause–effect model of how human activities impact biodiversity, and a clear indication of the role played by different indicators and monitoring approaches in assessing observed changes.

The first stage in the development of any management system is to establish an initial management plan based on clear goals and objectives, and incorporating a set of minimum practice standards, together with associated indicators and targets to verify performance. The definition of such a management plan will vary widely from place to place and is invariably determined through a combination of constraints set down by the relevant management authority (whether legal regulations, regional certification criteria, and/or economic, environmental and political concerns of local stakeholders) and any additional guidance that is available from existing scientific knowledge regarding the area and the type of forest under consideration. The performance targets that make up a standard may represent some minimum indicator value, range of values or threshold of acceptable deviation away from some baseline state (Ghazoul and Hellier, 2000; Wintle and Lindenmayer, 2008). The specific purpose of effectiveness monitoring is to assess the condition of a forest (as measured by selected indicators) against predetermined targets and determine whether compliance against a minimum level of performance has been achieved (Figure 8.4)

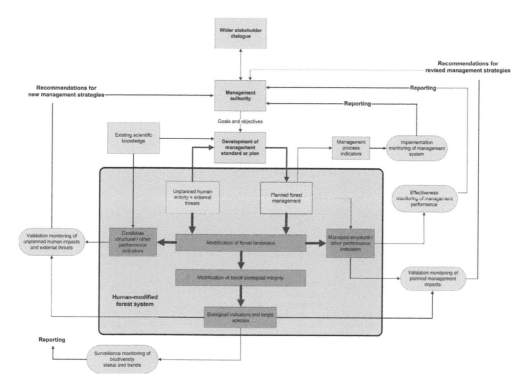

Figure 8.3 A general conceptual framework for a multi-tiered biodiversity monitoring system to help guide the development of an ecologically responsible adaptive forest management system

Note: Existing management standards are assessed through programmes of implementation and effectiveness monitoring that report on compliance against minimum practice requirements and performance targets respectively. Structural indicators, invariably provide more appropriate indicators of management performance than biological indicators but non-structural indicators such as disturbance processes (e.g. fire), and direct harvesting of non-timber forest products can also be important. Management standards can be iteratively improved following recommendations from validation monitoring (based on biological indicators) of the adequacy of existing performance standards in achieving progress towards long-term conservation goals. To effectively identify where the greatest opportunities for biodiversity conservation lie, validation monitoring should encompass the evaluation of both planned and unplanned human activities (right and left sides of diagram respectively). Because not all possible impacts on biodiversity can be evaluated at the same time, any validation monitoring programme needs to be underpinned by a research prioritisation exercise. See text for more details

Validation monitoring begins with the selection of a research objective (and set of associated indicators) that is focused on improving our understanding of a particular human impact, management practice or natural disturbance process. There are invariably a huge number of human activities and disturbance processes that could be evaluated and a key challenge lies in identifying those that are most deserving of priority attention (see Chapter 10). Monitoring is likely to make a more immediate contribution to improved management guidelines by evaluating the things that managers actually do as part of existing management plans, whether it is the logging of native forests, plantation management, the spatial design of multi-

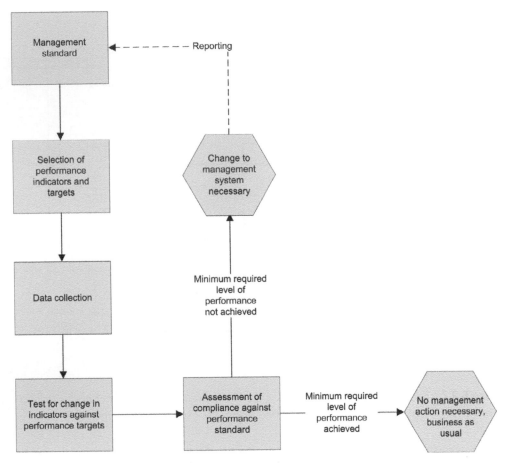

Figure 8.4 The process of effectiveness monitoring to assess compliance against minimum performance standards in forest management

ple-use landscapes, or the management of locally harvested non-timber forest products. Such activities represent potential threats facing biodiversity, but also potentially accessible leverage points for adjusting the established management system (Figure 8.2).

That said, it is important that biodiversity monitoring not only focuses on things that are actively controlled as part of a management plan, but also considers additional human impacts and potential conservation interventions that could become part of forest management in the future. Including additional objectives for monitoring is important because even in managed forest landscapes many human activities fall outside the remit of a formal management plan (e.g. extraction of non-timber forest products, as well as threats from outside the management area). Moreover, in poorly understood or poorly regulated systems, it is common that no management plan exists at all. Whether human activities in a landscape are entirely or only partially unplanned, it is worthwhile identifying a set of additional candidate indicators that are not part of any existing plan yet which could be included in a future management standard (Figure 8.2). With time, an increasing number of human impacts and potential conservation interventions can be evaluated through an ongoing programme of validation monitoring, and where necessary and appropriate incorporated into an increasingly more responsible and comprehensive management plan (Figure 8.2).

Once a research objective is established, the process of scientific inquiry then proceeds by drawing on our current ecological knowledge and information from past research to articulate multiple alternative hypotheses that represent competing ideas about how the management impact(s) may influence biodiversity (see Chapter 13). These hypotheses then form the basis for guiding the design and implementation of the data collection and analysis process (see Chapters 14 and 15) to improve our understanding of observed changes, inform recommendations for management and identify future research priorities (Figure 8.5). This represents a more scientifically rigorous approach to monitoring and research than that which characterizes many programmes to date, yet one that is crucial for the improved management of biodiversity and natural resources (Lindenmayer and Likens, 2009; Box 8.1).

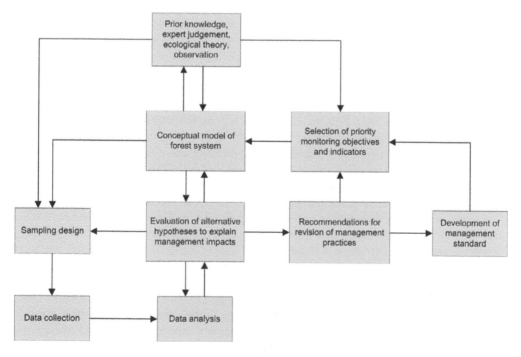

Figure 8.5 A conceptual framework of a validation monitoring programme as a research exercise designed to evaluate the adequacy of existing management standards and make recommendations for continued improvement in achieving progress towards long-term conservation goals

Box 8.1 The origins of a more scientifically rigorous approach to biodiversity monitoring and research

The science of biodiversity conservation has frequently been criticized as being both slow and uncoordinated (e.g. Peters, 1991), and research on the biodiversity of human-modified forest landscapes bears no exception (Lindenmayer and Fischer, 2006). In the case of monitoring, the popular adoption of a surveillance 'sit and watch' type approach has led to a reliance on inductive reasoning and retrospective analysis to elucidate the most likely causes for an observed change or decline in biodiversity. While such passive monitoring has its place (some of science's most important discoveries have been through serendipity), there is an urgent need for a more strategic and efficient approach to identifying more ecologically sustainable forms of forest use. In response to this need, the past two decades have seen an increasing number of calls for biodiversity monitoring to adhere more closely to something akin to the scientific method (Romesburg, 1981; Lindenmayer and Franklin, 2002; Rempel et al, 2004; Nichols and Williams, 2006; Lindenmayer and Likens, 2009).

In a seminal article published more than four decades ago, Platt (1964) described the scientific method as analogous to moving up an 'experimental tree'; alternative hypotheses are devised that relate to the question of interest, experiments are implemented to arbitrate between candidate hypotheses, and our confidence in these hypotheses is subsequently revised on the basis of empirical observations. Platt called this process 'strong inference'. However, ecology differs from the more experimental branches of science (e.g. physics, chemistry, cellular biology) in its ability to achieve the 'strong inference' of the scientific method as proposed by Platt. It is rarely possible, logistically, financially or ethically, to implement manipulative and fully replicated and randomized experiments in nature. By contrast, learning in applied ecology is accomplished through a gradual accumulation of evidence that varyingly favours alternative explanations (posed as hypotheses) about a particular real-world question or management problem (Hobbs and Hilborn, 2006). This distinction means that the interpretation of ecological monitoring data is often best served through approaches that explicitly accommodate uncertainty, and therefore often differ from more traditional approaches to data collection and analysis (Hilborn and Mangel, 1997; Hobbs and Hilborn, 2006; see Chapter 15).

In another important philosophical development, Holling (1998) called attention to two different 'cultures of ecology', noting that reductionist, experimental approaches that have seen such success in fields like molecular biology have important limitations when applied to complex, integrative and multi-scale ecological systems. Holling termed an alternative, although not mutually exclusive, approach – 'integrative ecology'. Integrative approaches are characterized by the explicit consideration of multiple competing hypotheses to explain human–ecological interactions, and acceptance that uncertainty and surprise represent integral parts of an anticipated set of adaptive responses. They are fundamentally interdisciplinary and combine historical, comparative and experimental approaches at scales appropriate to the issues at hand. The premise of this 'integrative ecology' is that knowledge of the system we deal with is always incomplete. There will rarely be unanimity of agreement among scientists and managers concerned with developing more ecologically responsible yet viable approaches to forest management – only an increasingly credible line of tested argument. Not only is the science incomplete, the system itself is a moving target, evolving because of the impacts of management and changing values of stakeholders concerned with managing forests for multiple-use.

CHAPTER 9

Setting Conservation Goals for Biodiversity Monitoring

Synopsis

- Conservation goals provide a reflection of societal values and political or institutional intent in management, create the entire context and sense of purpose of biodiversity monitoring, as well as the basis for selecting monitoring objectives and indicators.

- The concept of ecological condition, as measured by deviations from an appropriate reference condition or more desirable system state, provides an ecosystem perspective for management that is complementary to a limited focus on the conservation of individual species. The maintenance and restoration of ecological condition invokes a much broader conservation challenge than that which is focused on preserving a particular set of biodiversity elements, and incorporates notions of intactness and ecological resilience.

- Changes in ecological condition can be measured using ecological disturbance indicators that are classified according to their sensitivity to anthropogenic impacts.

- Identifying and measuring sites that can provide an appropriate reference condition is problematic but can be assisted by identifying reference sites based on an independent understanding of human impacts, and adopting a more realistic interpretation of a reference state that is not limited to undisturbed areas but also encompasses the notions of best available and best attainable condition for a given study region.

The importance of setting clear conservation goals for ecologically responsible forestry must not be underestimated. Goals provide a reflection of societal values and political or institutional intent in management, and create the entire context and sense of purpose of biodiversity monitoring, as well as the basis for identifying specific monitoring objectives and indicators (Noss, 1999; Tear et al, 2005, Nichols and Williams, 2006). However, in spite of its recognized importance, the process of defining management goals is rarely undertaken with sufficient care and attention (see Chapter 5).

The purpose of this chapter is to consider how concerns about biodiversity (e.g. as stated in the P&C of forest management standards) can be more clearly translated into conservation goals that can provide a meaningful guide to the ongoing process of AM. The focus here is on validation monitoring as the mechanism by which the adequacy of existing management standards for achieving long-term conservation goals is evaluated (see Chapter 8 for a full conceptual framework of the monitoring process). First, I briefly discuss how biodiversity values, which provide the basis for conservation goals, often vary among different stakeholders. Second, I outline a number of arguments for why the concept of ecological integrity or condition provides a useful overarching conservation goal, and encourages managers to look beyond a limited focus on the conservation of individual species. The final sections of the chapter address two related challenges in developing a biodiversity management and monitoring programme that places the restoration and maintenance of ecological condition as its underlying conservation goal: the selection of an appropriate set of indicators that can report on changes in condition; and the selection of a suitable reference condition to guide the interpretation of any observed changes.

STAKEHOLDERS AND THE VALUE OF BIODIVERSITY

Discussion of the value of biodiversity conservation and the role played by different stakeholder groups in justifying the value of conservation action is beyond the remit of this book. Nevertheless, it is very important that the scientists and managers who are responsible for implementing biodiversity monitoring programmes understand the different conservation values that are often held by different stakeholders or stakeholder groups. The term 'stakeholder' encompasses a diverse range of people affected by biodiversity, including those with health, welfare, intellectual, recreational, spiritual and financial interests who will be affected by biodiversity change within a given region (Green et al, 2005). Identifying such people can be very challenging. From a practical monitoring perspective, Green et al (2005) identify *interested parties* as a subset of stakeholders who specifically support a science-based measurement of biodiversity, and this is the group that is most relevant to the discussion in this book. They may be individuals but will more often be organized into institutions or associations, such as forest landowners, managers and communities, certification authorities, national forest management regulators, regional conservation organizations, research institutions and international forest governance groups. There will be other stakeholders who are likely to be influenced by changes to biodiversity and any decisions taken following the analysis of monitoring data, but who are not among this initial set of interested parties. These include those who may not wish to see any science-based assessment, do not know about it or understand it, or who lack the resources or political power to influence the evaluation (Green et al, 2005). Examples may include impoverished local people or those that live in areas distant to the landscape of concern yet nevertheless in someway influenced by changes to biodiversity (e.g. through existence and non-use values). Those responsible for implementing biodiversity management and monitoring programmes should endeavour to identify such groups and take their viewpoints into consideration, for example through the inclusion of additional biodiversity indicators, or in the assessment of trade-offs regarding the resources necessary for biodiversity conservation when reviewing management guidelines.

For those interested parties who already support and are involved in measuring the biodiversity impacts of forest management activities, broad recognition of the value of conserving biodiversity often already exists, either through management guidelines or voluntary certification standards (see Chapter 2). However, this value basis is often very vague and poorly articulated, and frequently emphasises unfeasible targets for conservation that may be of limited relevance to the actual ecological impacts caused by human activities.

MANAGING TO CONSERVE SPECIES AND MAINTAIN ECOLOGICAL INTEGRITY

Biodiversity is the buzzword that currently underpins the vast majority of conservation initiatives, and can be best defined as a multidimensional 'metaconcept' encompassing genes, populations, species, habitats and ecosystems (Noss, 1990; and see Chapter 5). In practice, many conservation projects are focused on protecting individual species, sets of species, or particular structural and compositional elements of biodiversity, such as vegetation types, that are of particular societal value or conservation concern. The second and related concept that, together with biodiversity, forms the conceptual foundation of modern conservation science is ecological integrity (Noss, 1990, 2004). Ecological integrity is defined broadly as the capacity of an ecosystem to support and maintain a community of organisms that has a structural, compositional and functional organization comparable to that of similar yet relatively undisturbed ecosystem, in the same region (Karr and Dudley, 1981; Karr, 1993). The maintenance of ecological integrity therefore invokes a much broader challenge for conservation than that which is focused on preserving a particular set of biodiversity elements, and incorporates notions of intactness, ecological resilience and the capacity to recover from disturbance, and the maintenance of evolutionary capacity (Angermeier and Karr, 1994; Noss, 2004; Folke et al, 2004). Changes to the ecological integrity of a system are reflected by deviations from an appropriate reference condition or more desirable system state, as characterized by relatively low levels of human impact. The terms 'integrity' and 'condition' are commonly used interchangeably, and I adopt the same approach. Ecological integrity is also closely related to the allied concept of ecological resilience (although the scientific literature on these two concepts is largely distinct), where high levels of integrity confer high resilience. This is defined by Walker et al (2004) as the capacity of a system to absorb disturbance and reorganize while undergoing change so as to retain essentially the same function, structure, identity and feedbacks (Walker et al 2004).

Distinguishing the concept of ecological integrity or condition as an ecosystem-wide perspective from a limited focus on individual species is important because the two perspectives underpin distinct approaches for management and monitoring. Although single-species management represents a valid and complementary element of a comprehensive approach to biodiversity conservation (Lindenmayer et al, 2007; Lindenmayer and Hobbs, 2007; Table 9.1), there are strong arguments for adopting an ecosystem-based monitoring approach that is focused foremost on the maintenance and restoration of ecological integrity.

Single species-based approaches to conservation management are often much easier to find in practice. In part, this is because it is much easier to draw attention to the needs of individual species; they frequently represent the cornerstone of efforts to mobilize interest and investment in conservation action (Box 9.1). Moreover, it is theoretically easier to develop prescriptive guidelines (and associated performance standards) for managing the resource and habitat requirements of individual species than entire ecosystems. Nevertheless, it is impossible to develop species-based approaches to satisfy the conservation requirements of all taxa, meaning that individual species monitoring will only ever make a piecemeal contribution to our understanding of what is needed to achieve truly ecologically responsible management. For most of the world, we have very little idea of the identity of those species that are most in need of conservation attention, let alone their specific resource requirements, or how they interact with one another. Furthermore, concerns about individual species vary widely among different landscapes making it hard to develop generic guidelines for good monitoring practice that can readily be implemented across multiple sites.

Box 9.1 The importance of single-species approaches in biodiversity conservation

Choices regarding the selection of individual species for biodiversity monitoring often reflect a tension between their ability to infer information on wider impacts of human activity for other species or ecosystem attributes (i.e. their indicator value) and their intrinsic value or importance to stakeholders. In many cases, this balance is clear-cut and individual species can represent a valuable focus for monitoring in their own right, including keystone species, pest species, invasive species and species of particular economic importance to local communities (see Chapter 12). In the case of species that are known to be highly threatened (yet where the nature of threats are poorly understood), monitoring work may be vital if we are to guarantee their survival. In other situations, the threat status of individual species, their intrinsic economic value or relevance to ecosystem function, may be unclear, casting doubt on the usefulness of any monitoring data in guiding improvements to management.

Biodiversity monitoring exists at the interface between science, policy and management and, in order to be most effective, the development and use of biological indicators or target species must be responsive to political and social values – at local, regional and global scales (Aldrich et al, 2004; Sheil et al, 2004; Turnhout et al, 2007). Irrespective of any scientific criticisms regarding the validity or cost-effectiveness of some single-species monitoring data, there may often be over-riding social and cultural justifications that make them particularly deserving of research and management attention. Species that encompass many of the more normative aspects of conservation can be important in helping to engage local stakeholders in the management process.

Alternative management approaches that focus on maintaining and restoring the integrity or condition of the overall forest system avoid this problem by

Table 9.1 Strengths and limitations of single-species and ecosystem management-based approaches to conservation research and monitoring

	Strengths	Limitations	Opportunities for cross-fertilisation
Single-species research	• Is easier to elucidate causal processes underpinning declines of individual species	• Detailed information on a limited number of species – other species may be overlooked	• Can highlight links with other species, especially if functional roles are studied
	• Can provide useful information for policy and management on threatened species, keystone species and invasive species	• 'Cute and cuddly syndrome' – not all species receive the same attention	• Research on keystone species benefits understanding of entire ecological communities
	• Has a strong history and methodological basis	• Many relationships are highly species-specific or site-specific, which can make generalisation difficult (or even detrimental)	• Invasive species research and research on disease-carrying species have ecosystem implications
	• Yields detailed understanding of relationships between a species and its environment; mechanistic understanding can sometimes provide insights applicable to other species ('the model system approach')	• Can be overly reductionist and produce insights of limited direct benefit	• Charismatic and highly endangered species may attract funding for conservation, including for activities that benefit many species simultaneously
		• May not address broad scale cause of decline and/or distract from more ultimate drivers of unsustainable resource management	• Charismatic species can act as focal or flagship species thereby fostering public interest and involvement in biodiversity conservation
Ecosystem-orientated research	• Encapsulates many species, so many species may gain from the research	• If done superficially, a lack of detail may hamper effective, targeted conservation	• Can identify areas where detailed single-species information is required while gaining general information about many species
	• Identifies key management priorities that may benefit many species simultaneously	• Specialised or wide-ranging species may be overlooked or undervalued	• Actions at the ecosystem scale may foster interest in particular species as indicators or focal species
	• Often involves species counts so it is immediately quantitative	• History of ecosystem-orientated management is relatively recent in the mindset of many scientists and practitioners	• Can highlight differences between species and scales if lower-level studies are effectively incorporated within a larger-scale study
	• Cost-effective to look at many species at one time if high level of detail is not required	• If done superficially, it can confuse correlation and causation	• Can stimulate improved ecosystem understanding, including the roles of and interactions among individual species

Note: Opportunities for cross-fertilization between single-species research and ecosystem-oriented research are also listed
Source: Adapted with permission from Lindenmayer et al (2007)

integrating data from a wide range of sources, including multiple disturbance-sensitive indicator taxa and structural and functional indicators, to provide a more complete understanding of the nature of human impacts. Efforts can then be made to maximize, across the entire managed landscape, the representation of compositional, structural and functional elements of forest biodiversity that are characteristic of less disturbed areas of forest for a given region. Because different human activities exert different impacts on biodiversity, achieving progress towards this goal is often best achieved by identifying spatial patterns of complementarity in conservation value across different areas of the managed landscape.

The notion of improvements to ecological condition as a broad-based conservation goal is well developed in the science and management of aquatic systems (e.g. Linke and Norris, 2003; Linke et al, 2007) and has its origins in the management of water resources in the US, where it has been used with tremendous practical success (Karr, 1991; Angermeier and Karr, 1994; Karr and Chu, 1999). By contrast, although the importance of ecological integrity is explicitly stated in many ecological criteria for responsible forest management (e.g. the FSC International Standard, 2002), there has been very little progress in translating the concept into terrestrial monitoring programmes.

A central requirement for monitoring the impacts of human activities on ecological condition is a mechanism for accurately measuring changes relative to a more desirable state or reference condition (also termed 'benchmark' or 'target state'). However, the accurate measurement of a reference condition, and deviations from it, as a basis for judging management performance is fraught with challenges that have confounded scientists working in natural resource management systems for decades (Woinarski, 2007; Lindenmayer and Hobbs, 2007; and see Chapter 6). Although there are no ready answers to this challenge, it is easier to identify opportunities for progress if we are clear about our goals and the limitations of the methods we use. We need to take a step back from some of the scientific discussions that surround the technical details of how to measure change in ecological condition, and first question what is necessary for monitoring to be useful for managers, as well as what is reasonable for most managers to be able to adopt and implement without losing sight of long-term sustainability goals (see Chapter 10). For example, having the capacity to accurately quantify the natural variability of a reference condition is not an essential prerequisite for monitoring to make a useful contribution towards guiding progress in management. Neither is it essential to have an exact understanding of the extent to which the abundance or diversity of individual forest attributes and species can be changed before the resilience of the system is irreversibly impaired. Both of these issues represent valid long-term research goals, but a lack of perfect scientific understanding should not be used as an excuse for inactivity in management – a tenet that lies at the heart of the AM philosophy.

Following on from this discussion, there are two main challenges to developing a validation monitoring programme that holds the maintenance and restoration of ecological integrity and condition as its primary conservation goal: the selection of an appropriate set of indicators that can report on changes in condition, and the selection of a suitable reference condition to guide the interpretation of any observed changes.

SELECTING INDICATORS TO VALIDATE CHANGES IN FOREST CONDITION

As defined above, the concept of ecological integrity and condition is very broad and encompasses a wide range of biological elements (e.g. populations, species, species functional groups), their trophic and other ecological interactions, as well as abiotic processes that are indirectly linked to different levels of the biological system. However, despite encompassing both processes and species, species are used more frequently than ecological processes as indicators of changing condition, because they are typically more sensitive to degradation, more fully understood, and less expensive to monitor (Angermeier and Karr, 1994). Species that serve this function, and are capable of providing a sensitive and reliable measure of disturbance impacts at a spatial scale that is relevant to management, can be termed 'ecological disturbance indicators' (Chapter 4; and see also Kremen et al, 1994; Dufrëne and Legendre, 1997; McGeoch, 1998; 2007; Howe et al, 2007). The information content of ecological disturbance indicators can be further enhanced when selected indicator groups are known to perform important and reasonably well understood ecological functions.

In order to be most useful for management, assessments of ecological condition and integrity should reveal information on the direction in which an area of forest is changing along a gradient of condition, rather than a simple snapshot assessment at one point in time, thus allowing monitoring to guide a process of continuous ecological improvement (Noss, 1999, 2004). It is therefore important to identify indicators that can be sampled across the full spectrum of an anthropogenic disturbance gradient. This enables the development of a reliable empirical relationship – consistent quantitative change – across a range of human influence (Karr and Chu, 1999; see also Oliver et al, 2007; Figure 1.2). Individual ecological disturbance indicator species are classified according to their sensitivity to anthropogenic impacts, with patterns of species replacement – from mature forest specialists to semi-tolerant species that inhabit modified forest systems and open-area and invasive species – serving to define the underlying gradient in ecological integrity (e.g. Canterbury et al, 2000; McGeoch et al, 2002). Thus, the responses of taxa associated with the undisturbed end of the gradient (i.e. positive indicators of ecological integrity) can be combined with those associated with open or structurally modified habitats (i.e. negative indicators) to give a more complete picture of overall changes (Brooks et al, 1998; Carignan and Villard, 2002; and see Chapter 12). Monitoring such replacement patterns can provide additional conservation information because species that depend, or partly depend on, undisturbed forest are often also regional endemics (e.g. Dunn and Romdall, 2006; Cleary and Mooers, 2006), locally endangered and especially susceptible to habitat fragmentation (Laurance et al, 2002; Tabarelli et al, 2004).

The use of species indicator groups to characterize the position of a given site along a gradient of increasing anthropogenic disturbance in this way is very common. Other terms that have been used to classify such species, and are synonymous with the notion of ecological disturbance indicators, include 'predictor species' (Basset et al, 2004a, 2008c), 'habitat assemblages' (Canterbury et al, 2000), 'management indicators' (Lindenmayer et al, 2000) or 'response guilds/groups' (Szaro, 1986) (see Chapter 12 for an extended discussion of the indicator selection process).

Because different species and species groups have distinct ecological requirements, and interact with their local environment at different spatial and temporal scales, combining multiple ecological indicator groups within a comprehensive monitoring programme can help in understanding a range of management impacts on multiple levels of ecological organization. Employing a combination or 'basket' of different indicators also helps to avoid the danger that once a single indicator is chosen it may no longer act as an independent yardstick of what it is supposed to indicate (i.e. tailor management to ensure high levels of a particular species or species group are maintained irrespective of underlying changes in other parts of the system (Landres et al, 1988)).

Recognition that species interact with their environment in diverse ways and at multiple scales also highlights the value of species-based indicators of ecological condition, as species experience a wide range and variation of environmental conditions through time. By contrast, structural indicators often provide only highly variable snap-shot measures that can easily misrepresent important underlying differences (Howe et al, 2007).

Monitoring of ecological disturbance indicators can be further supplemented with surveys of individual target species, such as sensitive keystone species, that make a disproportionate contribution to the structure and functioning of forest ecosystems (see Chapter 12), as well as the direct monitoring of ecological processes and functions themselves. Depending on the monitoring objective such measurements may be taken as covariates to enhance our understanding of changes in species-based indicators (see Chapter 13), or as targets in their own right that contribute to an overall goal of improving the integrity of the forest ecosystem. For example, the monitoring of soil and leaf-litter attributes and processes can provide particularly valuable data on changes in ecological integrity, as well as providing additional explanatory variables to help understand the biotic disturbance responses (Box 9.2). Raison and Rabb (2001) provide a comprehensive review of indicators of soil attributes and processes for forest management.

SELECTING A REFERENCE CONDITION TO GUIDE FOREST BIODIVERSITY MONITORING

A common difficulty facing monitoring programmes that are concerned with measuring changes in ecological condition following management is the choice and definition of the reference condition against which to interpret any observed change. Understanding what comprises a more desirable forest state is fundamental to guiding improvements in management, even if exact specifications are not needed to judge management performance. Deviation away from a desired reference condition is the arbitrator by which managers can distinguish the varying

---

Box 9.2 Evaluating the consequences of forest management on leaf-litter dynamics

Leaf litter represents a fundamentally important component of the overall structure and functioning of forests worldwide, contributing to the retention of nutrients and the maintenance of the chemical and physical properties of the soil that in turn are important for maintaining vegetation dynamics and diverse communities of soil biota (Vitousek, 1984).

Given its ecological importance, the quantitative measurement of parameters that describe the structure and function of the leaf-litter layer represent a powerful tool for capturing changes of far-reaching importance for the long-term viability and ecological integrity of managed forest systems. It is reasonable to assume that the maintenance of leaf-litter processes closer to the rates observed in minimally disturbed areas of forest will enhance the overall resilience of the system, helping to conserve those species that depend upon certain structural or functional attributes of the litter layer to provide key habitat requirements.

While the processes of litter fall and leaf-litter decomposition have been well studied across the world over recent decades, our understanding is limited for many forest types and management regimes, particularly disturbed and regenerating habitats in the humid tropics (Barlow et al, 2007b). Barlow et al (2007) employed a logistically straightforward and cost-effective method for measuring total leaf fall and decomposition rates, and showed how rates can then be compared between different levels of disturbance to illustrate the effects of management. In their case, it was shown that secondary forests regenerating on abandoned and degraded Amazonian land can be effective at restoring basic processes, such as leaf fall and decomposition. Further study will help illuminate the extent to which the litter layer in these secondary forests performs comparably to undisturbed forest for specific functions, such as the return of key nutrients to the soil and the maintenance of diverse invertebrate communities.

---

conservation benefits that are associated with alternative management strategies. Aquatic ecologists have made particular progress in devising effective reference-condition approaches to measuring human disturbances and the effectiveness of management, although very little of this has been translated into work in terrestrial systems. In the following sections, I outline five guidelines or considerations that can help in establishing a suitable reference condition for biodiversity monitoring.

Identifying reference sites based on an independent understanding of human impacts

Condition assessments are often subject to harsh and justifiable criticism when they are internally circular i.e., gradients of condition are determined entirely by changes in the indicator that is identified as defining condition in the first place. Sites that are classified as being of high condition are only described as such because they contain high levels of the selected indicator and nothing more – the indicator is just measuring a change in itself. It is perfectly acceptable to have a constrained definition of condition that relates to habitat quality for a species of particular conservation concern (e.g. availability of specialized roosting sites) but this cannot be conceived of in broader terms of forest condition or ecological integrity. Instead, it is necessary to first identify a gradient of disturbance based on our existing knowledge of human impacts, including the identification of a set of minimally disturbed reference sites, and then to subsequently employ a set of indicator taxa that can characterize the reference condition and quantify any deviations from it in ecologically meaningful terms. The initial identification of reference sites needs to be made on the basis of an agreed set of minimal-disturbance criteria (Bailey et al, 2007). This may include historical records of minimal human activity (e.g. as prescribed by protected areas), or areas that have made a substantial recovery from past disturbance (e.g. regrowth of mature secondary forests that were originally logged or clear-felled).

Accepting the problem of shifting baselines

In an ideal world some may argue that the concept of a reference condition would be founded in the idea of naturalness, and the ecological impacts of human activities would be measured against an intact or undisturbed state. However, such an approach is rarely, if ever, possible as anthropogenic impacts are evident in forests across the entire world, obscuring our ability to identify what is 'natural' (see Chapters 1 and 6). Even if an area of forest has been free from obvious human impacts, deciding how far back in time we need to go to define a natural benchmark for a given forest type is a subjective decision. Pre-arrival of European settlers? Pre-arrival of indigenous peoples? There are also strong philosophical arguments against such an approach as it controversially distinguishes humans as being outside 'nature', a position that is at odds with the growing body of research demonstrating the highly coupled nature of human–ecological systems. Clearly, a more flexible definition of what can constitute a reference condition is needed if progress is to be made.

We need to accept that in some places the baseline has shifted and that we have lost the chance to conserve (or even identify the loss of) the most vulnerable species. However, the characterization of even sub-optimal sites as a reference condition can still provide a valuable guide for managers working in even more biologically impoverished areas. This kind of logic introduces the notion of 'best-available condition' as the most appropriate, and indeed only, reference standard for many areas of the world. Stoddard et al (2006) defined a series of terms to describe different types of reference condition, with each type being associated with different opportunities for conservation (Table 9.2). While this typology was developed for stream systems, the concepts are directly transferable to forests.

Accepting that human disturbances are nearly ubiquitous features of forest landscapes worldwide is essential not only in framing how biodiversity monitoring can usefully proceed in degraded regions, but also in understanding what represents *a reasonable goal* for ecologically responsible management. This point is reflected in the concept of the 'best attainable position' as proposed by Stoddard et al (2006) (Table 9.2).

Table 9.2 A typology for defining the reference condition under varying levels of human disturbance

Term	Description
Reference condition for ecological integrity	Reserved for the traditional concept of the reference condition as a completely natural or intact state
Minimally disturbed condition	A measure of condition in the absence of significant human disturbance. The concept of a 'minimally disturbed condition' accepts that some level of disturbance is almost inevitable for most of the world, and provides what is often the best approximation of the reference condition for ecological integrity
Historical condition	This term describes the condition of a system at some point in its history. It may be an accurate estimator of true reference condition (for ecological integrity) if the historical point chosen is before the start of any human disturbance. However, many other historical reference points are possible (e.g. pre-industrial, pre-Columbian)
Least disturbed condition	Least disturbed condition is found in conjunction with the best available physical, chemical, and biological habitat conditions given today's state of the landscape. It is ideally described by evaluating data collected at sites selected according to a set of explicit criteria defining what is least disturbed by human activities (e.g. contemporary landscape assessments of structural complexity, and/or historical records of human land use and management). The specifics of these criteria will differ across ecological regions, as the characteristics of the landscape, and human use of the landscape, vary
Best attainable condition	Best attainable condition is equivalent to the expected ecological condition of least-disturbed sites if the best possible management practices were in use for some period of time. Sites in the best attainable condition would be places where the impact on biota of inevitable land use is minimized. This is a somewhat theoretical condition predicted by the convergence of management goals, best available technology, prevailing use of the landscape, and public commitment to achieving environmental goals. The upper and lower limits on the best attainable condition are set by the definitions of "minimally disturbed condition' (MDC) and 'least disturbed condition' (LDC) respectively. Best attainable condition will never be 'better' than MDC, nor 'worse' than LDC, but may be equivalent to either, depending on the prevailing level of human disturbance in a region.

Source: From Stoddard et al (2006)

Matching impact sites to the most appropriate reference site

Forest ecosystems are highly heterogeneous with respect to spatial patterns of variability in biophysical features (e.g. topography, hydrology and soil) and the impacts of natural disturbance regimes. As such it is important that evaluations of management impacts at a particular site are made with reference to a carefully selected reference site that is appropriate for providing an estimation of the ecological condition of the impact site in the absence of any human disturbance. This requires characterizing reference conditions across multiple independent sites from which the most appropriate site for comparison can be selected. The formalized 'reference-condition approach' from river assessments represents a powerful approach to achieving exactly this (Bailey et al, 2004). An array of reference sites are used to characterize the ecological condition of a region (using specialized indicators); a test site is then matched to an appropriate subset of reference sites using independent data on the physical and abiotic characteristics of the sites, and a condition impact assessment made. The particular strength of the reference condition approach is that it fine-tunes comparisons by matching impact sites with reference sites that have similar environmental characteristics, thereby allowing for ecologically meaningful comparisons.

Recognizing that forest ecosystems are naturally dynamic

Even relatively intact areas of forest can be highly dynamic. The composition, ecological structure and interactions of natural systems all vary in response to natural disturbances at multiple temporal scales, resulting in a variety of alternate states (Woinarski, 2007). This variability makes the definition of a reference condition problematic when it is based on short-term measurements from a single site. Consequently, wherever possible it is advisable to develop an understanding of natural variability in the reference condition by collecting information on long-term trends – or, where this is not possible, substituting time for space and collecting data across multiple sites that are likely to reflect different states in a natural disturbance or successional cycle (Woinarski, 2007). The extent to which these different forest states exhibit complementary conservation values can be evaluated by comparing them against disturbed sites.

Dealing with landscape context

Forests vary with respect to differences in structure, composition and function at both patch (i.e. individual forest blocks) and landscape scales. At the landscape scale,. ecological condition can be impaired when certain landscape features decline in abundance (e.g. riparian forest habitat), or when landscape patterns are altered to the point where flows of materials, energy or organisms are influenced (Franklin and Swanson, 2007). In regions where large areas of contiguous forest no longer exist, the process of selecting sites that can be used to provide a reference condition must take such considerations of landscape context into account. The science of designing human-modified landscapes for conservation is extremely complex, and thus far has generated few general principles for management (Lindenmayer and Hobbs, 2007; and see Chapter 10). However, we do know that different species experience changes to landscape context in very different ways (e.g. depending on their home-range sizes and dispersal capacities), and that depending on species life-history strategies there can be considerable lag-periods before the effects of human impacts (such as fragmentation) become evident. Consequently, when attempting to characterize a reference condition in landscapes that are already moderately degraded, it may be necessary to include data from even more reference sites than would be the case in undisturbed forest.

CHAPTER 10

Setting Objectives for Biodiversity Monitoring

Synopsis

- The process of setting objectives for biodiversity monitoring is an exercise in research prioritization that should seek to identify the management practices or threatening processes whose evaluation would make the greatest contribution towards improving opportunities for biodiversity conservation.
- Research priorities are commonly selected on a relatively *ad hoc* basis that is strongly influenced by the nature of funding opportunities, as well as differences in the personal interest of the researchers involved. Adopting a more systematic approach that compares alternative objectives with regard to their ability to generate real and practically relevant learning, could greatly increase the efficiency and cost-effectiveness of research.
- The first step in a more systematic approach to identifying research priorities is to identify a suite of existing and/or potential management practices or 'management control variables' that are candidates for evaluation. This review process can be aided by a series of assessments to summarize information on current management practices, the structure and composition of the study landscape, and the characteristics of reference sites.
- The second step is to employ a selection framework for identifying priority objectives from this candidate set. The three types of motivation that drive decisions about research priorities are: *the opportunity to learn* – what management practices are feasible to evaluate and monitor?; *the necessity to learn* – what management practices are associated with the greatest level of uncertainty regarding their implications for biodiversity conservation?; and *the value of learning* – what management practices, if adjusted or newly implemented, are likely to deliver the greatest benefits for biodiversity?
- The value of learning is determined not only by the estimated impact of management on biodiversity (e.g. as determined by a formal risk assessment), but also by the extent to which any findings can be extended to other landscapes and related management systems, and the likelihood with which any findings from monitoring will translate into policy change.

Once the overall conservation goals for monitoring have been defined (Chapter 9), the next step is to decide on specific monitoring objectives. The process of defining clear monitoring objectives is crucial because underlying design choices (indicator selection, sampling design, analytical approach, etc.) invariably depend upon the perspective and context set by the objectives. Furthermore, it is important to justify *why* certain monitoring objectives were chosen over any others. The *why* is important because, under the broad remit of improving the condition of a managed forest landscape for biodiversity (or even a small number of target species), there is a practically limitless number of questions that could, in theory, be asked. Hence, a key part of the process of defining specific objectives in biodiversity monitoring lies in prioritizing the use of limited resources, and understanding how research and monitoring can make the most cost-effective contribution to management.

Where management standards already exist, the objective of implementation monitoring is straightforward as the task of the assessor is simply to record whether certain agreed practices are being implemented or not. In the case of effectiveness monitoring, the setting of objectives is also straightforward as the requirement is simply to assess the status of a particular indicator against a given minimum management standard or threshold (see Chapter 8 for a discussion of the full conceptual framework for monitoring).

By contrast, the process of setting objectives for biodiversity validation monitoring represents an exercise in research prioritization – where it is necessary to identify those management activities (whether established within a standard or not) that are in most urgent need of scientific attention in the development of a more ecologically responsible management system. The problem of prioritizing objectives is the same for monitoring as it is for any other research-based exercise. In this chapter, I tackle the question of prioritizing research objectives by first reflecting on the diversity of ways in which conservation science has already contributed towards guiding more responsible management. What factors have helped define the research development process so far? What factors can be identified as acting to limit the transfer of research to management? Within this context, I then question the suitability and effectiveness of the current process by which most research priorities are selected for biodiversity validation monitoring programmes. In response to this problem, and to help guide a more effective approach to selecting objectives, I outline a framework of 'motivational factors' that should be considered when seeking to identify research priorities. This framework is relevant to both managed forests and applied ecological problems in general.

BIODIVERSITY CONSERVATION RESEARCH AND MONITORING IN MODIFIED FOREST SYSTEMS: AN ASSESSMENT OF WORK TO DATE

There are at least six major areas of research activity that are associated with different forms of direct or indirect human impact on the biodiversity and ecological integrity of forest ecosystems:

1   **Management of native forest**, including on-site management, such as harvesting

methods and schedules, as well as impact-mitigation techniques, such as structural retention systems.

2   **Plantation and restoration of cleared land**, including different types of commercial tree plantations, understorey and pest management, reforestation techniques, natural regeneration processes and afforestation.

3   **Spatial design of multiple-use landscapes**, including the composition and configuration of different forest and non-forest land-use types, and the value of forest patches, reserves, buffer zones, forest strips and riparian corridors.

4   **Disturbance regimes**, including fire, flood, pest-outbreaks and wind. Also the exacerbating effects of climate change on disturbance regimes and environmental conditions.

5   **Harvesting of non-timber resources**, including game-hunting, fruits and nuts, resins, latex, vines and other plants (either as a main economic activity or as an additional impact associated with the presence of local human communities in managed forest systems).

6   **Endogenous and indirect threats**, linked to harvesting of non-timber resources and including trophic cascades, disruption of so-called 'mobile link' species (e.g. pollinators, seed dispersers), and invasive species.

It is important to note that different human activities (whether in the form of threatening processes or conservation interventions) rarely impact biodiversity directly or in isolation, but instead operate through a diverse and interacting array of proximate mechanisms (e.g. changes to food and breeding resources, dispersal opportunities and physiological tolerance limits) and feedback effects from shifts in community composition and abundance structure (Gardner et al, 2009; and see Chapter 13).

Table 10.1 illustrates a small fraction of the empirical research conducted to date that has focused on evaluating the biodiversity impacts of particular forest and landscape management activities (including planted and native forest systems). I selected these studies from a pool of thousands in order to capture examples that encompass diverse objectives and research contexts. Although this is a small and biased sub-sample of the large body of work that has been conducted on this subject during the last few decades – and only a few of these studies represent multi-year monitoring programmes – it helps illustrate some of the key issues that characterize the contribution of biodiversity research to forest management to date and hence ways in which future research objectives can be improved.

Because the process of setting objectives is fundamental to the development of the entire programme, it is helpful to discuss how these issues relate to different elements of the wider monitoring framework (see Chapter 9). Here I focus on four such sets of issues: knowledge gaps in research; choice of biodiversity measurements; confounding factors in research design; and problems of disconnect between management and biodiversity research.

Table 10.1 Examples of biodiversity research and monitoring programmes in human-modified forest landscapes during the past decade

Study reference(s) and location	Research focus	Area of management control under investigation	Research objective	Research insights relevant to forest management and biodiversity conservation
Anand et al. (2008) Western Ghats, India	Spatial design of multiple-use landscapes	Management of stand diversity in shade-coffee, landscape design in locating coffee stands relevant to forest remnants	To evaluate the relative importance of local stand composition and landscape structure on the diversity of birds in shade coffee plantations	Increase in silver oak as the shade tree and increased distance from forest fragments were the primary factors driving declines in bird diversity and abundance, while increase in basal area of native trees in plantations also had a positive effect
Hawes et al. (2008) Brazilian Amazon	Spatial design of multiple-use landscapes	Maintenance and restoration of narrow forest strips and riparian buffers in plantation landscapes	To evaluate the value of *terra firme* (dry land) and riparian forest strips of varying lengths for birds in a *Eucalyptus* plantation landscape	Forest strips more deeply embedded within the plantation landscape exhibited impoverished bird assemblages characteristic of secondary forest. Riparian and *terra firme* strips harbour distinct groups of species
Lees & Peres (2006) Brazilian Amazon	Spatial design of multiple-use landscapes	Maintenance of native forest fragments in plantation landscapes, restriction of logging and fire in fragments	To evaluate the combined effects of fragmentation and forest disturbance from logging and fire on the composition and diversity of birds	Birds exhibited different responses to fragment size and patch degradation according to their level of dependence on primary forest habitat, yet patch size and shape accounted for the majority of spatial variability in species occupancy
Lees & Peres (2008) Brazilian Amazon	Spatial design of multiple-use landscapes	Maintenance and restoration of native forest corridors in cattle landscapes. Restriction of livestock access to riparian buffers	To evaluate the value of riparian forest corridors for native forest birds, and the relative importance of corridor width and structural integrity in maintaining diversity	Highly species-specific patterns of corridor use. Narrow and/or highly disturbed corridors were very depauperate while wide (>200m) and well-preserved corridors maintain an almost intact assemblage
Rondon et al. (2008) Peru	Spatial design of multiple-use landscapes	Use of strip clear-cutting as a silvicultural technique	To evaluate the recovery of forest structure, function and composition, including commercially important trees in areas that had been subject to strip clear-cutting	15 years after clear-cutting, trees in strips recovered approximately two-thirds of their basal area and two-thirds of their original species richness. Pioneer trees still dominated after 15 years, while silvicultural thinning of regenerating areas improved recovery of commercial trees

Table 10.1 Examples of biodiversity research and monitoring programmes in human-modified forest landscapes during the past decade (continued)

Study reference(s) and location	Research focus	Area of management control under investigation	Research objective	Research insights relevant to forest management and biodiversity conservation
Uezu et al. (2008). Atlantic Forest, Brazil	Spatial design of multiple-use landscapes	Establishment of agroforestry woodlots in cattle landscapes	To evaluate the contribution made by agroforestry woodlots to the avifaunal diversity of a cattle-dominated landscape, and to facilitate bird movement between native forest fragments	No positive effect found from agroforestry on total landscape diversity. There appears to be an 'optimal matrix permeability' at which stepping-stone features like agroforestry have a marked positive effect on landscape connectivity. This positive effect is greatly reduced when the matrix is either particularly resistant or particularly permeable
Lindenmayer et al. (2008). Australia	Plantation and restoration of cleared land, and spatial design of multiple-use landscapes	Maintenance of native forest fragments in plantation landscapes	To evaluate long-term changes in use of native forest fragments by birds following changes in surrounding landscape context (pasture to timber plantation)	Small forest fragments contained an appreciable number of species of bird. Patterns of bird occupancy increased across all forest fragments and control sites indicating landscape-scale (treatment independent) changes
Barlow et al. (2007), and references therein. Brazilian Amazon	Plantation and restoration of cleared land	Set-aside of fallow land in a plantation dominated landscape	To evaluate the multi-taxa biodiversity conservation value of *Eucalyptus* plantation and secondary forests	Secondary forests provided highly complementary species habitat to plantations in the wider landscape. Marked spatial turnover in species composition within each land use. Highly variable responses of different species groups to land-use change
Barlow et al. (2008). Brazilian Amazon	Plantation and restoration of cleared land	Clearance and management of understorey vegetation	To evaluate the relative importance of landscape configuration and plantation age, productivity and understorey structure and composition for the diversity of fruit-feeding butterflies in *Eucalyptus* plantations	Stand-level factors, rather than landscape context, had a dominant influence on spatial patterns of butterfly diversity, with species richness of understorey vegetation being strongly associated with butterfly richness
Eycott et al. (2006). England	Plantation and restoration of cleared land	Retention of mature plantation stands, stand clear-felling	To evaluate the relative importance of stand management, environmental and land-use history factors in determining understorey plant species composition in a plantation landscape	The majority of plant species recolonised stands following key phases of the plantation cycle, including after canopy closure and felling. Annuals dominated early growth stages while shrubs were dominant in mature stands. Maximizing representation of young plantation stands will help maximize plant diversity

Table 10.1 Examples of biodiversity research and monitoring programmes in human-modified forest landscapes during the past decade (continued)

Study reference(s) and location	Research focus	Area of management control under investigation	Research objective	Research insights relevant to forest management and biodiversity conservation
Aguilar-Amuchastegui & Henebry (2007). Costa Rica	Management of native forest	Logging intensity (trees felled per hectare)	To evaluate the relationship between different levels of logging intensity, forest stand structural heterogeneity, and dung beetle community structure and composition	Found a logging intensity threshold of four trees per hectare above which there is a significant shift in forest structural heterogeneity and loss in richness and diversity of associated dung beetle communities
Hutto & Gallo (2006) North America	Management of native forest	Reduction or prevention of salvage logging post wildfires	To evaluate the effects of post wildfire salvage logging on cavity nesting birds	All except one species (of a total of 18) nested at higher densities in unlogged areas. This effect was likely due to reduced food supply rather than reduction in cavity nest options
Lindenmayer (2007). Australia	Management of native forest	Application of a Variable Harvest Retention System (VHRS)	To evaluate the response of vertebrates and plants to the silvicultural VHRS that maintains a high retention of structural features and stand attributes during logging cycles	High variability in taxon responses. A significant number of small mammals and birds persisted in habitat retention islands within logging coupes, while reptiles exhibited a much more negative response
Mac Nally et al. (2001) Australia.	Management of native forest	Restoration of coarse-woody debris in degraded forests	To evaluate the consequences of varying levels of removal of coarse-woody debris (for fuel wood and fire management) on vertebrate diversity	Highly variable responses to loss of debris, with extreme sensitivity in the case of one native mammal, localised positive effects on birds, and no effect on frogs and reptiles. High spatial variance in debris restoration likely important for birds
Pearce & Venier (2005) Canada	Management of native forest	Management of forest stands (maintenance of dead wood, understorey complexity, rotation time).	To evaluate the adequacy of stand structural attributes in determining patterns of small mammal diversity and abundance	Coarse-scale relative changes in mammal species abundance could be tracked through changes in stand structural attributes but high levels of inter-annual variability confounded prediction of changes in minimum population size
Presley et al. (2008) Brazilian Amazon	Management of native forest	Implementation of RIL	To evaluate the impact of RIL and changes to forest physiognomy on species of bat	A high variability in species responses. Abundant species were resilient to impacts but up to 15 species of rare bat were negatively affected by RIL

Table 10.1 Examples of biodiversity research and monitoring programmes in human-modified forest landscapes during the past decade (continued)

Study reference(s) and location	Research focus	Area of management control under investigation	Research objective	Research insights relevant to forest management and biodiversity conservation
Parry et al. (2009) Brazilian Amazon	Harvesting of non-timber resources	Control of hunting in areas of secondary forest fallow land	To evaluate the capacity of secondary forests in a plantation landscape to support sustainable hunting of game animals	Secondary forests support lower levels of mammal off-take than neighbouring primary forest, and an individual smallholder would require more than 3km² of secondary forest to support a sustainable harvest
Terborgh et al. (2008) Peru	Endogenous and indirect threats	Prevention of hunting	To evaluate how severe hunting of large mammals influences plant recruitment processes	In heavily hunted areas, trees dispersed abiotically or by small birds and mammals were shown to be substituting those dispersed by large birds and mammals
Barlow & Peres (2004) Brazilian Amazon	Disturbance regimes	Prevention of wildfires escaping from agricultural land into neighbouring forests	To evaluate the impact of single and recurrent wildfires on forest understorey birds	Foraging guilds differed in their response to burn severity, although most guilds declined. There was almost no overlap in species composition between unburnt areas and areas of forest that had burnt more than once

Knowledge gaps and historical biases in research focus

There is no question that there are enormous biases in the prevailing choice of research questions and objectives that have defined biodiversity research in modified forest landscapes to date, with the result that we know much more about some management issues than others – and not necessarily in proportion to information needs.

For example, there has been a considerable amount of work done on the consequences of forest fragmentation (reduced area and increased isolation of forest patches), yet this has often been at the expense of understanding the conservation importance of the wider landscape matrix (Lindenmayer and Franklin, 2002; Franklin and Lindenmayer, 2009; see also Chapters 1 and 3). Despite this historical focus, the size and isolation of native forest fragments in managed landscapes may often be poor predictors of species distributions (e.g. see Prugh et al, 2008 for a global review). Instead, more subtle factors, such as the quality of the intervening, production landscape (Uezu et al, 2008; Lindenmayer et al, 2008a), the quantity of native forest at the landscape scale (Bennett et al, 2006), the interaction between habitat fragmentation and local disturbances (such as fire, logging and hunting of large mammals (Barlow and Peres, 2004; Lees and Peres, 2006; Peres and Michalski, 2006)), and cascading effects of species loss (e.g. Terborgh et al, 2008), may be more important in determining biodiversity prospects across multiple-use landscapes (Gardner et al, 2009). The fact that we are only just starting to grapple with many of these more subtle factors is in part due to the frequent simplification of research objectives to address landscape features and/or management impacts that are particularly conspicuous to the human eye (e.g. forest fragmentation). This approach carries the risk of masking important drivers of biodiversity change, including interaction or synergistic effects between different processes, and confounds attempts to establish reliable cause–effect relationships (Ewers and Didham, 2006). By more clearly dissecting different components of a given management problem, and being explicit about how much we already understand regarding the ecology of the system, we are likely to generate information that is more reliable and useful for management. Chapter 13 discusses how the development of conceptual models can facilitate this process.

Choice of biodiversity measurements

One conclusion that can readily be reached from the studies presented in Table 10.1 and the wider scientific literature is that different taxa and taxonomic groups invariably exhibit distinct responses to land-use change and forest management. Chapter 6 provides a detailed discussion of cross-taxon response diversity and the difficulties that this presents in selecting indicator groups for monitoring programmes. The same point is repeated again here because consideration of appropriate biodiversity response variables or indicator groups is central to the process of setting defensible research objectives. It is inevitable that our understanding of human impacts on biodiversity will always be contingent on the choice of study species. The answer to this apparent dilemma is not to try and identify all-encompassing biodiversity indicator groups (as almost all empirical and theoretical work to date indicates that this is not possible, see Chapter 6) but instead to adopt a transparent and defensible rationale for selecting indicator taxa that makes explicit the strengths and weaknesses of alternative choices. The key components of this indicator selection process are laid out in Chapter 12 and should be considered in parallel with the proposals made in this chapter.

Confounding factors in the detection of management impacts

In designing ecologically responsible management strategies, it is ultimately necessary to link a set of known management impacts through observable changes in forest structure and function to a set of known biodiversity responses. Our ability to make this connection is confounded by a number of both ecological and methodological factors that need to be considered during the process of setting research objectives. These factors are essential for establishing how much we already know about a given problem, and determining our capacity to implement any new programme of work.

As discussed in previous chapters of this book, the whole process of developing a monitoring programme, through each of the scoping, design and implementation stages, is rarely a linear activity but instead involves multiple and iterative feedbacks between component parts. The task of setting objectives and choosing appropriate response variables needs to take place together with thinking about the design and implementation of the sampling and data collection strategy. Recognizing the importance of confounding factors throughout the entire monitoring framework is therefore central to successfully linking research objectives with an effective monitoring design and implementation strategy. Chapter 13 provides a more detailed analysis of some of these ecological factors (e.g. interaction effects, synergistic and cascading impacts, time-lags and context effects), and elaborates on their importance in the development of a conceptual system model that is key to achieving this goal. Chapters 14 and 15 provide a more detailed consideration of various methodological and analytical biases and how they constrain our ability to draw reliable inferences from field data.

The disconnection between forest management and biodiversity research

Applied biodiversity research and monitoring can be linked to potential improvements in the management process through at least three classes of objective:

1   to evaluate the adequacy of existing management activities for achieving overall conservation goals;
2   to evaluate the potential conservation benefits of adjusting existing management activities; and
3   to evaluate the potential conservation benefits of implementing an altogether new type of management intervention.

All three classes of objective can be interpreted as types of validation monitoring, where researchers are seeking to evaluate the contribution of a particular management intervention or controlling factor against long-term conservation goals. As such, different objectives can all contribute towards the process of AM where the management regime is adjusted in the face of new information about the effectiveness of component, or potentially component parts. Nevertheless, these three approaches differ significantly in the type of information they gather and their relationship with existing management regimes.

In the first, most basic option, research provides a direct test of the adequacy of existing management standards, and can help identify areas of management that are responsible for the greatest impacts on biodiversity (e.g. comparing patterns of conservation value between different forest types in a multi-functional landscape (Barlow et al, 2007a)). However, short of calling attention to a problem, and indicating which types of human activity are more damaging than others and which should perhaps be stopped altogether, this form of evaluation provides limited guidance for how to develop a more responsible management regime. In contrast, the second and third types of objective evaluate the biodiversity benefits that may be possible from implementing alternatives to the existing regime. In the second option, the focus of research is on evaluating the possible benefits of adjusting current practices, for example, increasing the width of a riparian forest corridor (Lees and Peres, 2008b), reducing the distance between agroforestry plots and the nearest native forest fragments (Anand et al, 2008), or allowing the growth of understorey vegetation in plantation forests (Barlow et al, 2008). By contrast, the third option is concerned with evaluating the likely conservation benefits of introducing an additional component to the existing management regime, such as the restoration of degraded land or introducing key ecological features (e.g. coarse woody debris (Mac Nally et al, 2001)), or the control and prevention of stressors that are outside the management system (e.g. hunting (Parry et al, 2009)).

Although all three of these research approaches are capable of generating valuable conservation insights, they may vary significantly in their potential to contribute towards leveraging real change.

The first class of objective can only make a limited contribution to guiding future improvements as there is no evaluation of any alternatives for each management strategy (e.g. different sizes of corridor or logging intensity). By contrast, the third class of research objective, in focusing on the evaluation of novel management options that are not part of the existing regime, may be of limited practical value if there is significant political or economic resistance preventing the adoption of new responsibilities (for example, they may be excessively costly, such as restoration of degraded land or fire management, or socially unacceptable, such as preventing local people from subsistence hunting). That said caution is needed in limiting conservation research only to those factors that we currently view as being feasible to change (Box 10.1).

Despite the important role that science plays in generating new ideas, it is likely that many of the most pragmatic opportunities for improvements to forest conservation come from evaluating potential adjustments to the existing management regime (i.e. the second class of objective). The existing management regime provides a pool of 'control variables' that can, in many cases, be readily adjusted with potentially significant benefits for biodiversity (e.g. increasing the width of riparian corridors). There are many areas of existing forest management practice for which we still have a very poor understanding of the implications for biodiversity. One example of this is the biodiversity impacts of selective logging. RIL techniques are well established as the basis for ecologically responsible forestry, yet despite their widespread adoption, there are very few data that describe how differences in the level of key management variables (e.g. logging densities, harvest rotation schedules) affect biodiversity (see Chapter 2).

SELECTING HIGH-PRIORITY RESEARCH OBJECTIVES FOR BIODIVERSITY MONITORING

Research objectives are commonly selected on a relatively ad hoc basis that is strongly influenced by the nature of funding opportunities imposed by political agendas, and personal differences in the interest of researchers involved (Box 10.2).

Box 10.1 The role of science in providing a vision for management: Stretch goals and ambitious forest conservation initiatives

The evaluation of novel approaches to forest management can represent a fundamentally important contribution of research to management for two main reasons. First, many regions lack any formal or detailed management plan for researchers to connect to when they first set up a project. There may be little or no *a priori* guidance regarding the type and level of management, if any, that is currently conducted, or is socially and politically feasible. Such situations are arguably those that would benefit the most from biodiversity evaluation and monitoring. Second, even in forests that may already have a well-established management regime, the development of a genuinely ecologically responsible management system may require thinking 'outside the box' – that is, above and beyond what is commonly accepted as normal practice. It is important to be pragmatic but a key purpose of conservation science is also to shed light on what could be possible if societal expectations shifted.

Ambitious forest conservation and restoration initiatives will not happen by accident, or through the accumulation of small-scale, ad hoc efforts. They need a long-term vision to guide their progress. *Stretch goals* and *backcasting*, along with scenario planning and shorter-term milestones, offer potentially promising pathways for overcoming barriers to what otherwise may appear as overly ambitious initiatives. *Stretch goals* are well established in business, management and industry, and are a term used to describe highly ambitious long-term goals. They are intended to inspire creativity and innovation and to challenge those involved in a project to achieve something that currently seems impossible. A popular example is that of NASA's Apollo mission to the moon that occurred within the promised 10-year time frame pronounced by President John F. Kennedy. This particular example relates to technological barriers but similar benefits can be gained when confronting seemingly insurmountable social or economic barriers. *Backcasting* is a technique for identifying the key elements (e.g. technological, institutional) that are needed to achieve a desired yet ambitious objective. Backcasting, in contrast to forecasting, is not intended to predict how the future is likely to be. Rather, the aim is to work back from a desired objective to assess the feasibility and desirability of the end point rather than its likelihood.

There are many conservation objectives that could be applied to managed forest systems that may at first appear overly ambitious in a present day social and economic context. However, when assessed against the concept of stretch goals, such objectives may appear more realistic. One example is the Trees for Life project in the highlands of Scotland which aims to over 1500km² of Caledonian forest. Despite considerable initial scepticism, this ambitious objective has succeeded in driving a series of shorter-term milestones and, by 2005, over 500,000 trees had been planted. This example and the idea of stretch goals is important in highlighting that feasibility is frequently 'in the eye of the beholder'.

Source: From Manning et al (2006)

Intrinsic curiosity and personal interest of individual scientists can play a very important role in guiding research, and serendipity is often responsible for major breakthroughs in both pure and applied science. However, what I argue in this chapter is that in being more judicious in identifying and setting research objectives,

---

Box 10.2 Biases in the selection of research objectives

The process of setting objectives for biodiversity research and monitoring programmes is often unclear. Funding agency agendas frequently have a strong influence on research direction and focus, but in many cases the process of setting research objectives is equally influenced by circumstance – the personal interest and past experience of project coordinators and training, and scientific appeal (e.g. the generation of novel and topical results that can be important in facilitating a career in science). These factors often have the potential to lead to a gradual disconnection between fundamental and means objectives; the feeding back of research findings and recommendations to management is substituted for personal interest and academia performance criteria, with the production of science and scientific publications as an end-game in itself. Publication of the results of monitoring programmes in the peer reviewed literature is an essential part of monitoring work. Yet, as researchers become disconnected from the actual management process, the relevance and utility of the conclusions published in such papers to tackling real-world problems can be highly diluted.

Eroding this connection between fundamental and means objectives carries the risk that research becomes:

- wasteful of resources – not only money, but perhaps more importantly, time. Applied science that falls outside a priority research schedule inevitably passes up on important opportunities to secure conservation dividends. If left unheeded, this carries the risk of cultivating a 'crisis of credibility' as investors and policy makers become disillusioned with the contribution of science to problem-solving (see also Chapter 3).
- more superficial as academic rewards may often be greater when attention is shifted to other issues rather than dedicating time to resolving the detail of a particular problem.
- repetitive and redundant as unless we measure likely returns from new research based on their contribution to solving problems we implicitly fail to account for underlying trade-offs in learning versus doing. This risk runs counter to the previous point yet there is inevitably a diminishing return on the contribution of new research to the understanding of a problem. By failing to question the extent to which new research may be increasingly redundant, we risk stagnating the research process.
- idiosyncratic as failing to adopt something of an objective basis for setting research priorities will inevitably limit the extent to which different research endeavours are coordinated and complementary, and therefore able to coherently address broader-scale questions.

---

we can often greatly enhance opportunities for learning. In essence, we need to rationalize the decision-making process we use to decide on research direction, and develop a transparent rationale for conducting more defensible and effective conservation research. In doing so, we should generate an improved appreciation of the relative contribution of different research objectives to facilitating forest conservation, and an improved appreciation of the absolute limits of conservation research to facilitate change in management and policy.

There are five main, inter-related components to the process of setting priority research objectives for validation monitoring:

1 Define overall forest conservation goals (see Chapter 9).
2 Identify a suite of existing and/or potential management interventions (as related to specific threatening processes) that can be translated into research objectives.
3 Employ a clear and defensible selection process for identifying priority research objectives.
4 Decide on an appropriate set of indicators and target species as a focus for evaluation (see Chapters 11 and 12).
5 Develop conceptual models and associated hypotheses to articulate the objective(s) and guide the process of data collection and analysis (Chapter 13).

In the next two sections, I consider first the problem of identifying a candidate set of management interventions, and then the task of selecting priority objectives for research from this set.

Identifying management interventions as candidates for research attention and validation monitoring

If the purpose of biodiversity monitoring is to inform about the impact of different management practices and identify opportunities for continuous improvement, it is essential to start out with a good understanding of how forests are actually managed, or how they could be managed in the future. As discussed earlier in this chapter, opportunities for change to management come in three basic forms: (i) practices that could be discontinued and replaced with less ecologically damaging forms of management; (ii) practices that already form part of an existing management strategy yet which could be adjusted; and (iii) practices that are currently not part of any management plan yet if introduced could potentially yield significant benefits for biodiversity. To generate the greatest conservation benefits, biodiversity research and validation monitoring needs to focus on evaluating a blend of management options that encompasses both adaptations to existing strategies and the development of novel interventions.

The process of identifying a set of management practices that are candidates for evaluation can be facilitated by undertaking three systematic review exercises:

1 Management policy and process review

A management policy and process review is needed to identify all existing management strategies within a given management area. For example, in the case of native forest management, this would include rotation times, cutting schedules, harvest intensities and pre- and post-harvest silvicultural techniques. This will be an easy task in cases where a detailed management plan already exists (which will be true in the case of most certified forests). This same evaluation process can also encom-

pass additional types of human activity that may fall outside any formal management plan yet may also be susceptible to management. Examples here may include
features of forest ecosystems that are persistently ignored by management, such as
soil biodiversity and soil ecological processes, as well as forest management practices by local communities (e.g. natural resource extraction and hunting within
managed forests) and externally sourced threats, such as impacts from invasive and
pest species.

2   *Landscape structure and composition review*

A second worthwhile assessment exercise is an assessment of landscape patterns
to identify the spatial composition and configuration of key forest features that are
amenable to management in the study region (e.g. the area, shape and distribution of native forest fragments, the type and number of riparian corridors, density
of roads, etc.). Such a review could also detail any spatial variability in the type and
extent of historical disturbance effects in the study landscapes (e.g. arrival of
human settlements, changes in local population density, etc.). In some instances,
this task may be more challenging than an assessment of actual management activities as it requires technical capacity in geographical information systems, which
is unlikely to be available everywhere. It is also problematic because very few
forestry operations operate at the landscape scale (Leslie, 2004). Consequently
achieving an understanding of the composition of the landscape matrix that is
adjacent to each management area will require access to additional spatial data
(e.g. from third-party satellite imagery; Figure 10.1) and/or collaboration between
forest managers and neighbouring landowners and government departments.
Despite these challenges, understanding the wider landscape context is often vital
for comprehending the effectiveness of local management interventions
(Tscharntke et al, 2005).

3   *Forest condition review*

A third assessment exercise that could be useful in identifying priorities for
monitoring is a review of the current understanding of the structural and
functional elements that confer a high level of forest condition for the study region.
This could involve an assessment of potential reference condition sites as well as a
review of the relevant scientific literature. Take a recent study by Cantarello and
Newton (2008) example of this, where the authors identified dead wood volume,
tree structure and composition, tree regeneration and ground flora composition
as potential structural indicators for guiding responsible forest management (see
also Chapter 11).

In combination, these three assessment exercises help identify which management options or 'control variables' are available for evaluation in a given landscape.
Ideally, this kind of information would be collated at a regional scale, enabling
monitoring programmes to be coordinated across multiple landscapes to address a
set of standardized objectives.

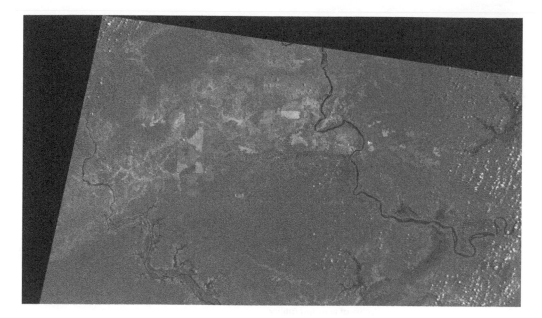

Figure 10.1 Landsat satellite image of the region between the Jari and Paru rivers in the north-east Brazilian Amazon

Note: The operations of a large plantation company are shown as lighter colours in the centre of the image, with extensive areas of adjacent primary forest

A selection framework for identifying high-priority research objectives

It is sensible to think about the priority of different research objectives in terms of their ability to facilitate learning. With this in mind, it is possible to identify three principal (yet partially overlapping) 'axes' of research motivation or choice that all influence the selection of a particular objective (Figure 10.2):

1　**The opportunity to learn** – What management practices are feasible to evaluate and monitor?
2　**The necessity to learn** – What management practices are associated with the greatest level of uncertainty regarding their implications for biodiversity conservation?
3　**The value of learning** – What management practices, if adjusted or implemented, are likely to deliver the greatest conservation benefits?

To be given priority, an objective should satisfy the criteria of all three axes i.e., it should be able to deliver results that are of high conservation importance, and concern a poorly understood problem that is practically feasible to study. Arbitrating between the relative priority of different research objectives is far from being an exact science yet explicit consideration of alternatives in the context of these three

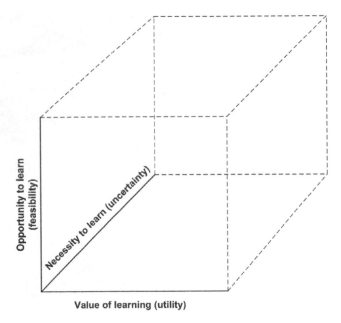

Figure 10.2 A selection framework for identifying high-priority research objectives

axes of motivation can be a very helpful exercise to improve the return on invest-ment in research. Establishing a common framework such as this is valuable in so far as it forces us to be explicit about the choices made when deciding what to do with limited resources. By being explicit, we can be more objective, defensible and transparent – and, as a consequence, foster more effective and meaningful science.

Research feasibility and opportunities to learn

The feasibility of a biodiversity research programme (whether as a snap-shot assessment or a long-term monitoring programme) is determined by both constraints in research design and cost. A feasible programme is one that has access to appropriate treatments, expertise and funding to ensure that the research is conducted with sufficient rigour. Identifying the key design features of a proposed biodiversity monitoring programme, and comparing these against what is available in the study landscape, is an important step in understanding the potential for effective research. A monitoring programme has two basic design elements: (i) a descriptive component that establishes a baseline condition within reference areas (or pre-impact time periods in cases where before and after comparisons are possible); and (ii) an experimental component that evaluates the effects of specific human impacts and management interventions ('treatments'). Experiments pro-vide the most convincing means of predicting ecological responses to environmental change (Sutherland, 2006), and as such some form of experimentation is desirable if we are to further our understanding of the consequences of anthropogenic stressors on natural systems. A true manipulative experiment capable of providing strong inferences (Platt, 1964) is defined by a random allocation of treatments to experimental units and replication of units within each treatment type. However, this kind of approach is rarely possible in large-scale and complex ecological systems. Some efforts have been made to set aside parts of a forest management system as 'demonstration' or 'model' forests in order to support AM and experimental research (e.g. North-West Forest Plan; Lindenmayer and Franklin, 2002), yet logistical, financial and political obstacles invariably combine to make such an option unfeasible or of only limited success (Stankey et al, 2003; Bormann et al, 2007).

Despite the many problems that exist with the practical implementation of AM (see Chapter 7), the idea of examining uncertainties and carrying out some form of experiment is still a sensible and realistic idea (Sutherland, 2006). An alternative approach to manipulative experiments is to overlay monitoring programmes on a natural experimental framework where existing management and disturbance regimes provide the treatments for evaluation (Box 10.3). The process of conducting a systematic assessment exercise of existing and potential management activities can therefore be invaluable in revealing the potential for natural experiments, including the number and type of treatments, their spatial distribution and level of replication.

Scientific uncertainty and the need to learn

Some types of human impact on biodiversity are far less well understood than others – for example, the consequences of different fire regimes or logging practices compared to the size of remnant forest patches (see Chapter 3). When screening candidate research objectives, it is important to identify such key areas of ignorance or scientific uncertainty. By focusing on areas of management that are poorly understood, researchers can increase the likely return on investment from any new monitoring programme.

Despite this logic, the *a priori* filtering of alternative research objectives with regards to differing levels of scientific uncertainty is a non-trivial task. The process can be facilitated by combining the initial assessment of existing and potential management activities (as discussed above) with a systematic review of the scientific literature (published or otherwise) and discussions with experts to identify:

- general areas of ignorance about the management and conservation of forest landscapes;
- what knowledge is available that relates to individual human impacts and management practices, and in what ways is this knowledge limited (e.g. geographically, taxonomically);
- which (if any) mitigation or restoration practices are routinely conducted that might not actually be effective;
- which human activities and management practices are routinely conducted with entirely unknown ecological consequences; and
- what potential management interventions exist that have never been evaluated?

Box 10.3 The use of natural experiments in biodiversity monitoring of managed systems

The concept of natural experimentation is not new (Diamond, 1986), and in the context of AM was first proposed by Walters and Holling (1990), who advised that in order to be successful, AM systems have to make effective use of existing opportunities for spatial replication and control. In realizing this objective Walters and Holling also observed that every management strategy is in fact a perturbation experiment (albeit usually with a highly uncertain outcome), and that an unambiguous assessment of the consequences of a given management intervention can often be achieved through the use of a careful experimental design to account for uncontrolled environmental factors (see also Walters and Green, 1997; Lindenmayer et al, 2000; Nichols and Williams, 2006). Treating stand- and landscape-scale forest management activities as experiments and overlaying well-designed monitoring and evaluation programmes on ensuing disturbance regimes, is therefore a very effective way of accumulating much needed information for guiding improvements in management.

A key issue here is that ecological research does not operate, of course, in a black and white world. While it is usually true that certain factors need to be present to dictate whether a particular research programme can go ahead (you cannot study forest corridors if there are not any available to study), other factors contribute more to the quality or power of the research design (e.g. availability of multiple independent treatment and controls sites), than its feasibility in any absolute sense.

In some cases, it may be possible to design a quasi-natural experiment where planned or ongoing management activities are adjusted in order to enhance the experimental study design of a research programme. A good example is the large-scale Variable Retention Harvest System (VRHS) experiment in the Victorian mountain ash forests of Australia (Lindenmayer, 2007). This ongoing experiment is comparing traditional clear-fell logging practices with the retention of large- and small-habitat islands in logging coupes, with six replicate blocks per treatment. Through the careful identification of study sites and 'piggy backing' on existing harvesting schedules, scientists and forest managers were able to develop a rigorously designed and powerful monitoring study at minimum cost.

Such a systematic approach to selecting areas of management and scientific uncertainty is rarely employed (see Sutherland, 2006) yet is likely to make a substantial contribution to identifying what are potentially the most fertile areas of research. The process of developing a conceptual model of the study system can also be very helpful in identifying knowledge gaps in the strength and nature of the relationship between different human activities and changes in biodiversity (see Chapter 13).

When evaluating different levels of uncertainty in management it is important to identify the point at which our understanding is sufficient to provide recommendations for improved practice. It is possible to dedicate too much time and effort to a single management objective. Continued research on an increasingly well-understood problem invariably prevents the same researchers focusing their time and resources on other problems – the solutions to which may be required with equal or greater urgency. Moreover, if resources for management and research come from the same source (as they may in the case of a national park authority or a commercial forestry company) then money spent on research and monitoring may result in less being available for implementation (Joseph et al, in press).

Research utility and the value of learning

In addition to concerns about feasibility and scientific uncertainty, the third axis of motivation that underpins the process of selecting research objectives is that of utility – what is the value of tackling one objective over another with respect to likely differences in the actual benefits for biodiversity conservation? Addressing this question is perhaps an even harder task than that found in distinguishing between different levels of scientific uncertainty or programme feasibility. However, as in other cases, progress can be made by encouraging a more transparent analysis and justification of the motives that underlie particular choices.

There are at least three principal factors that contribute to differences in the probable conservation value of researching different types of management impact: the likely effect of a given management action on biodiversity; the generality of any recommendations arising from the research; and the probability of research results being included in future management. An ideal case would be one where the likely impact of a new management action is very positive, widely applicable and of immediate appeal to management authorities.

Estimating management impacts on biodiversity

The immediate benefit of studying a particular research objective comes from learning about the relative effect of the size of a given human impact or management intervention – measured as the extent to which a given biodiversity response variable is expected to change following an intervention. One approach that can help in this task (as well as in the identification of areas of scientific uncertainty) is a systematic risk assessment which seeks to identify the relative risk to a valued attribute from different threats (Walshe et al, 2007).

Risk analysis is 'the consideration of sources of risk, their consequences and the likelihood that those consequences may occur' (Wintle and Lindenmayer, 2008). Inferring the relative risk of different management impacts involves the use of a matrix of consequence and likelihood. Outcomes depend upon the ability of expert analysts (i.e. those most familiar with the system), in consultation with stakeholders, to use subjective judgement to estimate the likelihood that an impact will occur (in the case of many forest management practices this will be a given) and the severity of its consequences. Priorities for research and management are those impacts that are considered higher risk and those for which there is considerable variation in perspectives among analysts (i.e. a high level of scientific uncertainty (Walshe et al, 2007)).

Risk analyses can be conducted with different levels of rigour. Walshe et al (2007) illustrated a simple and hypothetical example for the management of forest biodiversity (Table 10.2). In this example, the response of managers and auditors to this information may result in a preferential allocation of monitoring resources to the extent of regeneration failure, the incidence of fungal pathogen and the concentration of contaminants in streams. Lower risk impacts may be considered deserving of only low-level monitoring or none at all. While this grossly simplified approach has a number of limitations (Burgmann, 2005), it retains the advantage of being simple and quick, and provides a means for identifying and recording expert opinion on the probability and magnitude of management impacts.

Table 10.2 Hypothetical example of outcomes of a risk assessment conducted using the Australian standard

Hazard	Risk
Loss of soil nutrients	Medium
Reduction in hollow-bearing trees	Low–medium
Spray drift from aerial application of pesticides	Low
Loss of visual amenity	Low–medium
Intense insect attack	Medium
Remnant vegetation protection	Low
Regeneration failure	High
Spread of fungal pathogen	Low–high
Stream pollution	Low–high

Source: From Walshe et al (2007)

For more detailed risk assessments, there are powerful tools for assessing the likely biodiversity impacts of proposed management practices. For example, a recent study of the impacts of timber harvesting and plantation conversion on the Tasmanian wedge-tailed eagle (*Aquila audax fleayi*) used dynamic landscape metapopulation (DLMP) modelling techniques (Wintle et al, 2005) to assess how competing management options impact on future expected population sizes of the eagle (see also Wintle and Lindenmayer, 2008). However, quantitative methods like DLMP are only suited to well-studied individual species and not entire assemblages of the native biota that are the focus of the type of monitoring advocated in this book.

Despite the potential benefits there has been little attempt to reconcile the AM literature with research on formal prospective risk assessment methods, and as such formal assessments of risks faced by forest biodiversity from different human activities have been absent from the vast majority of management plans (Wintle and Lindenmayer, 2008).

Generality of management impacts

Learning about management impacts that are widespread will deliver a greater overall return on research investment than focusing on problems that are highly context-dependent and relevant to only a few sites. However, balancing questions of scale is difficult and the construction of any management standard necessarily reflects a trade-off between generic principles of good practice and the need to provide stand-level prescriptions that account for local differences in ecological, political and social conditions (Finegan, 2005; Lindenmayer et al, 2006; Lindenmayer and Hobbs, 2007).

There appear to be very few general management principles that can provide widespread yet practical guidance for the conservation of human-modified ecosystems (Lindenmayer and Hobbs, 2007). Instead, the generality of any management recommendation is commonly limited by a number of ecological and social factors, including: differences in landscape context and historical patterns of human disturbance between sites; differences between sites in the biological indicators or target species group(s) of interest (especially if monitoring is focused on species with limited distributions); and differences in opportunities for implementation – related either to physical limitations in available treatment options (e.g. pre-existence of large forest corridors) or socio-economic viability of different practices (see below). For many management practices, we still have a very poor understanding of the extent to which their conservation implications are determined by the nature of the practice itself, or the site-specific characteristics of the location where the practice takes place.

Despite the limitations of context dependency, it is naive to expect that all guidance necessary for developing ecologically responsible forestry can be derived solely from local, site-specific monitoring and research programmes. A balance is needed, and achieving this can be helped by focusing on those objectives that are neither overly specific nor overly general (Salafsky and Margoluis, 1999a). At one end of this continuum is a set of guidelines that are highly tailored to meet the specific requirements of a particular site, yet fail to take account of landscape- scale ecological and social heterogeneities, and are therefore of limited use or no relevance elsewhere (e.g. the maintenance of species-specific nesting requirements of a range-restricted and endangered bird). At the other extreme, are non-prescriptive generic guidelines that apply to all sites and are therefore, by definition, largely trivial and unhelpful to land managers (i.e. they are not amenable to testing). International or national scale ecological C&I for SFM are a very good example of the latter – and indeed they were not intended for implementation scale processes (Buck et al, 2003; McDonald and Lane, 2004; UNFF, 2006). The general yet non-trivial guidelines that lie between these two extremes may often be found by focusing, in the first instance, on the evaluation of research objectives that are common across multiple landscapes.

Uptake of management recommendations

Even if a particular management activity may be shown to have significant and widespread implications for biodiversity, other factors (political, economic and social) may act to limit the extent to which any new knowledge is transferred to policymakers and has the effect of encouraging behaviour change. Understanding such factors is very important yet perhaps represents the most difficult challenge of all in balancing the relative costs and benefits of different research objectives.

Predicting differences in research uptake is particularly controversial because, on the one hand, science needs to be pragmatic and look for solutions that are achievable (e.g. focus on testing minor adjustments to existing management standards), while at the same time work to stretch our imagination of what may be possible (Box 10.1). Big problems may ultimately require bold solutions and science cannot shy away from the task of presenting novel alternatives (Fischer et al, 2007).

We have a generally poor understanding of the ways in which different research findings contribute to conservation action – the so-called 'implementation gap' between knowing and doing as described by Knight et al (2006). However, careful reviews of previous studies may once again offer valuable insights regarding the types of information or research designs that have been historically effective in fostering real changes, and how the process of knowledge transfer varies among different institutional and governance frameworks, as well as among different countries.

Identifying high-priority research objectives for monitoring

The individual components of the selection framework for research objectives described in this chapter are not novel. When questioned, most researchers will admit to incorporating factors of value, uncertainty and feasibility in their decisions about what research objectives to prioritize, as well as idiosyncratic researcher-specific factors, such as personal interest, intrinsic curiosity and habit. In addition, it is common for researchers to have some level of interaction with managers to understand the array of management options that already exist and may be amenable to adaptation. However, this decision-making process is almost always implicit. Despite it inevitably being a somewhat subjective exercise, adopting a more transparent approach to setting research priorities is likely to greatly enhance the value of monitoring efforts, while also facilitating the external evaluation of existing monitoring programmes.

The process of evaluating a set of candidate research objectives across all three axes of utility, uncertainty and feasibility can range from a personal thought exercise to a semi-qualitative analysis based on ranking or vote-counting among experts and a fully quantitative approach using value functions to compare costs and benefits where data permit (e.g. Walters and Green, 1997). If applied more broadly the process of systematic review (Sutherland et al, 2004b; Pullin and Stewart, 2006) could greatly facilitate the evaluation process, although care is needed not to bias investigations towards only those questions that are well documented (and therefore susceptible to systematic reviews). To date, systematic reviews (whether quantitative or qualitative) have focused almost exclusively on evaluating only part of the 'utility' axis as described here – quantifying differences in the size and extent of known effects (e.g. through a formal meta-analysis). We very rarely perform analogous reviews of the extent to which opportunities for new learning vary among different research questions, or the degree to which different forms of research and data are more or less likely to precipitate real change.

Careful consideration needs to be given to the process of weighting between different motivating factors in the identification of research priorities. The contribution of different sources of research motivation is almost certainly not internally consistent when compared across a set of alternative objectives. For example, objectives that are only partially feasible (e.g. because they are costly or because there are few options for natural experimentation) may still be worthwhile if they are tackling important and poorly understood questions. On the other hand, questions that are relatively trivial and reasonably understood may still be worthwhile addressing if they are very cheap and easy to tackle.

Finally, it is important to remember that human-modified forest landscapes are dynamic phenomena. For this reason, biodiversity monitoring programmes need to plan for surprise and not be inflexibly tied to a limited set of objectives that are focused on the evaluation of landscape features of management practices that are unlikely to persist in the long term.

CHAPTER 11

Selecting Indicators of Forest Structure to Assess Management Performance

Synopsis

- As the most direct and conspicuous form of human impact on forest ecosystems is physical, indicators of forest structure often provide the most effective and appropriate basis for assessing management performance.
- Structural indicators link changes in management activities with changes in the biodiversity we wish to conserve – either directly through the provision of suitable species habitat resources, or indirectly via the conservation of key ecological processes that depend on the structural environment of the forest.
- Forest structural indicators can be distinguished into two basic kinds: (i) stand (e.g. measures of different vegetation strata and dead wood); and (ii) landscape (e.g. measures of forest extent and habitat composition and configuration) level indicators. In both cases, there are varyingly levels of empirical support regarding the relevance of different candidate indicators for different species groups and forest types around the world.
- A systematic approach to selecting structural indicators for forest management should follow a series of five assessment criteria: (i) the identification of a viable set of indicators based on the practical availability of necessary expertise and technical support; (ii) the 'responsiveness' of candidate indicators to management actions; (iii) the ease (cost and time) with which they can be measured; (iv) their relevance to changes in forest condition and biodiversity and/or individual target species; and (v) the generality with which they can be applied across similar management systems in other landscapes and regions.
- The initial selection of a set of structural indicators and appropriate indicator target levels for management is challenging, but can be assisted by drawing on expert knowledge and a review of management practices in surrounding region.
- Structural indicators alone are not sufficient to characterize the biodiversity potential of a given site and additional non-structural indicators (e.g. of disturbance processes, such as fire and grazing regimes) can often make a valuable addition to assessments of management performance.

The task of indicator selection lies at the heart of the overall monitoring framework as it is very tightly connected to other stages in the scoping, design and implementation process. Indicators provide the means by which management performance can be measured and evaluated against conservation goals. They are central to the definition of research objectives as well as the development of appropriate sampling strategies and data collection techniques. As discussed in Chapter 8, and depending on the purpose of monitoring, different types of indicator serve distinct and highly complementary functions.

Effectiveness monitoring assesses changes in selected management impacts that serve as indicators of performance, and compares any changes against minimum performance targets. As the most direct and conspicuous form of human impact on forest ecosystems is physical, indicators of forest structure often provide the most effective and appropriate method for assessing management performance, although non-structural indicators of disturbance regimes (e.g. fire and grazing) and levels of soil and water contaminants can also play an important role (Chapter 8).

Structural and other indicators are used in effectiveness monitoring work essentially to provide a link between changes in management activities and changes in the biodiversity we wish to conserve – either directly through the provision of suitable species habitat resources, or indirectly via the conservation of key ecological processes that depend on maintaining the structural environment of the forest, such as leaf-litter decomposition and micro-climate stability. In this sense, structural indicators should strictly be viewed as indirect indicators of true management performance as they rarely represent valued conservation attributes in their own right, and need to be validated against species data (i.e. through a process of validation monitoring) in order to be meaningful for biodiversity conservation (see Chapter 8).

Forest structural indicators can be distinguished into two basic kinds – stand and landscape level indicators (Lindenmayer et al, 2000). In both cases, a bewilderingly large number of measurements have been proposed in the scientific literature, with varyingly levels of empirical support regarding their relevance for different species groups and forest types around the world. One of the clearest conclusions to have emerged from work so far is that no single indicator can provide a satisfactory reflection of biodiversity change. Instead, a set of complementary indicators is necessary to adequately capture changes in overall forest condition. Nevertheless, there is an internal tension in deciding what represents a satisfactory number of structural indicators. The more indicators that are measured, the more likely it is that, in combination, they will provide a reliable indication of change in biodiversity. Yet, the addition of more indicators also makes it more difficult to evaluate the exact causal nature of any relationship, and hence where opportunities for cost-effective improvements to the management system exist (because of the masking effect of multiple, partially correlated variables). Hence, to be most useful, it is important to select structural indicators with care, drawing on existing knowledge to identify a set of ecologically important forest attributes that are both sensitive to management impacts and also amenable to reliable and repeatable measurement. An additional challenge to the task of establishing which structural indicators may be most appropriate is the need to identify quantifiable indicator levels or targets that can be used as a basis for setting auditable performance standards for management (e.g. minimum cover of native habitat, width of buffer zones, plantation understorey regeneration, maintenance of areas of regenerating habitat, etc. (Wintle and Lindenmayer, 2008; Villard and Jonsson, 2009)).

In this chapter, I briefly outline some of the key issues surrounding the selection of structural indicators to assess management performance. First, I distinguish between indicators of forest structure at stand and landscape scales. Second, I identify a number of basic criteria that can aid the selection process, namely the responsiveness of candidate indicators to management actions, the ease with which they can be measured, their relevance to changes in forest biodiversity and condition, and the generality with which they can be applied across similar management systems in other regions. Third, I present a general framework for selecting structural indicators, as well as discussing a number of important considerations in developing a set of indicators for monitoring, including the initial challenge of identifying candidate indicators, use of a common indicator classification scheme and multi-attribute indicators of forest structure. Finally I touch on some of the limitations in the use of structural indicators for understanding patterns of biodiversity change and the importance of other, non-structural indicators of management performance. Throughout the chapter, examples of specific indicators are given to be illustrative rather than prescriptive, and interested readers can look to key references (e.g. McElhinny et al, 2005; Newton, 2007; and see further references below) for more detailed guidance on individual cases.

INDICATORS OF FOREST STRUCTURE AT THE STAND SCALE

Stand-level structural indicators are commonly used to describe the key physical attributes of forest structure that are likely to be important in the maintenance of biodiversity. Yet, beyond this broad notion there is some confusion as to what actually comprises a structural attribute. Although they are often considered together in lists of management performance indicators (e.g. Allen et al, 2003; Smith et al, 2008), it is important to distinguish indicators of forest structure from indicators of floristic composition (herbaceous or woody). While the latter may be very important in explaining the distribution of animal species, it is much more difficult to clearly and defensibly link changes in plant composition to management impacts, than it is to link management to changes in forest structure. There is also often further confusion in comparing the structural and functional attributes of forest stands, as some attributes from one class may also be indicators for attributes in another – for example, dead wood, which has both structural and functional properties (McElhinny et al, 2005). Because this division is unclear, I follow McElhinny et al (2005) in classifying all such attributes as structural, while accepting that many of them also encompass important functional properties. Semantics aside, the most important criteria that defines a structural indicator is that it can be clearly linked to management impacts and is itself amenable to management. If it cannot, then it is unlikely to perform effectively as an indicator of management performance and should be reinterpreted as either a covariate for analyses (i.e. variables that are important in explaining patterns of biodiversity but which are not directly associated with management) or a valued attribute/response variable in its own right (and therefore, like all biological indicators, more appropriately viewed as an evaluator – rather than an indicator – of performance standards, such as floristic diversity, and key ecological functions, such as leaf-litter decomposition and nutrient cycling).

Stand-level structural indicators can be broken down into measurements that relate to the major structural canopy, understorey and herbaceous vegetation strata, as well measurements to describe stocks of standing and fallen dead wood. Depending on the type of managed forest and intensity of land use, the strength of association between management impacts and forest structural attributes will vary according to, the strata in question. For example, selective logging has a clear impact on the forest canopy and basal area of large trees while the management of plantation forests and agroforestry systems impacts primarily on the understorey and herbaceous layers. Within each of the different vegetation strata, stand structural attributes can be described in a variety of ways, including: measures of abundance (e.g. density of large trees, volume of dead wood); relative abundance (e.g. diameter at breast height diversity, basal area of a particular tree class, foliage height diversity); size variation (e.g. in tree diameter classes); and spatial variation (e.g. spatial distribution of large trees) (McElhinny et al, 2005). Attributes that quantify variation are equally important as those that quantify abundance, as measures of habitat heterogeneity at the stand scale can be highly informative for some species groups.

Combining across the different vegetation strata, together with the multitude of ways in which individual structural features can be measured and defined, generates an almost limitless number of forest structural attributes that could serve as structural indicators for forest management. In their recent review of forest ecology literature, McElhinny et al (2005) list some of those that have been used to date, highlighting the little consistency in the use of attributes among different studies and the relatively few empirical evaluations of structural indicators against biodiversity data (Table 11.1).

INDICATORS OF FOREST STRUCTURE AT THE LANDSCAPE SCALE

Measurements that describe the spatial pattern and characteristics of entire forest landscape mosaics can be divided into three main categories (Bennett and Radford, 2007; Newton, 2007):

1   The **extent** of habitat as referring to the sum of the spatial area of all patches that potentially provide habitat for a particular taxon, regardless of the size, shape or location of patches. For highly disturbance-sensitive species, this equates to the total area of remnant native forest.
2   The **composition** of a landscape as referring to the types of different landscape elements (land uses) present and their relative proportions.
3   The **configuration** of a landscape as referring to the spatial arrangement of landscape elements.

Table 11.1 Example attributes used to characterize forest stand structure

Forest stand element	Structural attribute
Foliage	Foliage height diversity; number of strata; foliage density within different strata
Canopy cover	Canopy cover; gap size classes; average gap size and the proportion of canopy in gaps; Proportion of tree crowns with broken and dead tops
Tree diameter	Tree diameter at breast height (dbh); standard deviation of dbh; tree size diversity; horizontal variation in dbh; diameter distribution; number of large trees
Tree height	Height of overstorey; standard deviation of tree height; horizontal variation in height; height richness class
Tree spacing	Clark Evans index; Cox index; percentage of trees in clusters; number of trees per hectare
Stand biomass	Stand basal area; stand volume
Understorey vegetation	Herbaceous cover and or its variation; shrub cover; shrub height; total cover of understorey; understorey stems per hectare
Dead wood	Number, volume or basal area of snags (by decay class or diameter class); volume of coarse woody debris; log volume by decay or diameter class; log length or cover; coefficient of variation of log density; litter biomass or cover

Source: Adapted from McElhinny et al (2005)

Composition metrics are only applicable at the landscape scale as they integrate data from across all patch types within a landscape. Measures include: the number of land-cover types present, as a percentage of the total possible number of cover types; the diversity, or evenness, of cover types across the landscape; and the dominance of individual cover types (as an alternative to diversity) (Turner et al, 2001).

The spatial configuration of a landscape refers to the arrangement of land-cover types within a landscape. A vast number of metrics have been proposed to describe the configuration of land-cover patches in fragmented landscapes, and are summarized in Table 11.2. Aside from the fact that there are so many options to choose from, there are at least three major problems that confound the interpretation of landscape-structure indicators. First, unless directly mapped with field data, landscape metrics are dependent on the resolution of the satellite imagery data used in their calculation. It is therefore vital that the same resolution data is used for any comparative analyses. Second, many of these metrics are highly correlated, largely because the majority are derived from a small number of primary measurements. Careful analyses are therefore needed to identify which metrics are most relevant for explaining observed biodiversity patterns, and to enable redundant information to be excluded from any analyses (Neel et al, 2004). Finally, and as is the case for stand-level indicators, relatively few of them have ever been validated against biodiversity data (McAlpine and Eyre, 2002; and see below). Banks-Leite (2009) showed that landscape-structure indicators can provide robust predictions of patterns of species composition of forest bird communities in modified forest landscapes in the Atlantic forest of Brazil (Box 11.1). This study is a powerful demonstration of the importance of landscape patterns in structuring assemblages of forest species. However, to improve its utility for management, the next stage is to disentangle the independent contribution of the individual indicators in explaining observed changes in biodiversity and identify appropriate management targets.

Table 11.2 Types of indicators available for measuring landscape configuration

Indicator group	Description
Area indicators	Measures describing area of patches of land-cover types, such as mean patch size, can be summarized for different land-cover types or for the entire landscape
Edge indicators	Measures of patch geometry that describe the length of edge between land-cover types, and can be used to assess the extent of total edge habitat
Shape indicators	Measures of land-cover patch shape, commonly represented as edge:area ratios or as a fractal dimension
Core indicators	Measures to describe interior of patches once an edge buffer has been removed, integrating patch size, shape and distance from edge into a single measure
Isolation/dispersion/ proximity indicators	Measures to describe whether patches are regularly distributed or clumped and how isolated different land-cover types are from each other
Contagion indicators	Measures of the tendency of land covers of a similar type to be spatially aggregated
Connectivity indicators	Measures of connection between patches, based either on strict adjacency, minimum distance between patches or even-sized buffers

Source: Adapted from Newton (2007)

All three landscape properties (extent, composition and configuration) are amenable to change through management. We currently have a relatively poor understanding of the implications of whole landscape properties for biodiversity and ecological processes because very few studies have replicated sampling across multiple landscapes. That said, it is not the case that differences in extent, composition and configuration are equally important for forest conservation and management. Numerous studies have demonstrated that the total extent of remaining native habitat is by far the most important factor in determining the biodiversity value of human-modified landscapes (Bennett and Radford, 2006;

Lindenmayer et al, 2008a; Gardner et al, 2009). Because different forms of forest use often support distinct sets of species (e.g. Pineda et al, 2005), differences in landscape composition can also have important implications for the total amount of biodiversity that can be supported in a given location. By contrast, we have a much poorer understanding of the importance of differences in landscape configuration for biodiversity, although it is difficult to reach firm conclusions from much of this work because of the complex nature of the metrics used.

SELECTION FOREST STRUCTURAL INDICATORS

The process of selecting a priority set of structural indicators from such a large number of options is an undeniably daunting task, and one that presents as much, if not more, of a challenge as the selection of biological indicators (despite having generally received less attention by scientists). Very few studies show evidence of having followed a systematic indicator selection process, and structural indicators are generally selected based on practical ease or historical precedent (i.e. what has been used in previous studies). Adopting a more judicious selection process provides the opportunity to improve our understanding of the connection between management activities and changes to biodiversity, and thus develop more defensible management performance standards.

Box 11.1 The use of landscape structure indicators to measure changes in the integrity of bird communities

Despite a considerable amount of research interest in the role played by different components of landscape structure in determining patterns of biodiversity in human-modified forest landscapes, relatively few robust empirical studies exist that have tested this relationship. In a recent study of forest fragmentation effects on bird communities in the Atlantic Forest of Brazil, Banks-Leite (2009) provides a powerful demonstration of the predictive power of a relatively small set of landscape structural indicators, with a combination of ten landscape measures explaining up to 99 per cent of the observed variability in the integrity of bird communities (Figure 11.1)

Landscape metrics were derived from the freeware Fragstats and were selected to represent variables known to be important to bird communities, including patch area, edge effects, connectivity and proportion of forest cover around sampling points. A set of 13 uncorrelated principal component analysis (PCA) axes were extracted to describe changes in landscape structure and compared against changes in bird community composition within and between three separate landscapes that were characterized by different levels of total deforestation (10, 30 and 50 per cent).

The results of this study are both novel and striking for at least three reasons. First, the relationship between landscape indicators and the bird community was tested across varyingly deforested landscapes and found to be strong in each case. Second, the transferability of landscape indicators was tested by applying indicators selected to explain the bird community in one landscape to other, independent landscapes – and the relationships still remained strong (Figure 11.1). Third, comparisons were also made between landscape-structure indicators based on high resolution (10m) and low resolution (30m) pixel data, and although the relationships were weakened (loss of explanatory power varied between 5 and 10 per cent), they still maintained a reasonable level of prediction (comparing between left and right panels in Figure 11.1).

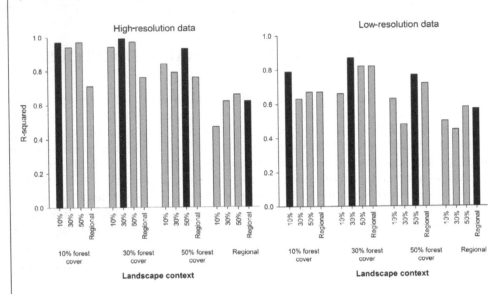

Figure 11.1 Sensitivity of selected subsets of landscape-structure indicators to changes in bird community integrity

Note: Results are compared between landscapes characterized by different levels of deforestation in three independent landscapes, as well as for the entire region. In each case, structural indicators are selected from one study landscape and applied to explain changes in the bird community across all others and the region, with comparisons derived from the same landscape illustrated as black bars. The left and right panels show the use of high- and low-resolution spatial data respectively
Source: Reproduced with permission from Banks-Leite (2009)

There are a number of basic principals that can help guide indicator selection at both stand and landscape scales for a given biodiversity monitoring and management programme. The first task (as for any type of indicator) is to contextualize the problem and clearly define the management-orientated monitoring goals and objectives that are being investigated (see Chapters 9 and 10). Selection can then proceed with respect to four principal criteria, namely: (i) the 'responsiveness' of candidate indicators to management actions; (ii) the ease with which they can be measured; (iii) their relevance to changes in forest condition and biodiversity and/or individual target species; and (iv) the generality with which they can be applied across similar management systems in other landscapes and regions (e.g. as part of a regional certification standard). As is the case for biological indicators, not all criteria are of equal importance and it is helpful to adopt a hierarchical approach to indicator selection which ensures that any final choices retain a common set of important characteristics.

The responsiveness of candidate structural indicators to management

In order to be useful in guiding the management process (i.e. most *fit for purpose*), the most important criteria that a structural indicator needs to satisfy is that it can be easily traced back to human impacts and management activity, and similarly that changes in management practice have a predictable and tractable impact on indicator values. Here, I term this reciprocal form of indicator sensitivity; 'responsiveness' to emphasize the fact that to be effective structural indicators need to be more than just sensitive to management impacts – the *explanation* for their sensitivity also needs to be sufficiently clear such that required changes to management can be inferred when the target level of an indicator is set (i.e. impacts need to both measurable and predictable). Establishing such a clear and well-understood connection between management and structural indicators allows indicators to perform effectively regarding both of their functions – as a transparent mechanism for assessing management performance, and a guide by which potential improvements in conservation value can be translated back into revised management standards. The selection of structural indicators that are appropriate for assessing management performance is therefore a much more stringent process than the selection of structural attributes (that may or may not be directly linked to management impacts) simply to provide explanatory variables in models of biodiversity.

The ease with which candidate structural indicators can be measured

The ability to make reliable and repeatable measurements of structural indicators is essential in ensuring that comparable data can be collected across different sites and years, and within the budget and logistical constraints of an individual monitoring programme. The question of cost-efficiency is especially important when management performance needs to be assessed across large areas.

Individual structural attributes can be measured directly in the field, or indirectly, using a variety of hand-held and remote sensor techniques, with varying levels of precision and accuracy (see Newton (2007) for a review of many techniques). New approaches are being adopted all the time to improve the efficiency with which field data can be collected, as well as the comparability of data within and among studies. A good example of this is in the use of digital photography and pixel analysing software (most of which is freely available; e.g. Gap Light Analyzer www.ecostudies.org/gla/) to reduce observer biases in measures of canopy cover (Figure 11.2). Proulx and Parrott (2008) recently extended this approach to the analysis of forest structural complexity in the understorey, and also found evidence that the complexity of light variations in digital images can be positively related to species richness. Lees and Peres (2008b) used digital photographs to develop a novel approach for indirectly estimating the structural complexity of forest corridors (height and uniformity of corridor profile) (Figure 11.3). At larger spatial scales, satellite imagery is being increasingly used to measure spatial variation in canopy structure and levels of degradation from disturbances, such as logging and fire, although as yet very few studies have linked these methods to changes in biodiversity (Box 11.2). When using any indirect measure of forest structure, it is essential to maintain a traceable connection between the indicator and changes to the management system that it is supposed to be indicating. This is achieved through a validation process that matches the surrogate measure (e.g. satellite image spectral class) with changes in the actual forest attributes that are impacted by management (e.g. canopy cover, forest basal area, etc.).

Measurement of landscape-level structural indicators is readily achieved through a combination of classified satellite imagery and freely available landscape-analysis software, such as Fragstats (www.umass.edu/landeco/research/fragstats/fragstats. html). Deriving indices of landscape structure and composition is more difficult in more finely graded landscapes where there are no clear boundaries among land-cover classes. The ease with which a large number of partially correlated landscape metrics can be generated by such programmes means that considerable care is also required not to include a large number of redundant and ecologically meaningless indicators.

Figure 11.2 Digital photographs taken to measure canopy cover in three major forest types – left to right, *Eucalyptus* plantation, secondary forest and primary rainforest

Source: Photographs courtesy of Jos Barlow

The relevance of structural indicators to biodiversity goals

The more ecologically relevant selected structural indicators are in providing a direct or surrogate measure of key habitat and resource requirements for bio-diversity, the more valid they will be in acting as genuine indicators of management performance, and the more effective they will be in guiding improvements towards long-term conservation goals. Despite the importance of making these links, extremely few studies have provided quantitative evidence connecting structural attributes with the provision of species habitat (either for stand- or landscape-level indicators, e.g. Mac Nally et al, 2001; McAlpine and Eyre, 2002; and see earlier in this chapter), and far more research is needed in this area. For example, a recent study by Smith et al (2008) evaluated the relationship between a suite of forest structural attributes and a range of different taxonomic groups. The authors demonstrated that while change in some attributes appears to exert a significant

Figure 11.3 Digital photographs of the vertical profile of forest corridors in a central Amazonian fragmented landscape

Source: Photograph courtesy of Alexander Lees

Box 11.2 Use of satellite imagery to measure spatial variation in forest degradation and canopy damage

Developments in remote sensing are rapidly expanding the toolkit of available methods for understanding human impacts on forest ecosystems (Chambers et al, 2007; DeFries, 2008). Whereas previously, satellite images were only useful in classifying discrete vegetation types and spatial patterns of forest loss, it is now possible, with the help of high-resolution images and sub-pixel spectral mixture analysis, to generate continuous-scale metrics of forest structure and canopy integrity.

Souza et al (2005) used spectral mixture analysis (SMA) to derive a normalized difference fraction index (NDFI) to effectively describe patterns of canopy disturbance – providing a single scale to evaluate disturbance intensity from bare soil (-1.0) to pure green vegetation (+1.0). Interpretation of the index is facilitated by a contextual classification algorithm, which enables accurate mapping of logging and fire-derived canopy damage.

Hyper-spatial imagery, like IKONOS and Quickbird, with resolutions as fine-scale as 1m have opened the door to even more detailed analyses of forest canopy damage, with analyses being possible at the scale of individual crowns, and the capacity to identify point disturbances from logging gaps, secondary roads, etc. (Figure 11.4). Cost and the difficulty of securing cloud-free images limit the current use of such high-resolution images, but with the advent of new technology and additional satellite providers their availability is improving every year. High-resolution radar, or LiDAR (light detection and ranging), offers one solution to the problem of cloud cover and can achieve an exceptionally detailed understanding of forest structure. LiDAR is based on the use of laser light that is emitted from a source (normally an aircraft) and reflected back to a sensor, providing proxy variables of the structural diversity of the forest, including the density of the canopy layer, forest gaps and elevation. The cost of LiDAR is out of the reach of most monitoring programmes but accessibility is expected to rapidly increase.

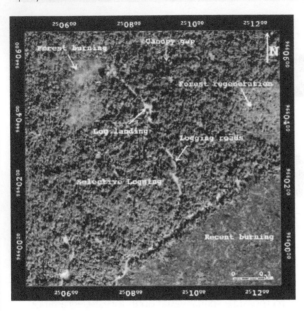

Figure 11.4 Ikonos image of a degraded forest showing log landings, logging roads, canopy damaged areas, burned forest, regeneration and forest cover

Source: Reproduced with permission from Souza and Roberts (2005)

Extremely few studies have attempted to link remote-sensing data on canopy structure and forest degradation to the impacts of forest management and changes in local biodiversity. One novel study by Aguilar-Amuchastegui and Henebry (2007) powerfully illustrates the potential value of remote-sensing imagery to help explain the impacts of selective logging on biodiversity. Using a vegetation index from SMA, the authors successfully linked changes in forest structural heterogeneity to logging intensity and changes in the composition and structure of dung beetle communities. In doing so, they also identified a logging intensity threshold of four trees per hectare as a transition point in forest structure, and the richness and diversity of associated dung beetles communities.

In another novel study, Müller and Brandl (2009) used LiDAR-derived variables to model both the species richness and composition of forest-dwelling beetles in a mountainous region of south-eastern Germany. This study is one of the first to attempt to link LiDAR and biodiversity data and the results were striking, with the LiDAR-derived data providing almost as much predictive power of beetle patterns as ground-based measurements of forest structure.

The use of satellite imagery to define structural indicators of forest degradation and management performance promises to be a very fertile area of research in the development of more cost-effective and larger scale forest biodiversity monitoring programmes. Progress will be facilitated by a closer integration of multi-disciplinary teams of remote-sensing and biodiversity scientists.

influence on a wide range of taxa (e.g. canopy cover), there is a high diversity of responses between species groups and species sensitivity classes (forest and non-forest species).

A particular difficulty in selecting a subset of structural indicators for monitoring lies in drawing a balance between stand and landscape-scale indicators. Landscape indicators are generally much easier to measure and map (as well as often being easier to manage). However, many species fail to conform to a traditional 'patch-centric' landscape model – as is implicitly presumed when applying patch-based indicators of landscape structure to try and explain species patterns (Haila, 2002). Instead, many species are known to perceive and respond to environmental heterogeneity across more continual environmental gradients (Manning et al, 2004), thus requiring stand-level attributes to improve our understanding of the distribution of species within individual forest patches that make up the wider landscape (e.g. McAlpine and Eyre, 2002; Barlow et al, 2008).

Where the goal of management is focused on the conservation of a single species, it is, of course, necessary to focus on improving aspects of forest habitat quality that relate to the ecological requirements of that particular species (e.g. the availability of specialized nesting and roosting sites). However, if the long-term conservation goal is to improve the overall forest condition, caution is needed to avoid circularity by only focusing on structural indicators that are tightly linked to the habitat of particular biological indicator groups. Instead, the selection of structural indicators (and associated target levels) need to be based on what we know about the consequences of changes to forest structure for biodiversity in general (including both stand- and landscape-level impacts), and our understanding of the best available reference conditions. Studies of reference sites (selected using independent knowledge of past human-impacts, see Chapter 9) provide an essential guide to the selection of indicators that characterize the structure of undisturbed or minimally disturbed areas of forest (e.g. density of large fruiting trees, canopy height and openness) and can help in measuring progress towards ecologically responsible management. Once a set of key structural attributes are selected, validation monitoring with biodiversity data can then be used to help qualify the relative importance of different structural indicators and indicator levels in contributing towards improving the condition of a managed forest. In highly modified and intensively managed systems, it may not be possible or appropriate to try and improve all aspects of forest structure to a state that more closely reflects an undisturbed system. Instead, the maintenance of different and complementary structural attributes (e.g. different age classes of plantation forest) in different parts of the landscape may provide the most effective and realistic means of conserving the entire forest biota (e.g. Eycott et al, 2006; Figure 11.5).

Generality in application

It is desirable if structural indicators of a particular management system or human land-use impact can be applied across a wide geographic area. Uniformity in management standards across different landscapes is essential to ensuring a fair and comparable system of environmental responsibility among different landowners

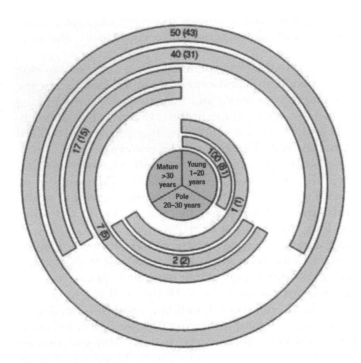

Figure 11.5 The total number of forest plant species recorded in different stages of the management cycle of a plantation forest in England

Note: Age classes are divided among young (before canopy closure, including thicket), pole (closed canopy) and mature (canopy opening, thinned stands). Circular bars show overlap of species across all three age stands, with only 50 species of a total of 217 recorded across all three stands. Numbers in brackets represent the number of angiosperm species
Source: Redrawn with permission from Eycott et al (2006)

(whether as the basis of a voluntary certification standard or as a legal obligation) while also providing a cost-effective and reliable means of assessing management compliance across a large area. Standards that vary significantly between regions are unlikely to be viable because of political and/or consumer (in the case of certification standards) resistance to perceived differences in the 'fairness' of assessments, and the high transaction-costs of implementing an assessment protocol that is tailored to the conditions of different sites.

However, despite such practical constraints, differences in the ecological (e.g. species composition, natural disturbance regimes) and biophysical (e.g. edaphic, hydrological) characteristics of different forest landscapes, as well as any interaction effects between varying environmental conditions and management impacts (see Chapter 13), make it inevitable that adjustments to the choice of structural indicators (and/or target indicator levels) are necessary in order to maintain comparable levels of environmental responsibility among regions.

Resolving this tension between pragmatism and context-dependency in the development of standards for forest management is extremely difficult for much of the world because we have a very poor understanding of spatial variability in biodiversity responses to forest modification. One way of tackling this problem is to improve the level of coordination among different biodiversity monitoring programmes so that it becomes possible to conduct cross-scale analyses of the landscape and regional dependency of different structural (or process) indicators and biodiversity disturbance responses (see Chapter 16).

BRINGING IT ALL TOGETHER: A GENERAL FRAMEWORK FOR SELECTING STRUCTURAL INDICATORS

In combining these criteria, it is possible to develop a general and hierarchical framework for selecting those structural indicators that are likely to provide the most rewarding measures of performance for a given forest management system (Figure 11.6).

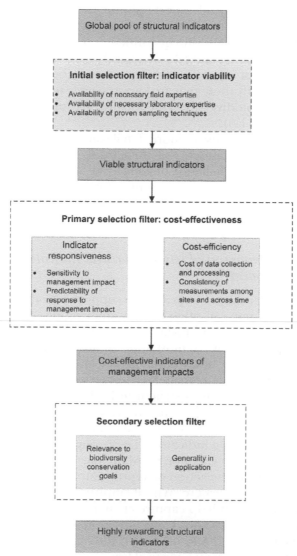

Figure 11.6 A general framework for selecting structural indicators that can provide a rewarding assessment of management performance with regard to a particular impact or disturbance

Note: Structural indicators are selected with respect to a particular management impact but also a particular conservation goal, which can be either ecosystem- or species-based. The framework is equally applicable for non-structural indicators of performance (see text for details)

From an initial pool of all possible structural indicators, a first selection filter is needed to identify those measurements that present viable candidates for monitoring and assessment based on the practical availability of necessary expertise and technical support (e.g. availability of geographic information systems (GIS) software, satellite imagery, digital camera for canopy photographs, etc.). The relative cost-effectiveness of the indicators in this set can then be evaluated with respect to indicator responsiveness to a given management impact (e.g. logging, clear-felling, forest fragmentation) and cost-efficiency. This evaluation process can draw on insights from previous studies and expert opinion, as well as comparative research on the structural consequences of different forest management activities (e.g. as demonstrated by Sist et al (2003), who quantified different measurements of stand-damage between conventional and RIL practices). Once indicators that provide a cost-effective assessment of management impacts have been selected, the second important filter is to identify those that are known *a priori* to be most relevant for the maintenance of forest biodiversity, and ideally, are applicable across a wide geographic area (Figure 11.6).

The following sections discuss in more detail some of the key considerations in the application and development of structural indicators for monitoring programmes, including: the challenge of identifying initial indicator sets; comparing structural indicators across studies; multi-attribute indicators; limitations of structural indicators; and the complementary importance of non-structural indicators for understanding biodiversity change.

The challenge of identifying initial sets of structural indicators and targets for management

The initial selection of structural indicators to support a biodiversity management and monitoring programme can often be very challenging. This is especially true in poorly studied areas of the world where little prior research exists to guide the identification of indicators that are most appropriate for particular types of forest or management systems. Many newly developed or interim management standards rely heavily upon indicators of management practices alone, rather than actual measures of performance (such as structural indicators). However, in order to ensure that real progress is being made, and to comply with many international standards authorities (e.g. ISO, FSC), it is vital to include a blend of both management process and performance indicators (i.e. through both implementation and effectiveness monitoring, see Chapter 8). Many parts of the world lack a strong empirical basis for the *a priori* selection of appropriate structural indicators, yet the process has to be started somewhere. Where empirical support is lacking, a defensible set of indicators that can help capture changes in forest condition can often be identified by drawing on the knowledge of regional experts.

For example, Oliver et al (2007) harnessed the expertise of both scientists and forest managers, and employed an analytical framework to derive a set of 'practical and defensible' structural indicators of vegetation condition as the basis for assessing the status of patch-scale forest biodiversity in south-eastern Australia. Using a form of multi-criteria analysis, the authors recorded and analysed the knowledge and opinion of 31 Australian forest experts on those condition indicators considered to be most important as biodiversity surrogates, as well as those that were considered most feasible to assess. A final indicator set was identified from an initial pool of 13 landscape-level indicators and 62 stand-level indicators. Experts considered that, in general, landscape-level indicators should contribute approximately one-third to an assessment of within-forest-type biodiversity, and stand-level indicators should make up the remaining two-thirds. Among the landscape-context indicators, *patch size, distance to nearest large patch* and *connectivity* were considered significantly more important than other attributes. In the case of stand-level forest condition, a set of ten structural indicators were identified, namely: *cover of native trees, shrubs and perennial grasses; cover of exotic shrubs, perennial grasses, legumes and forbs; cover of organic litter; recruitment of native tree/shrub saplings; native tree health;* and *evidence of grazing.*

Once a set of structural indicators has been identified by experts, it is then possible to initiate an ongoing process of evaluation and refinement against biodiversity data through a programme of validation monitoring and research. An often greater challenge than the selection of an initial set of structural indicators is to set appropriate indicator target levels that provide the basis for a management performance standard. This task is made particularly difficult by the need to compare indicators against an appropriate reference condition, together with a somewhat subjective decision about what constitutes an acceptable level of biodiversity loss for ecologically responsible management (accepting that few, if any, forms of forest management can be expected to incur zero loss; see also Chapter 9 for further discussion on issues regarding the reference condition). One way around this problem in the initial stages of the planning process is to draw on research findings from elsewhere in the region to inform local management standards. For example, Cantarello and Newton (2008) performed a comprehensive review of structural indicators of forest condition for two European forest sites (in England and Italy) and identified 17 indicators with associated target levels that may provide a starting point for measuring forest management performance (Table 11.3). Indicator choice and target levels can then be adjusted in response to local conditions following the onset of validation monitoring. In another example, Nilsson et al (2001) used general extinction models and some empirical work to recommend a target for 'sustainable forestry' in old-growth Scandinavian forests of maintaining at least 20 per cent of the original densities of large living and dead trees; this – translated into $20–30m^3/ha^{-1}$ of dead wood. While there is some evidence to suggest that this relation holds for some groups, much more research is needed to evaluate the wider validity of such a target (Nilsson, 2009).

Landscape scale indicators (e.g. fragment size, corridor design, etc.) obviously cannot be set against a reference condition as they are naturally anthropogenic in origin. In this case, appropriate levels need to be decided using available ecological research, taking care to set standards above possible threshold points that relate to potentially irreversible losses in biodiversity (e.g. fragments below a minimum size that are unlikely to provide viable habitat in the long term).

Table 11.3 Structural indicators used to assess the conservation status of forested habitats in Europe

Key factor	Structural indicator	Target or verifier	
		Foresta del Cansiglio, Italy	New Forest, England
Forest stand structure and composition		263	222
	Shannon–Wiener index for native trees	0.11	0.87
	Basal area (m² ha¹)	33	23
	Mean diameter (cm)	29	32
	Standard deviation of diameters (cm)	12	14
	Percentage of big trees	14	7
	Mean height (m)	24	17
	Number of total saplings (ha¹)	373	91
	Number of native saplings (ha¹)	373	91
Dead wood	Volume of downed dead wood (m³ ha¹)	61	26
	Downed dead wood decay class	0.5	0.5
	Volume of snag (m³ ha¹)	33	16
	Snag decay class	0.4	0.4
Tree regeneration	Number of total seedlings (ha¹)	25,000	63,219
	Number of native seedlings (ha¹)	25,000	63,219
	Shannon–Wiener Index for native seedlings	0.13	0.895
Ground vegetation	Number of ground vegetation species	34	33

Note: Proposed thresholds for two sites (Italy and England) are also presented
Source: Cantarello and Newton (2008)

Comparing structural indicators across studies

As already discussed few general and widely applicable methods for measuring structural indicators exist. The diversity of approaches available has meant that it is very difficult to make reliable comparisons of research findings between studies that employed different methods, or for newly initiated monitoring programmes to find clear advice as to recommended choices. Inevitably, different monitoring programmes will require somewhat different structural indicators as befits the type of forest and management system under investigation. However, even where indicators differ regarding some specific, it is possible to improve coordination among projects by adopting a common classification scheme. One such practical scheme has been proposed by Tews et al (2004) and classifies heterogeneity based on the type of structural variables, whether they are discrete or continuous measurements measurements based on a single or multiple study sites (Table 11.4).

Table 11.4 A general classification scheme for the measurement of heterogeneity in habitat structure

	Discrete variables (structural elements)		Continuous variables (structural elements)	
Study sites	Single	Multiple	Single	Multiple
Definition	Number of structural elements	Number and evenness of structural elements	Extent of structural qualities	Structural difference between various sites
Name	Structural richness	Structural diversity	Structural extent	Structural gradient
Measurement	Count of elements	Shannon's index of diversity	Measured structural quality	Gradient length, Euclidian distances
Example	Number of habitat types in a landscape	Diversity of habitat types in a landscape	Vegetation height or coverage	Difference in vegetation structure between sites

Source: From Tews et al (2004)

Multi-attribute indices of forest structure

One popular approach to dealing with the dual problem of an over-abundance of candidate structural indicators and a poor understanding of the relationship between individual indicators and biodiversity patterns, is to develop multi-attribute indices of stand and landscape structural complexity. For example, Spies et al (2007) developed an index of old-growth forest structure in the west coast of North America based on an average of separate measures of stand age, number of large trees, large snags (standing deadwood), volume of large snags and tree size diversity (itself an index). The combined index was scaled against attribute measures taken from reference forest stands over 200 years old.

In other examples, combined habitat and landscape indices are constructed based purely on multivariate statistics – particularly PCA that is used to identify independent axes of variability as inputs into a regression model of species patterns (e.g. Canterbury et al (2000) and Banks-Leite (2009) – Box 11.1 – for habitat and landscape-scale examples respectively).

Multi-attribute indices have both advantages and disadvantages. The principal advantage is that an index allows the collapsing of a large number of correlated variables, many of which are likely to be redundant, into a single tractable index that is often much more accessible to a non-scientific audience than the underlying data (Failing and Gregory, 2003). However, their main disadvantage relates specifically to the fact that changes across a large number of individual attributes and measurements are combined into a single, scaled index, where similar values can be generated through markedly different explanations. This masking effect can confound attempts to identify the relative importance of individual attributes in responding to, or driving, observed changes in the forest system.

Where index-based structural indicators are to be used, a balance is therefore needed between complexity and interpretability. McElhinny et al (2005) proposed the following simple four-point set of guidelines for the development of a relative index of stand structural complexity:

1   Start with a comprehensive set of structural attributes, in which there is a demonstrated association between attributes and the elements of biodiversity that are of interest.
2   Use a simple mathematical system to construct the index; this facilitates the use of multiple attributes and interpretation of the index in terms of real stand conditions.
3   Score attributes relative to the range of values occurring in stands of a comparable vegetation community.
4   Try different weightings of attributes in the index, adopting those weightings which most clearly distinguish between stands.

The Habitat Hectares approach (Parkes et al, 2003), developed for biodiversity compensation schemes in Victoria, Australia, is an index-based approach to measuring habitat 'quality' or condition that integrates both stand- and landscape-level attributes. The success of this method has been largely due to its intuitive and simple nature, although a number of problems remain (McCarthy et al, 2004; and see Box 6.2).

Limitations in the application of structural indicators for understanding patterns of biodiversity change

In discussing the selection of structural indicators for forest management standards, it is important to remember that no research project or monitoring programme will ever successfully identify a suite of structural attributes that is capable of characterizing the full biodiversity potential of a forest stand or landscape. There will always be a component of variance in the measured biodiversity response to a management impact that cannot be explained through changes to structural indicators (Smyth et al, 2002). This discrepancy in our understanding of the biodiversity consequences of forest management is due to at least six sets of reasons.

1   Management impacts are invariably conditioned by background environmental heterogeneities, including spatial variability in historical disturbances and the distribution of interacting non-focal taxa (e.g. plants, predators, etc.), which confound attempts to extrapolate patterns of cause and effect. It is important to try and account for these confounding factors in the analysis of monitoring data (see Chapter 13). Of particular importance is the fact that the same structural attributes may have distinct implications for biodiversity when they are found in landscapes that differ strongly with respect to historical patterns of human-use. This is especially likely to be the case in Europe, where the structure and biodiversity of some landscapes have co-evolved with humans over centuries or millennia.

2   We will always pass over an unknown number of additional structural attributes that are both sensitive to human activity and reflect important species resources. To account for this uncertainty, it is often useful to include the management treatment itself as a factor in analyses, and then employ post hoc pattern analyses to try and explain observed responses with respect to individual structural attributes.

3   The importance of certain structural indicators for biodiversity can depend on the condition of the surrounding forest. For example, tree holes can be seen as providing important habitat for birds in many forests, but the same holes can also be exploited by invasive Africanized honey bees (*Apis melifera scutellata*) in the New World.

4   Many of the ways in which human activities impact forest systems are not adequately captured by changes in structural forest attributes. Consequently, it is often necessary to incorporate additional, non-structural indicators into assessments of management performance (see below). Such indicators can then be evaluated against patterns of biodiversity in the same way as structural indicators.

5   Human activities can exert indirect impacts on biodiversity through changes to key ecological processes and functions (e.g. vegetation dynamics, nutrient cycling, decomposition rates, etc.), as well as secondary extinctions following the loss of functionally important taxa (e.g. from hunting or the harvesting of keystone plants). Discussion of management impacts on ecological processes is largely beyond the scope of this book but because they form a key part of the overall ecological integrity of the forest ecosystem they, like biological indicators, can contribute important additional response variables to the process of validation monitoring. Subsequent findings from this research can also then provide additional explanatory variables for understanding patterns of biodiversity.

1   Finally, our ability to connect management impacts to changes in biodiversity via changes in forest structural and functional attributes is limited by the imperfect nature of field studies and problems of sampling bias that often make it difficult to collect reliable and standardized measurements across different sites or time periods (see Chapter 14).

It is important to recognize that any failure to explain biodiversity patterns and link them to particular current management impacts can itself be a valuable result, and help identify important insights regarding the role of additional factors, including historical disturbance legacies, additional non-focal threats, complex indirect processes and sampling limitations and biases.

Non-structural indicators of management performance

Beyond management interventions that are focused on avoiding, mitigating or restoring the impact of human activities on the physical structure of forest ecosystems, there are a large number of additional impacts that need to be considered as part of any wider strategy for responsible forest use. Examples include disruption to natural disturbance regimes, introduction of invasive species and chemical pollutants, and direct harvesting of non-timber forest products (including game animals) (see also Chapter 10). Wherever possible, such impacts should be incorporated as part of an overall standard for ecologically responsible management in the form of minimum practice requirements, performance indicators and associated validation monitoring programmes. Non-structural performance indicators may take the form of the frequency of fire or pest outbreaks, occurrence of particular exotic species, the level of a given pesticide or herbicide in the soil, or the amount of harvest off-take of a particular non-timber forest product or game animal.

The disruption of natural disturbance regimes can have a particularly strong impact on the biodiversity value of modified forest ecosystems (Bengtsson et al, 2000; Nilsson, 2009). Moreover, because changes to disturbance processes are often tightly linked to local human or forest management activities, they represent an opportunity for securing significant benefits. For example, in sub-tropical, dry, temperate and boreal forests, fire plays a dominating role in shaping the structure and function of the ecosystem. Maintaining a fire regime that successfully reflects natural patterns of frequency and intensity is essential for the conservation of many forest species that depend directly on fire impacts or benefit from the novel environments it creates (Lindbladh et al, 2003; Nilsson, 2009). By contrast, fire is a novel and highly damaging source of disturbance in tropical wet forests and can result in significant losses of biodiversity (Barlow and Peres, 2004). Other important disturbance regimes include: changes to physical and chemical properties of water in local streams and rivers as a consequence of forestry activities (Roberts, 2001); grazing and browsing by both native herbivores (especially in temperate and paleotropical forests and woodlands (Nilsson, 2009)); and introduced species (e.g. sheep, cattle and goats (Allen et al, 2003)), and the dynamics of natural pest species that can be altered by human activity (such as the spruce bark beetle (*Ips typographus*) that operates as a keystone species in many northern European forests (Muller et al, 2008)).

CHAPTER 12

Selecting Biological Indicators and Target Species to Evaluate Progress Towards Conservation Goals

Synopsis

- Indicator taxa that are particularly valuable for monitoring are those that are both sensitive to management impacts and cost-efficient to sample.
- Because different species and species groups have distinct ecological requirements, and interact with their local environment at distinct spatial and temporal scales, combining multiple ecological indicator groups, each encompassing individual species that exhibit a wide range of responses to disturbance, can help in understanding a range of management impacts on multiple levels of ecological organization.
- Accounting for differences in survey cost among taxa requires a detailed audit of the cost of field and laboratory work, followed by a comparison of standardized costs that recalibrates sampling effort into units of money or time.
- Once a set of viable and cost-effective candidate indicator groups has been identified, additional considerations can provide further guidance for identifying the species groups that will provide the most useful information for management, including: the level of prior ecological knowledge to help the interpretation of sample data; the extent to which an indicator group can be applied to other regions and management systems; and the extent to which individual groups succeed in capturing the disturbance responses of other taxa.
- Individual target species can provide complementary information on management impacts to that which can be derived from indicator groups, and include those species that play particularly important functional roles in the forest ecosystem, are highly threatened, or are of particular economic or cultural value to stakeholders. Species of economic and cultural importance can include those that are harvested by local people (e.g. for game), as well as invasive and pest species. Other species may be valuable to include in monitoring because they are effective at garnering public support for conservation.

As laid out in Chapter 8, species data make their most useful contribution to forest management through validation monitoring – that is, the process of linking changes in management practice to changes in the actual conservation value or ecological condition of a forest via changes in measured and manageable forest structural or functional attributes. Depending on the choice of study species, there is a diverse number of ways in which this link can be made, with approaches grouping into two main types: ecosystem-wide, condition-based and individual species-based . As laid out in Chapter 9, there are powerful arguments for adopting an ecosystem-wide goal of improving forest condition as the primary basis of a biodiversity monitoring programme for responsible forestry. The first section of this chapter deals with the problem of selecting whole assemblages of ecological disturbance indicator species (see Chapter 9) that are capable of delivering on this task and providing meaningful and cost-effective information to guide management. I present a simple indicator-selection framework that can assist in identifying high-performing species groups that are capable of providing both reliable and cost-efficient sample data. The final section of the chapter addresses the role of individual 'target' species that may be of particular ecological, economic or cultural importance (including keystone, invasive, directly harvested, threatened and flagship species) as important complementary components of an overall monitoring programme.

A FRAMEWORK FOR SELECTING ECOLOGICAL DISTURBANCE INDICATOR GROUPS

A major challenge in the design of any biodiversity monitoring programme concerned with detecting and evaluating changes in forest condition lies in identifying the most appropriate taxonomic groups for study. The process of selecting groups of species that can provide reliable information on the ecological impacts of human-induced environmental disturbances has attracted considerable research attention during the last two decades (for reviews see: Landres et al, 1988; Noss, 1990; Kremen, 1992; Pearson, 1994; Stork et al, 1997; McGeoch, 1998; Spector and Forsyth, 1998; Caro and O'Doherty, 1999; Hilty and Merenlender, 2000; UNEP, 2000; Dale and Beyeler, 2001; Guynn et al, 2004; Hagan and Whitman, 2006). Yet, despite this concerted research effort, there remains considerable confusion and controversy among scientists and practitioners concerning the choice of appropriate indicators (Hagan and Whitman 2006; and see Chapter 6). One explanation for this confusion stems from the proliferation of different selection criteria that have been used to justify particular indicator choices, ranging from sensitivity to disturbance and natural environmental heterogeneity, distributional aspects, niche and life-history characteristics and criteria concerned with sampling methodologies (Hilty and Merenlender, 2000).

An initial task is therefore to *prioritize those selection criteria that must be satisfied in order to make any given species group a valid candidate for monitoring.* Here, I argue that priority should be given to those taxa that are: (i) sensitive to management impacts (and can therefore provide a reliable measure of changes in ecological condition –

that is, ecological disturbance indicators, (see Chapter 4)); and (ii) can be sampled feasibly with available resources. Gardner et al (2008a) classify such species groups as 'high-performance indicator taxa', describing them as having a high strategic value for biodiversity monitoring in providing both meaningful and cost-efficient information on the ecological consequences of human activity. Despite the logic of balancing measures of indicator utility (as related to a particular monitoring goal) and sampling feasibility to identify cost-effective indicator taxa, there are extremely few examples of where such an approach has been employed in practice. In the remainder of this section, I present a framework for selecting such high-performance indicator taxa. I emphasize the value of conducting empirical assessments of indicator performance, and present effective yet simple methods for measuring cross-taxon differences in sensitivity to particular disturbance and management impacts, as well as standardized survey costs. I also discuss important additional criteria that can help in making the final selection of high-performing indicator groups for a given monitoring programme. Many of these ideas draw upon the work by Gardner et al (2008a), who presented the first multi-taxa evaluation of the cost-effectiveness of biodiversity surveys for a forest ecosystem.

The value of conducting *a priori* multi-taxa assessments of ecological indicator performance

A priori multi-taxa field assessments are needed to develop an evidence basis for selecting high-performance indicator taxa for biodiversity monitoring programmes (Niemi and McDonald, 2004). Previous research has recommended the use of post-hoc pilot studies to validate the response of a preselected indicator taxon to a particular environmental gradient (e.g. McGeoch, 1998; Caro and O'Doherty, 1999). Yet, while indicator validation is extremely valuable, field trials of candidate indicator groups should ideally be conducted as part of the selection process itself (Rohr et al, 2007). There are a large number of studies that have independently recommended individual taxa for biodiversity monitoring work, based on the information value and/or sampling efficiency of that particular taxon (e.g. birds and plants (Pereira and Cooper, 2006); tiger beetles (Pearson and Cassola, 1992); carib beetles (Niemelä, 2000); and dung beetles (Halffter and Favila, 1993; Spector, 2006); and see Chapter 6). Yet very few studies have compared how different taxa perform under different management situations or in different forests around the world (see Lawton et al, 1998; Basset et al, 2004a; Schulze et al, 2004; Barlow et al, 2007a; Basset et al, 2008a; Smith et al, 2008).

The use of field trials to inform the process of indicator selection needs to be conducted within a framework that balances both economic and ecological reality. For example, it is unfeasible to conduct a comprehensive multi-taxa selection process for every monitoring programme. In some places, it may be possible to employ pre-existing multi-taxa data that were collected for another purpose, and apply them retrospectively to the indicator-selection framework. In other cases, it may be possible to draw upon resources from a number of different interested parties (e.g. research institutes, universities, forest management authorities) to implement a collaborative field campaign that can provide guidance on indicator performance for an entire region (as well as generating many other valuable research findings).

In planning the design of a multi-taxa indicator evaluation study, there are at least four important considerations to bear in mind:

1   Include as many taxonomic groups as possible in order to encompass a wide range of potential ecological response patterns (Barlow et al, 2007a; Basset et al, 2008a; Pardini et al, 2009). While it is impossible to sample all taxa, the choice of viable candidates for monitoring is often automatically limited by the availability of taxonomic and field experts, as well as proven and cost-efficient sampling techniques (Basset et al, 2004b; Rohr et al, 2007). Bockstaller and Girardin (2003) classify this initial stage in the indicator selection process as 'design validation' – the process of employing expert opinion to identify taxa that satisfy such basic requirements, thereby minimizing the wastage of resources on field testing of non-viable groups. In order to be sustainable, it is essential that any species group which may form the basis of a long-term monitoring programme is supported by a reliable team of local experts from the study region (rather than depending upon links with particular individuals in foreign institutions).

2   Samples should be collected across the fullest possible extent of a land-use intensity gradient, encompassing control areas, and areas of varyingly degraded native forest and modified systems (i.e. planted and regenerating forest) (e.g. Figure 1.2). This enables the demonstration of a reliable empirical relationship – consistent quantitative change – across a range of human influence (Karr and Chu, 1999; and see Chapter 9). It is, of course, impossible to conduct a multi-taxa assessment exercise across all the different forms of management that may be in need of evaluation. However, identifying those taxonomic groups that encompass a wide range of individual species responses to a broad disturbance gradient can often serve as a valuable proxy for indicating which groups are most likely to encompass species capable of discriminating more subtle and finer scale changes in habitat structure (Gardner et al, 2008a).

3   The sampling design should include sites that represent the least disturbed forest areas that are available in a region in order to provide a reference point for evaluating management impacts. As discussed in Chapter 9, reference sites do not have to be pristine in order to be useful, but rather just provide a general guide to measuring improvements in conservation value. It is important to include multiple reference sites in order to incorporate natural patterns of turnover in species composition.

4   Sampling of candidate indicator groups should be conducted at temporal and spatial scales relevant to actual management practices. Failure to link the scales at which ecological research is conducted to the scales at which management happens is one of the key explanations for the disconnect between the 'learning' and 'doing' of biodiversity conservation (Guynn et al, 2004; and see Chapter 10).

The Land-Use Change and Amazonian Biodiversity project in Jari (north-eastern Brazilian Amazonia) provides a good example of a multi-taxa biodiversity study that was designed to evaluate the conservation implications of land-use change (conversion of primary forests, establishment of *Eucalyptus* plantations and natural regeneration forests, Figure 12.1), as well as assess the performance of different indicator groups, and has generated tangible benefits for guiding long-term monitoring (see Barlow et al, 2007a; Gardner et al, 2008a).

Figure 12.1 The Jari landscape is comprised of large areas of secondary forest and *Eucalyptus* plantations embedded within a large area of almost intact primary rainforest

Source: Photographs by author

Evaluating the sensitivity of ecological indicators to disturbance

Ecological disturbance indicators provide a crucial link between the direct consequences of forest management and the underlying ecological condition and integrity of the ecosystem. The basic requirement for such an indicator therefore, is that it is sensitive to human-induced disturbance. A species group that performs well as an ecological disturbance indicator taxon is one that is comprised of individual species that are sensitive to human activities, but which also exhibit a wide range of responses to disturbance thereby allowing the evaluation of a wide range of management impacts. Species groups that are extremely sensitive to disturbance are often of limited practical value as they quickly disappear from all but the most undisturbed areas. Similarly, species groups that are so resilient as to be found almost everywhere are also of limited use. It is arguable that, at some level, most forest species will be sensitive to some kind of structural or functional modification to the forest environment. However, to be effective, indicators need to exhibit a clear response within the constraints of the sampling regime (e.g. spatial and temporal scale, and level of sample replication) that provides the data for evaluation, and at scales that are relevant to management. Taxa for which there are no standardized sampling methods or which exhibit significant lag-periods in their responses to disturbance will not provide reliable indicators for monitoring, even if they may ultimately decline in the face of human impacts. The same is true for rare species – those that naturally occur in low numbers everywhere or which are highly specialized on resources with a very discrete spatial or temporal distribution. In either case, it can be very difficult to collect sufficient data to draw reliable inferences about the distribution and abundance of species.

A large number of statistical techniques have been employed to help establish the relationship between an indicator species group and disturbance factor or management variable of interest. Depending on the treatment of the biological data, approaches range from univariate analyses (e.g. regressions of abundance and species richness) to a wide range of multivariate statistics, including various types of cluster and ordination analyses (e.g. Kremen, 1992; Davis et al, 2001a; Hausner et al, 2003; Basset et al, 2004b; Malcolm et al, 2004; Rohr et al, 2007). Although multivariate techniques can provide a powerful means of relating changes in entire species assemblages to changes in their local environment, they are of limited value in selecting ecological disturbance indicators as they mask information on individual species responses. An arguably more useful approach is to analyse species-specific responses (or levels of 'indicator value') in turn, and then combine this information to allow comparisons of overall sensitivity and response diversity between taxonomic groups. By providing additional information on within-group variability in disturbance responses, an individual species-based approach also provides investigators with information on which individual species are associated with different levels or types of impact.

The indicator value of individual species can be quantified by examining patterns of species abundance and occupancy. Species that perform well as indicators are those that exhibit marked and consistent responses to a particular type of disturbance, and are readily sampled in the field. One of the most significant methodological advances for the calculation of species indicator values is the *IndVal* method of Dufrêne and Legendre (1997). The IndVal method has seen increasing application in biodiversity monitoring and evaluation work (e.g. McGeoch et al, 2002; Pohl et al, 2007; McGeoch, 2007; da Mata et al, 2008; Gardner et al, 2008a; Smith et al, 2008), and has a number of advantages over other methods (Box 12.1). The method combines a measure of habitat specificity of a species to a particular level of disturbance or ecological state, with its fidelity (or frequency of occurrence) within that state (Dufrêne and Legendre, 1997). Species with a high level of specificity *and* fidelity for a particular disturbance state will have a high indicator value for that state. Dufrêne and Legendre's (1997) random reallocation procedure of samples within sample groups (e.g. disturbance classes or land-use types) can be used to test the significance of the IndVal measure for each species.

The IndVal technique can be applied at two levels of biological organization.

First, the analysis can be used to compare the generic indicator value of different taxonomic groups during initial field trials. Gardner et al (2008a) adopted this approach to identify which taxonomic groups from a candidate set of 14 (including vertebrate, invertebrate and plant groups) encompassed the greatest proportion of significant indicator taxa (Figure 12.2). The authors calculated the average percentage of significant indicators for each taxon across all three forest types. Comparing these percentage values across all taxonomic groups provided an effective means of identifying which groups were best able to discriminate changes in structural integrity across the sampled disturbance gradient (in this case, primary – secondary – plantation forests). An important step here is to check that variations in indicator performance are not simply a result of differences in sample representation (poorly sampled taxa cannot be expected to perform well as indicators).

Second, once certain indicator species groups have been selected the IndVal method can then be used to discriminate the potential roles played by individual species in evaluating management impacts. For example, the IndVal method allows the identification of both 'characteristic' species (i.e. those species with a high specificity and fidelity to a particular state) and 'detector' species. Detector species are those with only moderate specificity levels, whose relative changes in abundance across a disturbance gradient may provide an indication of the *direction* in which change is occurring (McGeoch et al, 2002; McGeoch, 2007; Figure 12.3). In combination, characteristic and detector species can provide complementary information on changes in ecological integrity. McGeoch et al (2002) demonstrated the value of identifying the indicator value of individual species using dung beetles as indicators of habitat conversion from closed canopy forest to open, mixed woodland in South Africa, while da Mata et al (2008) employed the same approach but using Drosophilid flies to evaluate the degradation of gallery forests in the Brazilian Cerrado, a vast Tropical savannah ecoregion.

Like any analytical technique, the IndVal approach comes with its own limitations, and more widespread application and analysis is needed to understand the generality of its properties, such as the formal relationship between the fidelity and specificity components (McGeoch, 2007). One well-known limitation is that it is highly dependent on the way in which samples are clustered. In the Gardner et al (2008a) study, the sample classification was very clear cut (primary, secondary and

---

Box 12.1 Advantages of the IndVal method for selecting ecological
disturbance indicators

Apart from its simplicity, the IndVal technique has a number of technical advantages over other methods. First, species selected by the IndVal analysis have to explicitly pass *both* the tests of habitat specificity and habitat fidelity. That is to say, selected indicators not only have a high information content regarding changes in their local environment, but also have a *high probability of being sampled* within a given monitoring period i.e., they are reliable as indicators of a given level or type of disturbance (Dufrêne and Legendre, 1997; McGeoch and Chown, 1998). This means that the analysis is robust to variation in the inclusion of rare taxa that may be present in samples as ecological vagrants, or because they are insensitive to common sampling techniques rather than being truly 'rare' (Basset et al, 2004a). Such an advantage is critical for indicator species to be of practical utility for cost-efficient monitoring because most classical sampling techniques perform poorly for sampling truly rare species (Yoccoz et al, 2001). While ordination and cluster analyses are theoretically able to incorporate information on differences in both site association (specificity) and occupancy (fidelity), the two factors can be confounded, and the explanation behind the position of a given species within an ordination plot or on a cluster diagram is not always clear. Second, the IndVal technique has the distinct advantage that the results for each species are calculated *independently* from the other species in the assemblage, meaning that comparisons can be made between unrelated species and species from different communities (McGeoch and Chown, 1998; Gardner et al, 2008a). In contrast, multivariate techniques are context-dependent and position different species on the basis of the distribution and abundance of other species in the assemblage. Finally, there are a number of additional advantages that lend the IndVal approach to being applicable to a wide variety of different situations, including the fact that it does not rely upon any hierarchical data structure, and can be generalized to any classification of sites, based either on species distributions or on *a priori* defined categories (Dufrêne and Legendre, 1997). This is extremely useful in refining sets of indicator taxa to evaluate particular management questions (e.g. edge effects, frequency of rotation, age of regeneration, etc.). Another advantage that could be very useful in certain situations is the ability to create an indicator value of species *absence* (McGeoch and Chown, 1998). For example, this could be used to incorporate exotic or invasive species into a multi-species index (see below) where positive values of the combined index represent higher levels of ecological integrity.

---

plantation forests), but the same is not true for situations where ecological disturbance gradients are more subtle or continuous. To get around this problem, Pohl et al (2007) presented an alternative classification approach that accumulated forest stands from older to younger along a chronosequence gradient. In another study, Howe et al (2007) presented an alternative method for calculating indicators of ecological condition based on species-specific logistic functions to represent the probabilities of finding individual species across a reference gradient of land-use intensity. They demonstrate, a successful application of the technique using bird assemblage from the Great Lakes region of the US.

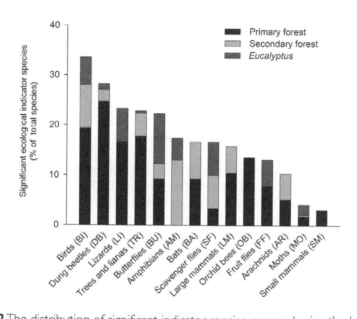

Figure 12.2 The distribution of significant indicator species, assessed using the IndVal technique, for primary, secondary and plantation forests for 14 taxa sampled across a large forestry landscape in Brazilian Amazonia

Source: Redrawn with permission from Gardner et al, (2008a)

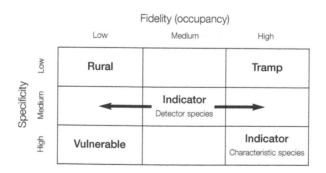

Figure 12.3 Species characterized by a combination of their degree of environmental specificity and fidelity, and classified on this basis as either characteristic (high specificity and fidelity) or detector (medium state) indicator species or rural, tramp and vulnerable

Source: Redrawn with permission from McGeoch et al (2002)

Irrespective of the technique used to calculate the sensitivity to disturbance of individual species or groups of species, it is highly desirable to validate indicator responses using independent datasets so as to understand the degree of confidence in which any chosen indicator(s) may be used in future work (McGeoch et al, 2002; da Mata et al, 2008). Although this is not always possible, validation should be viewed as an ongoing process of improvement in our understanding of the information content that can be derived from sample data. Finally, it should be emphasized that irrespective of the statistical technique used, analyses of indicator data should always be interpreted alongside pre-existing information on the species ecology and conservation status to add further meaning to results.

Identifying cost-efficient indicator groups

Despite the recognized importance of selecting indicators that can be sampled cost-efficiently, traditional guidelines provide few recommendations on how this should be done. Hagan and Whitman (2006) suggest that issues of practicality are usually only addressed by default, such that only indicators that can be afforded are selected. However, there is ample evidence to suggest that indicators are commonly selected mostly as a function of personal experience or preference of individual scientists, managers or auditors, and with little consideration of expense.

A number of researchers have evaluated the sampling efficiency of different inventory and monitoring designs (e.g. Longino and Colwell, 1997; Fisher, 1999) and more recently of sampling different taxonomic groups (Rohr et al, 2006). However, such analyses have assumed equal sampling and identification time and cost for different species groups (Rohr et al, 2006), despite the fact that such assumptions are rarely met (e.g. Lawton et al, 1998). In developing a new approach, Gardner et al (2008a) explicitly accounted for differences in the cost of surveying different taxa in the Jari project by recalibrating measures of sampling effort from 'individuals' to 'dollars per individual'. If the recording of data on survey costs was incorporated into biodiversity monitoring as standard (and given the rigorous accounting procedures demanded by most funding mechanisms this should not represent a lot of work), we would rapidly improve our understanding of the costs of biodiversity research and monitoring – and thus help to identify ways in which efficiency improvements can be made. The generic cost-standardization procedure adopted by Gardner et al (2008a) encompasses two principal stages: auditing of survey costs and standardization of survey costs.

Auditing of survey costs

A first step in any evaluation of survey cost is to conduct a detailed audit of the costs of work conducted for each taxa – in both the field and the laboratory (Figure 12.4). Cost should be measured in terms of both time and money and ideally calculated across all sampled treatments. It is particularly important to conduct a separate assessment for the control sites as they are generally the most diverse and difficult (expensive time consuming) to survey. To deal with project idiosyncrasies of who conducted what surveys (e.g. an MSc student sampled butterflies while a PhD student was responsible for the frogs) and to allow for direct comparison among taxa, salary costs for any field assistants, students or experts should be set at a standard rate – and only the minimum skill level necessary assigned to survey each taxon (i.e. don't include a PhD stipend rate for the frog surveys if they could perfectly easily have been sampled by an MSc student should one have been available – or better still by locally sourced field assistants). A subjective decision needs to be made as to whether to include non-perishable equipment costs, and it is often instructive to repeat the analysis with and without. If the objective of the analysis is to provide guidance for biodiversity work elsewhere, then transport and accommodation costs should be excluded as these are site specific. It is normally appropriate to exclude capital overhead costs, such as support from museum curation staff, this support is normally provided free of charge if there are mutual benefits from the work. Time costs should be divided into work by different skill levels and separated into field and laboratory time. In the Jari project, Gardner et al (2008a) found that total survey cost and time were closely correlated, and therefore based their main analysis on monetary cost alone.

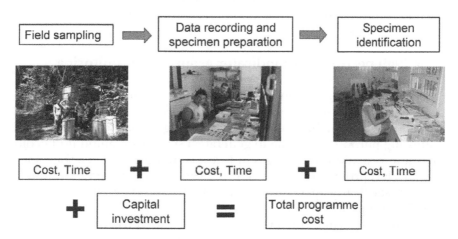

Figure 12.4 Auditing a multi-taxa biodiversity study

Source: Photographs by the author and Jos Barlow

Standardization of survey costs

It is inevitable that the amount of survey effort invested (in terms of money or time) will vary significantly between taxa, leading to differences in sample representation in the raw survey data. Hence, to allow direct comparisons, the challenge is to standardize costs between taxa that were surveyed with varying levels of effort. Gardner et al (2008a) presented a five-stage solution to this problem. First, individual-based rarefaction curves are constructed for each taxon. Individual-based curves are necessary because the units of sampling effort differ for each taxon.

Second, to account for differences in the total species richness of each taxon the y-axis scale should be recalibrated so that it represents the proportion of the estimated total number of species for each level of sampling effort (i.e. number of individuals). Where the estimated 'true' species richness is calculated using one of a number of species richness estimators (Gardner et al, 2008a), use the average of three non-parametric statistical estimators – ACE, Chao 1 and Jack 1 – to account for taxon-specific biases in estimator performance. Third, the total cost of surveying each taxon is calculated and used to estimate the survey cost per individual or encounter. Fourth, to account for differences between taxa in the total number of individuals or encounters, the x-axis of each rarefaction curve is recalibrated to represent survey cost (e.g. in US dollars) rather than number of individuals. Finally, to calculate the standardized survey costs for all taxa, the cost of surveying each higher taxon is rarefied to equate to the point at which the sample representation was equivalent to that of the least effectively sampled taxon. This point provides a measure of survey cost at a standardized level of sample coverage – or the 'standardized survey cost' (Figure 12.5). In cases where the estimated sample representation is very low for some taxa, it may be necessary to set an arbitrary threshold of representation (e.g. 50 per cent of the total number of expected species) and discard those taxa that do not reach this minimum standard.

Selecting high performance indicator groups for monitoring

Once an assessment has been made of candidate indicator groups regarding both their relative sensitivity to disturbance and standardized survey costs, it is a trivial step to then identify which, if any, qualify as having a relatively high performance for monitoring purposes i.e., able to generate a significant amount of ecologically meaningful information on changes in forest condition for a relatively low investment. In the Jari project, Gardner et al (2008a) demonstrated that dung beetles and birds are set clearly apart from the remaining 12 vertebrate, invertebrate and plant species groups and are deserving of being labelled as high performing (Box 12.2). Of course the same framework can also be used to compare standardized survey cost against other metrics of indicator value or information gained from surveying different taxa. For example, Gardner et al (2008a) also conducted an analysis comparing survey costs against biodiversity indicator values (congruency in cross-taxon disturbance responses), although in this case no clear pattern of indicator performance was produced.

Box 12.2 Selection of high-performance ecological disturbance indicators for the Jari project

Following a comprehensive multi-taxa field evaluation of the indicator performance of 14 species groups in the 1.7 million ha Jari landscape in north-east Brazilian Amazonia, Gardner et al (2008a) identified dung beetles and birds as being both highly sensitive to changes in forest structure and cost-efficient to sample (Figure 12.6).

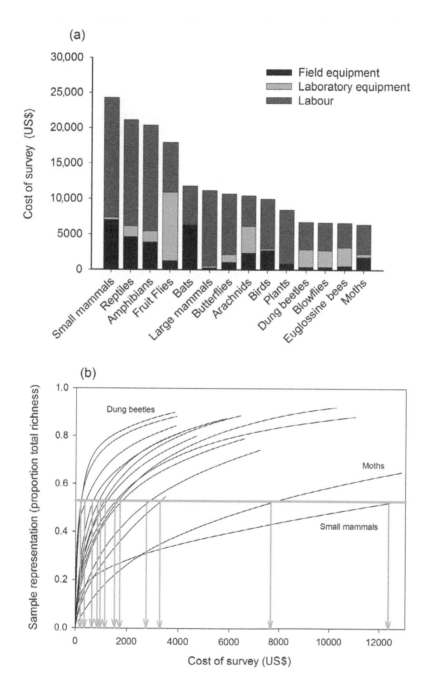

Figure 12.5 Raw and standardized survey costs (top and bottom panels respectively) for the 14 taxa surveyed in the Jari project

Source: Redrawn with permission from Gardner et al. (2008)

In addition to the results of the empirical cost-effectiveness assessment, previous work on both taxa provides further support as to their suitability for monitoring work. Dung beetles are a globally distributed insect group (Hanski and Cambefort, 1991), form a dominant component of the tropical insect fauna (Peck and Forsyth, 1982), play key functional roles in tropical forest ecosystems (e.g. seed dispersal and nutrient cycling, Nichols et al, 2008), and are used as effective indicators of environmental change (Spector, 2006). In addition, studies of both community (e.g. Davis et al, 2001a; Nichols et al, 2007; Gardner et al, 2008a) and species-level (e.g. McGeoch et al, 2002) responses of dung beetles to habitat loss and modification, collectively demonstrate a high degree of sensitivity to both local- and landscape-scale changes in vegetation physiognomy. Compared to invertebrates, few vertebrate taxa actually perform well as ecological disturbance indicators. In addition to often being very difficult and expensive to sample, many vertebrates are highly mobile and thus exhibit weak or delayed associations with local environmental impacts that are difficult to detect until it becomes too late for proactive management (Landres et al, 1988; Kremen et al, 1993; Hilty and Merenlender, 2000). However, birds provide a notable exception (Bibby, 1999; Carignan and Villard, 2002; Venier and Pearce, 2004), and have been shown to respond to environmental changes over many scales (Cushman and McGarigal, 2002), thus offering a compromise between small bodied insects and large mammals (Carignan and Villard, 2002). In addition, birds, like dung beetles, are known to perform a wide range of important ecosystem functions (e.g. seed dispersal and pollination, Sekercioglu, 2006). Finally, both dung beetles and birds can make a particularly valuable contribution to biodiversity monitoring and evaluation programmes due to the large volume and high quality of pre-existing ecological and natural history information (Halffter and Matthews, 1966; Bibby, 1999).

These two taxa have now been adopted as the core element of a long-term biodiversity monitoring programme in the Jari region, and are being used to address a number of research objectives on ecological implications of plantation management and RIL forestry.

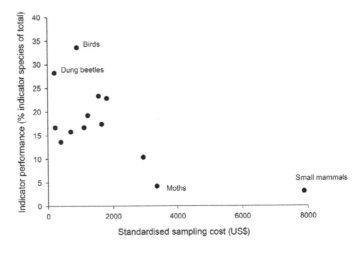

Figure 12.6 Comparing patterns of indicator value and standardized survey cost across 14 taxa sampled in an area of primary rainforest in Jari, Brazilian Amazonia

While the indicator selection framework described here represents a significant improvement from the status quo, it is not in itself sufficient for making final choices about which taxonomic groups should form the basis of a programme to monitor changes in forest condition. Additional, secondary considerations can provide further guidance for identifying the taxa likely to deliver the greatest amount of useful information for management. These considerations include recognizing the importance of prior ecological knowledge, our understanding of the role played by different taxa in the maintenance of key ecological functions and processes, the extent to which an indicator group can be applied to other regions and management systems from where it was selected, and the extent to which individual groups succeed in capturing disturbance responses of other taxa.

Prior ecological knowledge of indicators

Prior ecological knowledge of a species group is necessary to help understand patterns of abundance and occupancy in sample data, link observed changes in species patterns to measured forest stand and landscape attributes, and ultimately to develop a reliable understanding of cause–effect relationships (Landres et al, 1988). A fundamental requirement of any indicator taxon is that it provides data that are *interpretable*. Even if they are very sensitive to disturbance and cheap to survey, there is limited value in monitoring changes in a species group for which we have very little understanding of its resource and breeding requirements, or interaction with other species and ecological processes. That said, it is important that we avoid the danger of a positive feedback cycle – where we focus only on monitoring a small number of currently well-known taxa (e.g. birds), such that the more we learn about them the more valuable they become for monitoring and less likely we are to include other taxa (a research trap that has been referred to by Pawar (2003) as 'taxonomic chauvinism'). Given our poor understanding of man's long-term impact on the natural world, it is beholden upon conservation scientists to not only be cost-effective but also push the boundaries of knowledge – even if this may sometimes require a significant investment of time and money (see also Chapter 10 on the need to sometimes think 'outside the box'). For example, there have been extremely few studies on the consequences of land-use change for below-ground biodiversity yet we know that soil organisms play an essential role in nutrient cycling and soil structural engineering. Another major knowledge gap relates to the consequence of forest loss and degradation on the biodiversity of aquatic systems. Despite the fact that terrestrial and freshwater systems are inextricably linked both physically and functionally, there are extremely few, if any, biodiversity monitoring programmes that have combined sample data from both terrestrial and aquatic surveys.

Known functional importance of indicators

The value of any data collected on the response of an indicator group to a particular management impact is greatly enhanced when the indicator is *known to play an important functional role in the wider forest ecosystem*. Understanding how changes to an indicator group can have a cascading effect on key ecological processes, such as seed dispersal, pollination or nutrient cycling, or on biotic interactions with other taxa, can greatly enhance our ability to predict the functional consequences of biodiversity loss. In this sense, the greater the functional diversity exhibited by different individual species of an indicator group, the more valuable any monitoring data will be. In addition, taxa that are functionally linked to a wide range of other species or ecological attributes can provide important information on some of the indirect effects of human impacts (e.g. the cascading effects of mammal hunting on resource-dependent dung beetle assemblages and their associated ecological functions, Nichols et al, 2009). Of course, our understanding of the relative functional importance of different taxa is dependent on the level of prior ecological knowledge, and while it is inevitable that all taxa play some role in maintaining the ecological integrity of forest ecosystems, greater returns can be had by focusing initially on those for which we have a clearer understanding at present.

Generality in application of indicators

The more the same indicator group can be applied to different management regimes and different regions or forest types, the more valuable a contribution it can make to forest monitoring programmes in general. Employing the same indicator taxa across a network of coordinated monitoring programmes for an entire region would greatly increase the value of sample data – enabling multi-site analyses of the importance of landscape-scale and context-dependent biodiversity processes. Although not focused on forests one example of such an initiative is the GLOBENET programme, which is an attempt to develop a coordinated network of studies on biodiversity in urban landscapes, using ground beetles (Carabidae) as a common ecological disturbance indicator group (Niemelä, 2000). More such initiatives are urgently needed if we are to be effective in scaling up the results of individual monitoring programmes to tackle questions at larger scales.

Understanding the extent to which an indicator group selected for monitoring in one location or management regime can be recommended for use elsewhere or for another type of management involves a number of considerations, including: the extent to which species-specific differences in environmental sensitivity (indicator value) are a function of taxon-wide properties rather than particular characteristics of locally endemic species; variability in indicator sensitivity towards different forest and land-use types and disturbance regimes; and variability in access to necessary taxonomic and field expertise.

Cross-taxon representation

An ideal indicator group would provide information not only on changes in overall forest condition (as inferred through the loss or gain of a limited set of disturbance-sensitive species), but also on the disturbance response patterns of other, non-sampled taxa. As discussed in Chapter 6, there is only weak empirical support for the concept of cross-taxon disturbance response indicators, and no general theoretical argument why changes in one taxon should be representative of changes in any other. However, species groups or guilds that depend upon similar resources (e.g. leaf litter invertebrates for ground dwelling frogs and lizards, or fruit-bearing trees for primates and some birds), or that show similar levels of dependency on native forest, may exhibit similar responses to particular management impacts (e.g. Pardini et al, 2009). In accounting for differences in cross-taxon congruency (or lack thereof) in disturbance responses during the indicator selection process, the most important consideration is to maximize the extent to which sampling multiple species groups can provide *complementary information* on the ecological implications of forest management. In this sense, it is better to choose taxa that exhibit distinct, rather than congruent, responses to management impacts (Barlow et al, 2007a, and see below).

Bringing it all together: A general framework for selecting ecological disturbance indicators

By combining across this hierarchy of selection criteria it is possible to develop a general framework for identifying those indicator groups that are likely to deliver the highest level of performance for a given monitoring programme (Figure 12.7). From a pool of all possible species groups, an initial selection filter is needed to identify which taxa are viable for monitoring based on availability of necessary expertise and sampling techniques. The relative cost-effectiveness of this set of candidate groups can then be evaluated using field data by comparing differences in both indicator values (the sensitivity and reliability of responses to disturbance, where possible, including an empirical validation based on independent data) and cost-efficiency (Gardner et al, 2008a). The final choice(s) can then be made from this set of cost-effective candidates by taking account of background information on what is known on the ecology of each group and their capacity to deliver on additional agendas (Figure 12.7).

As discussed elsewhere in the book (e.g. Chapters 6 and 8), different species and species groups often exhibit a diverse range of responses to different forms of forest disturbance and management, and as such, no single ecological disturbance indicator group can provide all the information necessary to interpret the response of an entire system (Caro and O'Doherty, 1999; Carignan and Villard, 2002). Given the diversity of impacts that human activities exert on natural systems, we would ideally employ an integrated framework of biological indicators across multiple levels of organization, and at multiple spatial and temporal scales (Noss, 1990; Angermeier and Karr, 1994; Soberón et al, 2000). The adoption of such a 'bet-hedging' strategy can help to improve the robustness and sensitivity of monitoring programmes, and is a recommendation that has received wide support from the scientific community (e.g. Landres et al, 1988; Noss, 1990; Karr, 1991; Kremen, 1992, 1994; Hilty and Merenlender, 2000; Roberge and Angelstam, 2004; Basset et al, 2004a; Rohr et al, 2007). A difficult question therefore facing all monitoring programmes is to decide how many different indicator groups are sufficient to comprise an effective monitoring programme, bearing in mind that studying few groups well is almost always preferable to studying many groups superficially. The

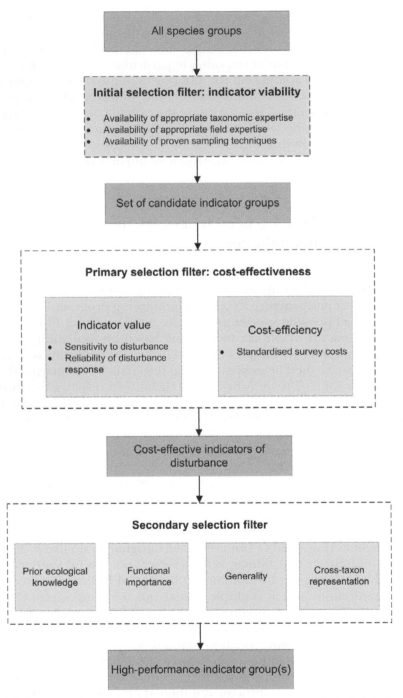

Figure 12.7 A general framework for selecting high-performance ecological disturbance indicators for monitoring changes in ecological condition

Note: The framework can be applied across different levels of taxonomic organization (see text for details)

degree of information provided by a variety of indicator species will depend upon the variability in their microhabitat use, niche breadth, ecological function and scale of response to environmental change (Terborgh, 1974; Kremen, 1992; Carignan and Villard, 2002; Smith et al, 2008). In the absence of detailed ecological knowledge on individual taxonomic groups a high diversity of disturbance responses in chosen indicator groups, can often be achieved by selecting groups from a broad taxonomic spectrum and across a broad range of spatial, temporal and organization scales (Landres et al, 1988; Lindenmayer and Burgman, 2005). The selection of complementary groups can then be balanced against the resources that are available for monitoring.

In addition to basic patterns of species occupancy and abundance, biodiversity data also provides information on species-specific life-history traits (e.g. body size and reproductive success) that are often associated with different levels of environmental disturbance, and can help in attempting to understand the mechanisms that underlie observed changes (see Chapter 14). Differences in individual responses to disturbance can also provide complementary information to that which can be gained from studying species and population-level responses alone, including differences in individual physiological and morphological condition and behavioural patterns (Leary and Allendorf, 1989; Wikelski and Cooke, 2006). For example, Delgado-Acevedo and Restrepo (2008) found that both the body size and shape of two endemic Puerto Rican frogs, *Eleutherodactylus antillensis* and *E. coqui*, varied between areas characterized by different levels of deforestation. Individuals of both species that were sampled in highly disturbed areas (<20 per cent forest cover) were smaller and had more asymmetrical body forms than individuals sampled in either intermediately (20–70 per cent forest cover) or little-disturbed (>70 per cent forest cover) landscapes. This kind of fine-scale information on differences in the state or condition of individuals sampled across different levels of human disturbance can be particularly important for quantifying species response patterns across short time scales (e.g. weeks–months), thereby providing an early warning system of changes in environmental stressors before they have a serious impact on population viabilities. Finally, and although they are given little attention in this book, environmental indicators (i.e. species that are used to measure a specific environmental state, see Chapter 4) can make a valuable addition to a wider monitoring programme in situations where there is concern about the impact of certain environmental pollutants. For example, gastropod molluscs can provide an effective means of measuring environmental exposure to endocrine-disrupting compounds (e.g. pesticides) that may be commonly used in forest management (Hall et al, 2009).

THE CONTRIBUTION OF INDIVIDUAL TARGET SPECIES TO BIODIVERSITY MONITORING

Species-level approaches to research and management can be highly complementary to more system-based approaches that focus on sampling whole biological assemblages (see Chapter 9). Additional management goals are usually necessary to accommodate the conservation requirements of individual species that either play particularly important functional roles in the forest ecosystem, or are highly threatened, or are of particular economic or cultural value to stakeholders (both local and global). Although such species generally fall outside the traditional remit of 'indicators', many of them nevertheless provide some kind of indicator function in contributing to our overall understanding of human impacts. In some circumstances the presence of certain species can itself represent a direct threat to biodiversity (e.g. invasive and pest species), or a direct consequence of human management (e.g. decline of non-timber forest products). Some species, such as those of particular conservation (e.g. IUCN Red List species) or societal importance (e.g. economically valuable, and flagship species) are required by law, or voluntary management standards, to be included as part of a biodiversity monitoring programme (e.g. Box 6.3). Finally, and especially where resources for monitoring are limited, individual species may be chosen for monitoring because they are particularly conspicuous and easy to survey.

The following non-mutually exclusive categories highlight some of the areas where individual or target species could make particularly important and complementary additions to ecological disturbance indicators as part of an integrated biodiversity monitoring programme. By definition, the process of selecting individual species for monitoring is highly dependent on context- and location-specific ecological and social factors.

Strongly interacting species

The idea that some species play exceptionally important roles in maintaining and organizing the diversity of ecological systems was first introduced by Paine (1966) to describe the phenomenon of 'keystone species' – starfish in this case – in structuring intertidal communities. This concept has now been extended to all ecosystems and is used to highlight the need to research and manage those species that exhibit particularly strong interactions with their surrounding environment (Mills et al, 1993; Soulé et al, 2005). Mills et al (1993) argued that the complexity of ecological interactions, and our daunting ignorance of them, mean's that there is little practical value in attempting to define individual species as 'keystone' or 'non-keystone'. Instead, they recommended the study of interaction strengths more generally, and how they can vary across environmental gradients and within different spatial and temporal contexts. There are a number of different categories of possible 'strongly interactive' species, including predators, prey, plants, link-species (e.g. seed dispersers and pollinators) and habitat modifiers. Some of the most famous examples of strongly interacting species in forest ecosystems include the cascading effects on lower-level trophic groups and vegetation of the removal of top-predators (Terborgh et al, 2001), and the impact of 'ecosystem engineers' – species that have a powerful impact on the physical structure of the forest, such as the role played by beavers in creating forest wetlands (Jones et al, 1994).

A major difficulty in incorporating strongly interacting species into a biodiversity monitoring programme lies in identifying, *a priori*, those species that are genuinely deserving of attention. In principle, a given species should receive special attention for monitoring and management if its loss or population decline precipitates cascading, dissipative transformations in ecosystems, including alterations or simplifications in ecological structure, function or composition (Soulé et al, 2005). To help in identifying such situations Soulé, et al (2005) recommended asking a number of key questions concerning the potential importance of candidate focal taxa (Table 12.1).

Table 12.1 Key questions that can assist in identifying strongly interacting or keystone species

Question	Example
Does the absence or decrease in abundance of the species lead directly or indirectly to a reduction in local species diversity?	The absence of coyotes from arid ecosystems can lead to a reduction in bird diversity via mesopredator release, or a reduction in rodent diversity via competitive exclusion (and large population increases of a small number of dominant taxa)
Does the absence, decrease in abundance or range contraction of the species directly or indirectly reduce reproduction or recruitment of other species?	The number of forest tree species that successfully reproduced on predator-free islands in a Venezuelan reservoir fell from about 65 to about 10 due to an extreme of herbivores
Does the absence or decrease in abundance of the species lead directly or indirectly to a change in habitat structure or composition of ecosystems?	High levels of herbivory by elk on willow trees in Rocky Mountain and Yellowstone National Parks in the US have been implicated in the disappearance of the beaver and loss of wetlands
Does the absence or decrease in abundance of the species lead directly or indirectly to a change in productivity or nutrient dynamics in or between ecosystems?	Prairie dog colonies shape nutrient cycling, soil chemistry and physical structure, and the productivity and nutrient content of vegetation through their burrowing and grazing activities
Does the absence or decrease in abundance of the species change an important ecological process in the system?	Beavers have a profound effect on stream dynamics, water-tables, flooding and the extent of wetlands
Does the absence or decrease in abundance of the species reduce the resilience of the system to disturbances, such as fire, drought, flood or exotic species?	Loss of the dingo in some regions of Australia has had an indirectly-led habitat degradation effect because dingoes play a critical role in limiting herbivory by rabbits, red kangaroos and other herbivores

Note: Examples relating to each question are also given although not all are from forest environments
Source: Modified with permission from Soulé et al (2005)

Answering such questions and identifying candidate strongly interacting species may require a combination of existing ecological knowledge and theory, expert judgement, comparative analyses, modelling and historical data reconstructions (Soulé et al, 2005). Ultimately, however, the identification of such species will remain a partly subjective, context-dependent and dynamic exercise – and one that should be subject to constant revision within an adaptive monitoring and management framework. Two specific sub-categories of 'strongly interacting' species that are deserving of particular attention for forest biodiversity and monitoring programmes are invasive and pest species – see below.

Invasive species

In many forest ecosystems, conservation managers are primarily concerned with the conspicuous impacts of habitat loss and degradation. However, the more cryptic or indirect impacts of invasive species can often have devastating consequences for local biodiversity. For example, in the case of tropical forests, one of the starkest lessons of the potential ecological consequences of a single, successful invasive species is that of the brown tree snake (*Boiga irregularis*) on Guam (Mortensen et al, 2008). Since the introduction of the snake around 1950, most of the native forest birds, nearly half of the native lizards and two of Guam's three bat species have gone extinct. These losses have also led to reduced recruitment of plants that depend on flower-visiting birds for pollination. In another example, Srbek-Araujo and Chiarello (2008) found in two years of monitoring that dogs were the fourth most abundant of all mammals encountered, and the most abundant carnivore, in a reserve in the Atlantic Forest of Brazil. The cascading effects of such high abundances of an exotic carnivore are poorly understood yet are thought to have serious consequences on the native forest biota (e.g. through predation of small mammals, competition with native carnivores and as vectors of disease (Srbek-Araujo and Chiarello, 2008)). There are multiple examples of where invasive plants have had a marked impact on forest biodiversity. Island systems that have become dominated by alien plant species have been particularly well studied, with Puerto Rico (Lugo and Helmer, 2004) and Hawaii (Asner et al, 2008) representing particularly stark cases. In the majority of cases, our understanding of the mechanisms that determine such impacts remains weak.

In addition to the challenge of unravelling the mechanisms by which invasive species impact native biodiversity, an effective forest monitoring programme should also encompass a targeted surveillance or 'early warning' component to identify when and where invasive species arrive. In well studied areas, some high-tech methods exist that can facilitate the recording of known invasive species. For example, Brockerhoff et al (2006) describe the use of pheromones to detect exotic insects, and report that moth sex pheromones were successfully employed to detect, delimit and monitor several recent arrivals to New Zealand, including the white-spotted tussock moth, the gypsy moth and the gum leaf skeletonizer.

Pest species

Alongside invasive species native pests are another class of strongly interacting species that can have major ramifications for the structure and function of forest ecosystems and the maintenance of local biodiversity. Perhaps the most conspicuous example of this is in the case of bark beetles whose infestations regularly have massive ecological and economic impacts on boreal forests of Europe and North America. Of course, outbreaks of native pests represent natural disturbance events and therefore play important engineering roles in the maintenance of biodiversity and ecological processes (Raffa et al, 2008). For example, Muller et al (2008) found that the spruce bark beetle (*Ips typographus*), which is a serious economic pest of mature spruce stands throughout Eurasia, can have a strong positive effect on the diversity of native saproxylic beetles that thrive in canopy gaps created by beetle outbreaks.

Monitoring work can be vital in helping to develop management strategies that are effective at limiting economic damage from pest outbreaks, while also ensuring that functionally important disturbance regimes are maintained. This is particularly important in situations where intensive forest management for economic production can limit the scale and intensity at which natural disturbance regimes would otherwise occur (the same problem occurs with respect to fire). In other situations, human activities can work to exacerbate the impact of pest outbreaks, and increase the risk of fundamental regime shifts. For example, Raffa et al (2008) illustrate how human activities can mediate the interaction between bark beetles and conifer trees, showing how reductions in forest heterogeneity, coupled with climate change, can increase the likelihood of landscape scale outbreaks in North American forests. The authors also show how decades of careful monitoring and observational work has contributed to the development of a predictive framework to help guide more sustainable management strategies.

Economically important species

Aside from the main production species (i.e. plantation trees, hard wood, plantation coffee, cocoa, etc.), other elements of forest biodiversity may be of considerable economic importance, especially to local people. Such species encompass a wide range of 'non-timber forest products', including both plants (e.g. bamboo, palms, fruit, nuts) and animals (e.g. game meat, skins) that are harvested from areas of native forest both for local consumption and commercial gain by millions of people worldwide (Godoy and Bawa, 1993). The harvesting of individual forest species is a vast subject in its own right, and accordingly has received considerable attention from scientists with regards to both plants (reviewed in Ticktin, 2004), and animals (e.g. the impacts of game hunting on tropical forest mammals (Bodmer et al, 1994; Peres, 2000; Milner-Gulland and Rowcliffe, 2007)). Even entire journals have been dedicated to the subject (e.g. *Economic Botany*). Although a review of this literature is beyond the remit of this book, it is important to recognize both the need to monitor such species and the particular requirements that are necessary for their assessment (Hall and Bawa, 1993). There are at least four separate arguments for including economically important, non-timber species in biodiversity monitoring programmes for managed forest systems:

1   To help develop management strategies that can ensure sustainable harvesting of natural resources (including co-management systems of both timber and non-timber resources).

2   To directly engage local people who use forest resources with the wider monitoring and management system (e.g. Danielsen et al, 2007, and see Chapter 16).
3   To evaluate the cascading effects of species harvesting on other species and
ecosystem processes. This is particularly important in the case of species that
are strongly interacting with other components of the forest (e.g. seed-dispersing primates, the loss of which can have serious, yet poorly understood, indirect
effects on the composition of tree species, e.g. Terborgh et al, 2008)
4   To better understand possible interaction and synergistic effects between
human-induced impacts on forest patch and landscape structure and direct
species harvesting (e.g. Peres and Michalski, 2006).

Threatened and endangered species

If the overall goal of ecologically responsible forest management is to conserve and
restore biodiversity, it makes logical sense for species that are threatened with extinction or severe population declines to be given priority attention in both management
and monitoring work (e.g. Meijaard et al, 2006). Nevertheless, there are a number of
serious practical and theoretical problems associated with monitoring threatened
species, which in combination often serve to limit the value of such data for
management (for a critique see Chapter 6). Aside from the difficulties of sampling
species that are often found in low numbers, as well as limited and patchy distributions,
one of the greatest challenges lies in identifying threatened species in the first place.

Threatened species are invariably identified through the use of Red Lists,
whether the IUCN system (a global standard for the classification of extinction risk)
(Mace et al, 2008; Box 12.3), or national equivalents. However, while these protocols
have been highly successful in attracting significant conservation attention to the
plight of individual species, they suffer from a number of severe practical limitations,
especially in poorly studied areas of the world. For example, Red List protocols have
been criticized by many as being too simplistic, relying upon poorly understood
assumptions (e.g. the idea that naturally rare species are more extinction-prone),
while lacking any serious consideration of deterministic threats to individual species
persistence (Gaston and Fuller, 2007). They are also subject to considerable
measurement error as well as semantic uncertainty (Akçakaya et al, 2000), and are
frequently scale-dependent (Hartley and Kunin, 2003). Long-term survey data are
often required to overcome these problems and generate reliable species
assessments, and a lack of adequate data has been considered a strong impediment
for applying Red List protocols in developing countries where few biodiversity data
are available. Although Red Lists can be a powerful tool for evaluating the status of
relatively well-studied species at the national or regional level, such limitations
greatly reduce their utility for landscape-scale conservation planning. There may
either be very few listed species in a given area or so many that they become difficult
to choose from. The listings themselves may be questionable, and there is invariably
a large number of species that are threatened with local population declines or
extinction yet are not listed (sometimes because of scientist concerns about
repercussions for research and collecting on listed species).

Box 12.3 IUCN Red List classification of endangered species

The IUCN Red List of Threatened Species was initially established in the 1980s to assess the conservation status of species for policy and planning purposes. This use stimulated the more formal development of a new set of quantitative criteria for listing species in the following categories of threat: critically endangered, endangered and vulnerable. These criteria are part of a broader system for classifying threatened species that was fully implemented by IUCN in 2000 (see www.iucnredlist.org for the full listing of threat criteria). The system and the criteria have been widely used by conservation practitioners and scientists and now underpin one indicator being used to assess the CBD 2010 biodiversity target. The threat criteria at the foundation of the IUCN Red List refer to fundamental biological processes underlying population decline and extinction. Because of major differences between species, the threatening processes affecting them, and the paucity of knowledge relating to most species, the IUCN system is both broad and flexible in order to remain applicable to the majority of described species. The system was designed to measure the symptoms of extinction risk, and uses five independent criteria relating to aspects of population size, population loss and decline of range size. A species is assigned to a threat category if it meets the quantitative threshold for at least one criterion. The criteria and the accompanying rules and guidelines used by IUCN are intended to increase the consistency, transparency and validity of its categorization system, but it necessitates some compromises regarding species of uncertain status that affect the applicability of the system and the species lists that result.

The IUCN Red List relates to global extinction risk, and uptake of the IUCN Red List criteria has also been extensive at national level; at last count, 76 countries were using them. However, while the relationship between sub-global to global extinction is clear, the relationship of global to regional threatened status is unfortunately much more complicated. Certain species may be assessed to be relatively secure within a country but nevertheless be at risk globally, whereas other species that are relatively secure globally may be highly threatened within a particular region, for example, at the edge of their geographic range. The distinction between the IUCN Red List and individual national lists is important because conservation actions take place locally. National lists of threatened species frequently form an important part of national management standards. Continuing work is underway to improve the consistency between the data and assessment methods used nationally and globally, but extinction risk within individual countries is a complex concept because of the interdependence of many species on adjoining nations (or states in federal nations with multiple lists), and there will inevitably be variability among countries in assessment methods and uses of these lists.

Source: Text based on Mace et al (2008)

Despite such serious problems in both identifying and monitoring threatened species, they often retain a vital role in forest biodiversity monitoring programmes. In part this is because many species that are listed as threatened are indeed genuinely threatened with extinction, and in need of urgent conservation attention. In addition, the laws of many countries demand that species listed as threatened be protected – and such regulations are often one of the most powerful tools available to conservation (e.g. the US Environmental Protection Agency listing, which has significant powers to halt or limit development plans). Where additional and targeted monitoring of threatened species is necessary, care should be taken in selecting those species that are most deserving of research attention. A prioritization schedule is essential because individual species monitoring can be very costly and some form of triage is necessary to maximize the return on any investment in research (Possingham et al, 2001). Regan et al (2008) present one such strategy that takes into account the spatial and temporal extent of known threats and can be used to help prioritize species for monitoring and management in multi-species conservation plans (Box 12.4).

Box 12.4 A strategy for selecting priority at-risk species for monitoring and management in multi-species conservation plans

Regan et al (2008) presented a two-tiered system for selecting priority at-risk species for monitoring, and illustrated the approach with a case study of 85 plants and animals in San Diego county, California. The first tier is based on at-risk and focal species prioritization recommendations from Andelman et al (2001) and the broader ecological monitoring literature:

1   Apply an at-risk species classification using established risk rankings (i.e. national or IUCN Red List criteria).
2   For each at-risk group, allocate species to categories based on the nature of the risk factor.
3   Using information on home ranges (or a surrogate, such as body size or known distributions), further classify species in each group according to their spatial scale of response to risk factors.
4   Using information on life span or plant functional group (as a surrogate for life span), further classify species in each group according to their temporal scale of response to risk factors.
5   Select one or more priority species from each group that best represent the rest of the species in the group.
6   Stop when each discrete vegetation community type is represented by at least one priority species or when all risk factors have been associated with at least one priority species.

The pertinent pieces of information needed to prioritize the covered species according to steps 1–6 above are: (i) at-risk category (based on applicable ranking systems); (ii) threats and the degree and spatial extent of threats across a species' range; (iii) broad-habitat associations of species; and (iv) temporal scale of the impact of threats. In compiling this information, researchers can draw upon published and non-published data and the opinion of relevant experts.

In order to prioritize the species that are likely to be at greatest risk, Regan et al (2008) also recommend an additional selection tier whereby species are grouped according to their at-risk-ranking into Risk Groups, which are determined by the number of high-level threats and the temporal scale over which they occur.

Source: Modified with permission from Regan et al (2008)

Flagship species

Flagship species, as the name suggests, are species that act to draw attention to a conservation area and facilitate the engagement of stakeholders (Caro and O'Doherty, 1999). In the same vein, Lindenmayer and Fischer (2003) have also termed such species 'social hooks' in light of their often important role as primary motivators of political and community engagement in difficult conservation problems. Flagships species are invariably charismatic and conspicuous taxa that are easy 'to sell' (e.g. tigers, wolves, rhinos, etc.). There are few criteria for selecting a flagship species and it is often the case that they are also endangered and/or operate as keystone species in maintaining key functional processes – thereby providing multiple scientific, social and legal arguments to justify their inclusion as part of a wider monitoring programme. Perhaps the most important example of the use of a flagship species in efforts to improve the ecologically responsibility of forest management is the case of the northern spotted owl (*Strix occidentalis lucida*, Figure 12.8) and the North West Forest Plan (Lindenmayer and Franklin, 2002). The northern spotted owl was listed as threatened under US law in 1990 and formed the basis of one of the world's most ambitious conservation plans, helping to reserve 77 per cent of the 11.1 million hectares of federal forest land from timber harvest within the range of the owl (Lindenmayer and Franklin, 2002).

Figure 12.8 Northern spotted owl (*Strix occidentalis lucida*) – the primary flagship species behind the North West Forest Plan on the west coast of the US

Source: Photograph from US Fish and Wildlife Service's digital library collection

Focal species

The concept of individual focal species (*sensu* Lambeck, 1997), as those species whose ecological requirements serve as an umbrella for the conservation of other taxa, has received considerable criticism due to a common lack of sufficient ecological information for implementation (Lindenmayer et al, 2002; and see Chapter 6), and has yet to gain significant traction in the conservation community.

Nevertheless, the concept is not without merit. In well understood systems, it is possible that examining the disturbance responses of species with very different life-history strategies can provide valuable insights into the variability of ecological impacts from forest management. For example, McComb et al., (2007) used spatial simulation models to evaluate current and potential future habitat availability over a 100-year period for three focal species (Pacific fisher (*Martes pennanti*), pileated Woodpecker (*Dryocopus pileatus*) and Warbling Vireo (*Vireo gilvus*)) in the managed forest landscapes of Oregon, north-west US. These species were chosen because their habitats represent a broad range of spatial scales and forest types. Simulations showed that the area of habitat for Pacific fishers and pileated woodpeckers is predicted to increase over time under current forest land management policies, while that of warbling vireos is predicted to decline (see Chapter 15). The difficulty lies in linking the relevance of these results to other species or the ecological integrity of the system as a whole – unless the focal species are known to perform key functional roles (and therefore also operate as keystone species). One of the main benefits of monitoring focal species may come from their capacity to act as a 'social hook' and engage local stakeholders to think about conservation problems (Lindenmayer and Fischer, 2003). In this sense, the concept of focal species blurs with that of flagship species, although well-studied yet non-charismatic species may still hold considerable didactic value in facilitating stakeholder dialogue about the dynamics of the forest ecosystem and impacts of management.

In conclusion, the concept of focal species has remained difficult to grasp by many researchers, and comprehensive empirical assessments are lacking. Focal species may be most valuable for monitoring programmes when they simultaneously act as ecological keystone species and/or some kind of flagship species that strengthens the value of inferences that can be made from sample data, and facilitates engagement with stakeholders.

CHAPTER 13

Making Assumptions Explicit: The Value of Conceptual Modelling in Biodiversity Monitoring

Synopsis

- To successfully tackle the complexity of human-modified forest ecosystems biodiversity monitoring programmes need to make explicit the logic and assumptions that lie behind design choices and stated associations among individual system components.

- Conceptual frameworks and models are powerful tools that can help in addressing these challenges. The process of developing conceptual models is as important as the product itself as it forces us to think about how systems work.

- Conceptual frameworks are a means of categorizing variables and indicators and identifying which are most appropriate for assessment, while conceptual models are a method for articulating causal relationships and dynamics of the study system.

- Well-conceived conceptual models can facilitate the development of biodiversity monitoring programmes in at least five inter-related ways by helping to: (i) contextualize the relative importance of evaluating different management impacts on biodiversity; (ii) make explicit our assumptions about the nature of human impacts on biodiversity; (iii) develop a basis for constructing research hypotheses; (iv) design robust sampling strategies for monitoring; and (v) provide a valuable aid to communicating biodiversity monitoring and research.

- A validation monitoring programme needs to consider the factors that mediate the biodiversity consequences of a particular human impact or management intervention. An important function of a conceptual model is to help clarify these mediating factors, including the importance of interaction and synergistic effects as well as time-lags, thresholds, landscape context and stochastic dynamics in interpreting observed biodiversity patterns.

- Conceptual models that encompass not only ecological factors but also the social processes that relate to the structure and function of forest management can be a highly effective way of integrating the monitoring process with the wider management system

Human-modified forest ecosystems are complex and dynamic phenomena. First, the background ecological processes and functions that serve to maintain biodiversity are themselves highly complex and inter-related, and play out across multiple spatial and temporal scales. Second, human-induced impacts, from physical landscape change through to the introduction of exotic species, contribute yet further complexity and confound our ability to interpret the drivers of observed changes in species distributions and abundance. Finally the entire ecological system is tightly, coupled to the broader social system by a myriad of linkages and feedbacks that are governed by complex and multi-scale market, regulatory and cultural factors, relating to both economic and conservation activities.

In a very general sense, the purpose of monitoring forest biodiversity for conservation is to track changes in forest biota, and disentangle the contribution of human impacts to observed patterns, while at the same time delivering this information to the relevant authorities and stakeholders in a manner that is effective in facilitating improvements to management. In light of the complexity that characterizes human-modified forest systems, understanding and conceptualizing how this process can best be achieved is far from a trivial exercise.

To be effective in tackling complexity, it is essential that biodiversity monitoring programmes make explicit the logic and assumptions that lie behind design choices and stated associations among individual system components. This is true with respect to both the design and functioning of the overall monitoring process itself, and the ways in which we interpret data on the impact of human activities on forest ecosystems and biodiversity. Only when choices and assumptions are made explicit does it become possible to:

- generate defensible arguments in support of particular recommended changes to management practice;
- identify the greatest areas of uncertainty in the management system; and
- justify the relevance of monitoring to stakeholders and interested parties who may not be directly involved in the design and implementation process.

Conceptual frameworks and models are powerful tools that can help in addressing these challenges. They function by encouraging the adoption of a more transparent and systematic approach to programme development, and help to identify key factors and relationships to measure and analyse (Failing and Gregory, 2003; Stem et al, 2005; King and Hobbs, 2006; Lookingbill et al, 2007; Margoluis et al, 2009). Specifically, they help make explicit assumptions regarding how the monitoring process is able to link human activities to observed changes in biodiversity, and hence its capacity to assess and evaluate management performance. While we will never fully understand all the relevant variables and relationships, conceptual models can greatly facilitate our understanding of complexity, and help in identifying those issues that are of most practical relevance, as well as those that require more research attention.

The term 'model' as used here is divorced from the idea of a complex computer-based exercise with quantitative parameters and the capacity to predict change. Instead, conceptual models are exactly what the name suggests: models of concepts.

In essence, they provide a simplified and explicit overview of what we understand about the structure and function of a system, and serve as an important first step in defining and clarifying the relationship between science, monitoring and management (Figure 13.1). The process of developing conceptual models is as important as the product itself as it forces us to *think about how systems work – or, in the case of a management system, how we would like it to work if given sufficient incentive for change.*

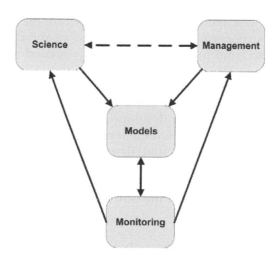

Figure 13.1 Conceptual models provide a linchpin that helps connect science, management and monitoring

Note: By working together, science and management can use conceptual models to inform monitoring that contributes to both scientific understanding and management decision-making. Models do not have to be complex in order to be useful
Source: Adapted from Lookingbill et al (2007)

Conceptual models can take multiple forms, including diagrams, pictures, graphs, tables and words.

However, arguably the most valuable approach is to build a form of pressure–state–response or 'influence diagram' that allows a visual and conceptual interpretation of the relationship between management practices, indicators and monitoring objectives (Stork et al, 1997; Newton and Kapos, 2002; King and Hobbs, 2006).The purpose of this chapter is to highlight the utility of conceptual modelling in the development and implementation of biodiversity monitoring programmes. First, I deal with some of the semantics that relate to the definition and purpose of conceptual models, and separate conceptual frameworks and models as being distinct and complementary aids to research and management. Next, I outline the advantages of developing conceptual models to make explicit the cause–effect relationships that are central to detecting and understanding human impacts on biodiversity. Finally, I highlight key considerations in the building of a conceptual model.

DISTINGUISHING THE ROLE OF CONCEPTUAL FRAMEWORKS AND MODELS IN BIODIVERSITY MONITORING

There is considerable uncertainty in the terminology used to define different types of conceptual models (Newton and Kapos, 2002). While there is no need to become too distracted by semantics, the use of clear terminology and definitions can be very helpful in identifying which methods and approaches are the most appropriate for a given task. For the purposes of biodiversity monitoring and the proposals made in this book, I distinguish conceptual models from the related notion of conceptual frameworks by defining the former as a method for articulating cause–effect relationships, and the latter as a way of categorizing variables and indicators and identifying which are most appropriate for assessment (see also Newton and Kapos, 2002). Both are required in the development of a biodiversity monitoring programme and serve highly complementary functions. For example, it is often useful to nest conceptual models that describe the causal and conditioning relationships of different parts of an ecological or management system within a wider conceptual framework that identifies how different types of monitoring information relate to each other and the wider management context.

Conceptual frameworks in biodiversity monitoring

Conceptual frameworks have been employed extensively in earlier chapters of this book to help articulate the purpose and design of biodiversity monitoring programmes. For example, Figure 8.2 provides an overall conceptual framework of how the coupled management monitoring process should ideally operate in order to foster progress towards more ecologically responsible systems of forest use. In clearly distinguishing the complementary role played by different monitoring approaches and types of indicator, this conceptual framework, and accompanying text, provides a basis for translating monitoring data into assessments of management performance, as well as recommendations for how practices can be revised within an AM system. This simple organizational framework also provides a template for further elaborating on the nature of cause–effect relationships within individual components of the system (see below). While the logic behind this framework may seem clear and self-explanatory, it contrasts with the way in which monitoring programmes are frequently conducted, where the analysis of ecological indicator data is largely disconnected from the management process rather than being placed in the context of specific interventions and performance assessments (see also Stem et al, 2005).

Elsewhere in the book, I have used more focused conceptual frameworks to help guide prioritization and selection processes within key parts of the overall monitoring framework. These can be very simple constructs, yet by making important choices explicit and visually intuitive they can provide very useful aids to decision-making. For example, Figure 10.2 illustrates how the choices involved in selecting

priority research objectives for evaluating the effects of a given management practice can be split into considerations of utility, uncertainty and feasibility. In adopting
a more systematic process for selecting monitoring objectives, we are more likely to
identify those questions that will deliver greater returns on researcher investment
(see Chapter 10). Chapters 11 and 12 tackle the related problems of selecting indicators for assessing and evaluating management performance (mostly structural
and biological indicators respectively). Indicator selection is an involved exercise
that is tightly coupled with the purpose, goals and objectives of monitoring while
also being central to designing the process of data collection and analysis. In keeping with this, Figures 11.6 and 12.7 present simple conceptual frameworks for
selecting structural and biological indicators that help to identify which indicators
are both cost-effective and most fit for their respective purposes.

THE VALUE OF CONCEPTUAL MODELS IN ARTICULATING CAUSE–EFFECT RELATIONSHIPS FOR BIODIVERSITY MONITORING PROGRAMMES

The process of developing an effective biodiversity monitoring programme can be
greatly facilitated by taking stock of existing knowledge and ecological theory. This
understanding can then be expressed in the form of conceptual models and associated research hypotheses that make explicit the causal and conditioning
relationships that exist between different indicators and other system elements.
Well-conceived conceptual models can facilitate the development of biodiversity
monitoring programmes in at least five inter-related ways. Conceptual models:

1   can help contextualize the relative importance of evaluating different manage-
    ment impacts on biodiversity;
2   can help make explicit our assumptions about the nature of human impacts on
    biodiversity;
3   provide a basis for developing research hypotheses;
4   can help in designing robust sampling strategies for monitoring; and
5   can provide a valuable aid to communicating biodiversity monitoring and
    research.

Conceptual models can help contextualize the relative importance of evaluating different management impacts on biodiversity

We will never have sufficient scientific knowledge or resources to evaluate all possible impacts of human activity on biodiversity. Yet, through a judicious approach to
identifying priority research objectives, it is possible to identify where the greatest
potential conservation returns on research investment are likely to be (Chapter 10).
By visually depicting the major human-related factors that are considered to be
responsible for driving biodiversity change in a given landscape, and placing them

in context of each other and additional indirect or external factors, conceptual models can be of considerable assistance in this task.

Information on the relative importance of multiple drivers of biodiversity change is particularly important for at least two reasons. First, if monitoring efforts focus on evaluating a limited set of management factors whose impacts are correlated with other unmeasured factors that are actually responsible for part (or even all) of observed changes, then any analyses of sample data can produce misleading recommendations regarding where management practices need to be adjusted. Second, multiple drivers of biodiversity change can often interact in important ways. For example, the conceptual modelling of possible interaction effects may highlight the need for particular monitoring attention to be given to human activities that are responsible for precipitating cascading effects on biodiversity (e.g. hunting of large, seed-dispersing mammals) or that operate synergistically with other impacts or threatening processes (e.g. logging and fire) (see below). In this way, the actual process of model building can be invaluable for any *a priori* research assessment, helping to reveal factors that were previously not considered worthy of study (Fazey et al, 2006).

Conceptual models can help make explicit our assumptions about the nature of human impacts on biodiversity

Our expectation of the ways in which human activities may alter patterns of biodiversity invariably draws upon some combination of ecological theory, empirical observations and personal experience. Much of the time, the contribution of these different sources of information to the ways in which we perceive and interpret biodiversity data remains implicit in our thinking. As a consequence, it can often be difficult to identify the assumptions that underpin inferred cause–effect relationships, or know the extent to which they are supported by established theory or evidence. By helping make explicit our understanding of the mechanisms that link human impacts to changes in biodiversity, and the ways in which the expression of such mechanisms is dependent on the context of a given forest landscape, conceptual models can greatly enhance the capacity of a validation monitoring programme to provide a reliable interpretation of sample data. Researchers differ in their personal experiences and theoretical perspectives. For example, different landscape ecologists and managers often rely on different conceptual models of species persistence in human-modified forest landscapes – ranging from a purely patch-centric fragmentation models to a species-individualist habitat continuum model (Fischer and Lindenmayer, 2006; Table 13.1) – to simplify management problems and guide decision-making. Although the choice of landscape model can have profound influence on how species-sample data are interpreted (Burgman et al, 2005), alternatives are rarely discussed, and theoretical perspectives remain implicit.

Implicit knowledge is frequently marginalized in science, although there is growing awareness that it can have a profound influence on the way in which research is conducted and management decisions are made (Fazey et al, 2006).

Table 13.1 Assumptions made by the 'fragmentation model' (derived from the island biogeography theory) regarding how species persist in human-modified landscapes, and how they are relaxed in the 'continuum model', which allows species-individualistic responses to continuous environmental gradients

	Habitat fragmentation model	Habitat continuum model
Landscape pattern	Assumes clear contrast between patches and areas outside patches	Allows landscapes with gradually changing patterns
The notion of 'patches'	Requires human-defined patch boundaries to correspond closely with animal-perceived patch boundaries; patches are assumed to be internally homogeneous	Human-defined patches are not of primary interest, and no assumptions are needed about their internal homogeneity
Identity of species	Restricted to single species or multiple species with similar requirements	Allows consideration of multiple species with vastly different requirements
Species distributions	Requires species to be restricted to patches, ideally as metapopulations	Species can be distributed through space in complex and continuous ways
Ecological processes	Assumes that landscape pattern is a reasonable proxy for a multitude of interacting ecological processes	Attempts to study ecological processes directly

Note: Different landscape conceptual models are varyingly appropriate in different circumstances and the fragmentation and continuum models represent two extremes along a gradient of alternatives
Source: Based on Fischer and Lindenmayer (2006)

Conceptual models provide a basis for developing research hypotheses

Hypotheses are identifiable and testable elements of a broader scientific or management paradigm (Hilborn and Mangel, 1997). More specifically, the function of a hypothesis is to provide a more general explanation of nature than is possible from the limited set of observations used in its development (Williams, 1997). Consequently, if monitoring programmes are to succeed in allowing robust inferences to be made from sample data (and therefore make a practical contribution towards AM) then they require the construction of explicit and testable hypotheses. In identifying possible cause–effect relationships between human impacts and biodiversity data, conceptual models are essentially comprised of a set of hypotheses (or just a single hypothesis) that reflect our understanding of how an ecological system may respond to change. Developing these hypotheses into formal statements and predictions regarding the nature of key causal and conditioning relationships (e.g. the shape and strength of a functional relationship between different variables) provides an important step in constructing research objectives that can be used to test the relative effectiveness of different management strategies. The testing of a single hypothesis (e.g. a null hypothesis) very rarely provides the most appropriate means of evaluating whether a particular management intervention is effective or not, and why. A much more powerful approach is to identify multiple working hypotheses (or alternative sets of hypotheses), that represent alternative explanations and theories (i.e. alternative related conceptual models) for an observed change (Box 13.1; and see Chapter 14).

Box 13.1 The importance of multiple working hypotheses

Much of traditional natural resource management was founded on a desire for certainty in choosing between alternative strategies, and many scientists and managers have been trained to solve problems rather than develop alternatives (Franklin, 1995). Associated with this desire for certainty, was the development of a reductionist approach towards evaluating the consequences of a particular management intervention that relies on the testing (and subsequent acceptance or rejection) of a single hypothesis (Williams, 1997). Either something works or it doesn't. While such an approach is ideal for evaluating the results of fully replicated experimental manipulations, unfortunately it is unsuitable for most problems in ecology, which are generally characterized by a complex interaction of multiple complementary factors (Hilborn and Mangel, 1997; Williams, 1997; and see Chapter 8). In response to this problem, the last decade of ecology has witnessed a growing appreciation of the inadequacy of single hypothesis testing, recognizing instead the need to explicitly articulate scientific uncertainty in the form of alternative working hypotheses (Burnham and Anderson, 2002; Johnson and Omland, 2004; Hobbs and Hilborn, 2006). In fact, this idea is not new and was eloquently put forward by T. C. Chamberlain (1890) more than a century ago, as well as being re-emphasized in Platt's (1964) seminal paper on the importance of the scientific method in developing strong inferences.

Chamberlain (1890) proposed his Method of Multiple Working Hypotheses as a means of reducing the chance that particular, favoured theories are able to dominate in science. It is instructive to summarize Chamberlain's original thesis in his own words:

'The moment one has offered an original explanation for a phenomenon which seems satisfactory, that moment affection for his intellectual child springs into existence and as the explanation grows into a definite theory his parental affections cluster about his offspring and it grows more and more dear to him… There springs up also unwittingly a pressing of the theory to make it fit the facts and a pressing of the facts to make them fit the theory… To avoid this grave danger, the method of multiple hypotheses is urged… It differs from the simple working hypothesis in that it distributes the effort and divides the affections. Each hypothesis suggests its own criteria, its own means of proof, its own method of developing the truth, and if a group of hypotheses encompass the subject on all sides, the total outcome of means and of methods is full and rich.'

This concern is as valid today as it was 100 years ago and bears direct relevance to the dangers of holding onto certain simplified theories of how species can exist in human-modified landscapes (e.g. the theory of island biogeography and its legacy of fostering a patch-centric view of fragmented forests (Laurance, 2008)) and the need to consider alternative explanations that accommodate the potential for site-specific differences in landscape structure and processes (e.g. the importance of edge effects, forest degradation and neighbourhood effects from the surrounding matrix), as well as differences in the ecology of sampled species or species groups.

Despite this slow acceptance, the process of creating and evaluating a range of alternatives is now widely recognized as the most effective method of defining research objectives (Tear et al, 2005; see also Chapter 10), as well as the most effective approach for integrating research into forest policy (Franklin, 1995). All management plans and conservation strategies have properties that can be stated as alternative hypotheses, derived from a combination of empirical observations and ecological theory (Murphy and Noon, 1991; Irwin and Wigley, 1993). Nevertheless, the formulation of alternative hypotheses (and their expression as alternative conceptual and statistical models) that are both ecologically meaningful, and feasible in the context of available data remains perhaps the most significant challenge in ecology (Burnham and Anderson, 2002; Nichols and Williams, 2006) and forest management (Failing and Gregory, 2003), as well as the one that is most readily trivialized (e.g. through the generation of 'silly' null hypotheses or 'straw men' (Anderson et al, 2000; see Chapter 15))

Conceptual models can help in designing robust sampling strategies for monitoring

The process of conceptual modelling can be extremely valuable in developing a robust strategy for monitoring. For example, visually depicting the relationship between particular management impacts and monitoring indicators can help in the process of identifying and defining both the indicators themselves (see Chapters 11 and 12) and additional environmental variables that may be important in interpreting the cause of any observed changes (i.e. covariates for data analysis, see final section of this chapter). Many of studies that have documented biodiversity responses to land-use change are purely pattern based, or collect only a limited and standardized subset of environmental variables that can help explain observed changes in biodiversity (Gardner et al, 2007). The selection of environmental factors, whether as indicators of management performance or covariates for analysis, that can help explain biodiversity responses to habitat change is a fertile area of conservation and ecological research, and one that would benefit significantly from a more explicit conceptualization of hypothesized cause–effect relationships. Because almost all biodiversity monitoring programmes are resource limited, one of the most useful features of conceptual models for monitoring is that they can help determine not only what to measure, but also just as importantly, what not to measure (Margoluis et al, 2009).

Conceptualizing the relationship between forest management activities and biodiversity can also help in guiding the appropriate spatial and temporal scales over which monitoring needs to occur in order to distinguish different effects. This is particularly true regarding the importance of human-induced threats or natural ecological processes that operate at scales much larger than a given management area, yet which may play an important confounding role in determining local patterns of biodiversity change (see end section).

Conceptual models can provide a valuable aid to communicating biodiversity monitoring and research

By providing a lot of information in a simple and concise format, conceptual models represent a powerful means of visually communicating the design and results of a monitoring programme (Margoluis et al, 2009). Scientists and managers of monitoring programmes can use them to communicate with non-specialist audiences, which may include landowners, local communities, politicians and funding agencies. Similarly, anybody involved in an external evaluation or audit of forest management systems can use conceptual models of cause–effect relationships to help explain and support recommendations they may have reached, and where individual strategies can be improved and why (Margoluis et al, 2009).

Revealing the main assumptions associated with our current understanding of an ecological system, and explaining them in simple terms to non-specialists, can itself be a powerful technique for identifying previously hidden limitations and ecological processes (Fazey et al, 2006). It can also provide a valuable confidence-building tool for improving dialogue between different stakeholders. To enhance communication with non-specialist stakeholder groups, it may often be necessary to simplify technical terminology. For example, Schiller et al (2001) developed a set of *common language indicators* to translate a regional ecological condition assessment by the US Environment Protection Agency to the general public, and found higher levels of engagement where indicators were linked to general aspects of the environment that are valued by society.

Finally, conceptual models can aid communication of biodiversity monitoring data by providing a realistic expectation of the timing of results (Margoluis et al, 2009). All too often, conservation researchers are expected to identify measurable changes soon after a monitoring programme has commenced, and quickly draw conclusions regarding recommendations for management. Improving appreciation about the complex and multi-scale nature of biodiversity responses to human impacts could be of significant help in securing the long-term investment that is needed for many conservation projects.

BUILDING CONCEPTUAL MODELS FOR BIODIVERSITY MONITORING

Figure 13.2 presents a generalized conceptual model of the impacts, processes and feedbacks that determine patterns and trends of biodiversity in human-modified landscapes (Gardner et al, 2009), and highlights three general observations that are true of all biodiversity monitoring programme. First, far more is currently understood about the first-order human impacts of landscape change (correlative factors, such as forest fragmentation) than the ultimate causative mechanisms and second-order cascading effects, yet understanding the mechanisms that drive observed changes is key to generating reliable recommendations for management. Second, inferences from biodiversity studies are constrained by differences in regional and landscape context, as well as the context imposed by study-specific differences in research design and focus. Third, changes to modified forest landscapes over time (e.g. land-use intensification, land abandonment, new conservation interventions and internal cascading effects of altered biodiversity processes) can lead to feedbacks that alter the relative importance of different ecological processes and context-dependent factors.

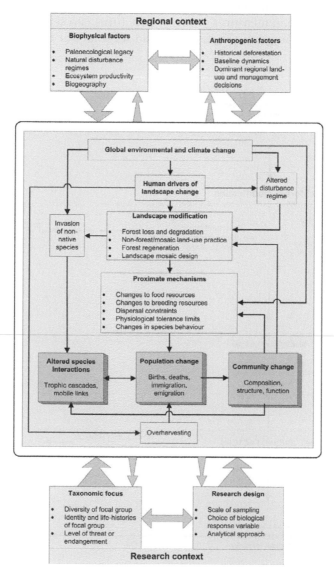

Figure 13.2 A conceptual model of the impacts, processes and feedbacks that determine patterns and trends of biodiversity in human-modified forest landscapes

Source: Redrawn with permission from Gardner et al (2009)

Building a conceptual model that is able to provide a useful guide for monitoring requires moving from theoretical generalizations, such as Figure 13.2, to a tailored model structure that explicitly links indicators and measurements in the context of a particular research objective (or set of objectives), management system and location. Because conceptual modelling is essentially about making explicit the cause–effect relationships that link human impact(s) with changes to biodiversity, it can make a significant contribution to the development of validation monitoring programmes concerned with evaluating the extent to which indicators of management performance (e.g. measures of forest structure) can secure progress towards long-term conservation goals (Figure 13.3; and see Chapter 8).

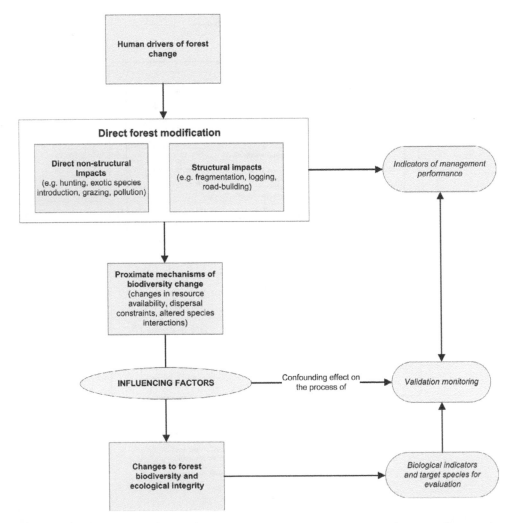

Figure 13.3 A simplified conceptual model illustrating the hypothesized cause–effect relationship between human impacts and changes to biodiversity and ecological integrity, and how the process of validation monitoring serves to articulate this relationship

There are six basic and non-exclusive steps to building a conceptual model:

1   Define the spatial and ecological scale of the programme with respect to the size of the study area and type(s) of vegetation.
2   Define the scope of the programme with respect to overall conservation goals and values (see Chapter 9).
3   Identify the main threatening processes (whether managed or unmanaged human impacts, external threats or natural disturbance regimes) that are likely to affect conservation values, and identify priority processes or impacts that can provide a focus for evaluation (see Chapter 10).
4   Select indicators of management performance that can reflect measurable changes in the level of human impact (see Chapter 11).
5   Select biological indicators and/or target species that can provide a means of evaluating the adequacy of current management practices (and associated performance indicators) to fostering progress towards long-term conservation goals (see Chapter 12).
6   Identify the influencing or confounding factors that condition the nature of perceived cause–effect relationships between threatening processes and changes to biodiversity (Figure 13.3).

The model-building process is not linear, and the identification of threatening processes, indicators and other environmental measurements should ideally occur together. The process can be simplified in situations where the conservation goal represents a very specific target of maintaining or restoring a single species or species group, such that management can be tailored to achieve that target. However, where the goal is broader and concerned with conserving and maintaining overall forest condition, it is helpful to separately articulate both the goal and the selection process used to identify indicate sets that are appropriate for evaluating progress towards that goal (e.g. ecological disturbance indicators; see Chapter 12). The contribution of existing knowledge to the formulation of a conceptual model may come from general ecological theory, previous studies of the same system, or if the similarities are sufficiently strong, from the results of studies on other systems (Green et al, 2005). While conceptual models are most useful if they are developed during the initial implementation of a monitoring programme, they can still provide valuable guidance for the interpretation of monitoring data when developed retrospectively. In this case, the model-building process can often act as a valuable consensus building and negotiation tool between different stakeholders in identifying the most likely causes of past changes. Regardless of how and when they are built, conceptual models should be interpreted as dynamic constructs. Lessons learnt during the model-building process, and via the subsequent collection and analysis of sample data, can feedback to an improved understanding of the ecological dynamics of the study landscape and management system, as well as the process of selecting research objectives and indicators.

In the rest of the chapter, I focus on some of the confounding factors that are central to the development of any conceptual model.

Identifying factors that influence cause–effect relationships between human impacts and biodiversity in modified ecosystems

Once the basic ingredients of monitoring are in place with respect to overall conservation goals, research objectives and indicators, the key challenge to building a conceptual model that can facilitate programme design and implementation lies in articulating hypothesized cause–effect relationships between selected human impacts and components of biodiversity. An effective validation monitoring programme cannot just focus on the outcome, but also needs to consider the factors that mediate or influence the consequences of a particular human impact or management intervention (Salafsky and Margoluis, 2003; Stem et al, 2005). While every monitoring programme has its own peculiarities, there are a number of general issues that should be given consideration. These issues include the potential for: interaction effects between background environmental variables and human impacts; synergistic effects among multiple drivers of change; cascading effects from the disruption of species interactions (trophic or otherwise); time-lags and legacy effects from historical human impacts; threshold effects and regime shifts; landscape context effects from the influence of land-use change in neighbouring areas; and stochastic or unpredictable events.

By giving explicit and systematic consideration to these various types of influencing factor, it is possible to build a conceptual model that is much richer in structure and therefore more useful in guiding the design of monitoring programmes and analysis of sample data (Figure 13.3). These factors are not necessarily exclusive and may be difficult to distinguish in reality, but to facilitate conceptualization are given separate consideration in the following sections.

Environmental interaction effects

The impacts of human activities, whether local, regional or global, negative or positive, on the biodiversity of any particular site are invariably conditioned by the natural environmental and biophysical characteristics of that site. Thus, accounting for such factors in the design and analysis of monitoring programmes is essential for generating a reliable and unbiased understanding of the causes of biodiversity change. The more reliable our understanding of the drivers of biodiversity change, the more confidence we can have in transferring recommendations from research in one area to other landscapes and time periods.

There are at least three difficulties to overcome in accommodating key environmental interaction effects in the analysis of monitoring data: (i) identifying the key environmental factors that condition human impacts; (ii) encompassing a representative level of spatial variability in each factor during sampling; and (iii) adequately integrating the factors into data analyses in order to identify the independent and conditional effects of management.

Environmental interaction effects vary from being very subtle to quite obvious. For example, Eycott et al (2006) found that plantation stands in south-east England had a greater and more variable number of understorey plant species when on higher pH soils, moderating the effect of clear-felling and thinning. In a more obvious example, Hawes et al (2008) found that the diversity and composition of bird communities in native forest corridors in a Brazilian plantation landscape changed significantly depending on whether the corridor included a stream or not. There can also be important interaction effects between local biodiversity responses to forest management and large-scale (e.g. regional or global) environmental change processes. The obvious example here is climate change, the effects of which have been shown to interact in critical ways with human land-uses (e.g. through fire and altered patterns of precipitation). Integrating factors that are outside a manager's control into a visual conceptual model can go a long way towards facilitating the interpretation of monitoring results, and understanding the limitations of any inferences that can be drawn from sample data (Margoluis et al, 2009).

Synergistic effects

While a particular impact may be identified as the ultimate driver of an observed change in biodiversity, the proximate drivers are often the result of complex and synergistic interactions between different stressors and reciprocal system responses. Over-simplified and reductionist approaches to conceptualizing and investigating biodiversity patterns can attribute change to the wrong factor, or result in a failure to fully understand situations where seemingly distinct drivers of change are intrinsically linked or highly interdependent, for example in the case of fire and logging (Barlow and Peres, 2004), habitat modification and species invasion (Didham et al, 2007), habitat loss and fragmentation (Koper et al, 2007), fragmentation and logging impacts (Hillers et al, 2008) and area and edge effects (Ewers et al, 2007). Such shortcomings are particularly problematic in situations where synergistic effects are ultimately responsible for driving population changes, as is the case in forest fragmentation (Tabarelli et al, 2004; Figure 13.4).

Ecological cascades and the indirect consequences of biodiversity change

Biodiversity research and monitoring has predominantly focused on evaluating the direct impacts of human activity, while comparatively little attention has been given to the cascade effects and the indirect consequences of biodiversity change (Brook et al, 2008; Tylianakis et al, 2008). However, despite the difficulties inherent in untangling indirect drivers of biodiversity change, understanding the consequences of altered species interactions and the potential for co-extinction is fundamental to determining the future of biodiversity in modified terrestrial ecosystems (Koh et al, 2004; Laurance, 2008; Tylianakis et al, 2008), and extending our perception of how biotic communities respond to landscape modification (Figure 13.2). There are three main ways in which human activities can have indirect yet potentially very serious consequences for biodiversity: (i) via the disruption of trophic cascades; (ii) disruption to 'mobile-link' species (e.g. species with key connective functions in the forest such as pollinators and seed dispersers); and (iii) by facilitating the introduction of invasive species (Gardner et al, 2009). In all of these cases, a failure to give explicit recognition to the importance of indirect human impacts in the

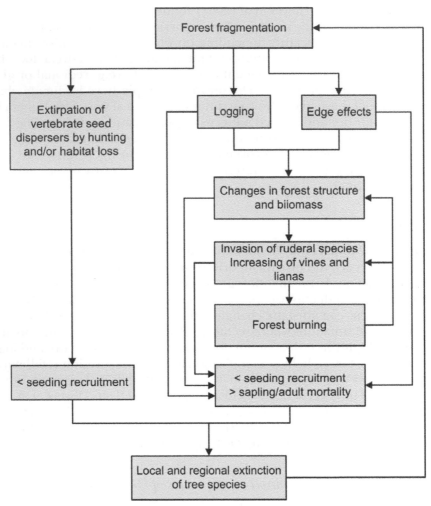

Figure 13.4 Forest fragmentation and its synergistic relationship with other threats and ecological processes that drive local and regional extinction of tree populations

Source: Redrawn with permission from Tabarelli et al (2004)

design and implementation of a biodiversity monitoring programme can greatly limit our ability to understand the proximal drivers of any observed changes.

For example, regarding the disruption of food webs following land-use change, Terborgh et al (2001) showed how the cascading effect of vertebrate predator removal from recently isolated forest islands in Lago Guri, Venezuela, resulted in highly elevated densities of herbivores (including rodents, howler monkeys, iguanas and leaf-cutter ants), contributing to declines in seedling and sapling recruitment of nearly every plant species present (Terborgh et al, 2006), as well as the loss of some bird species (Feeley and Terborgh, 2008). In the case of mobile-link species the decline of loss or even a single keystone species can instigate a cryptic cascade of 'downstream' extinctions among dependent taxa, with dramatic implications for subsequent patterns of community structure and ecosystem function. Such a scenario has been proposed for dung beetles following the decline of resource provisioning mammal populations due to over-hunting and landscape change (Nichols et al, 2009; Figure 13.5). Finally, the changes precipitated by invasive species can also be cryptic yet severe (see Chapter 12).

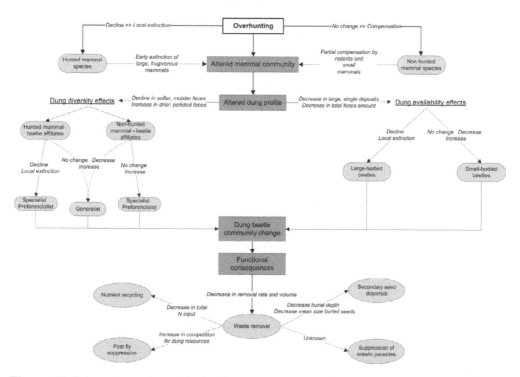

Figure 13.5 A conceptual model of pathways by which overhunting in tropical forests impacts coprophagous dung beetle community structure and dung beetle ecosystem function

Note: Text along arrows describe the potential direction or mechanism driving each effect
Source: Redrawn with permission from Nichols et al (2009)

Time lags and legacy effects of past disturbance

Legacies of human impacts on natural systems are remarkably persistent, and the constraints imposed by differences in site history are varyingly embedded in the structure and function of all forest ecosystems. A major challenge in understanding the future of forest species in human-modified forest ecosystems lies in reconciling the mismatch in temporal scale over which human impacts and ecological processes are acted out. Only by doing so is it possible to isolate the relative importance of current management strategies from the legacy of historical disturbances that have been inherited by today's landowners. Achieving this capacity is vital both for tailoring management strategies to the conditions of a particular site, but also for understanding how management standards may need to be flexible among landscapes with different disturbance histories in order to maintain comparable levels of performance.

Encapsulating the importance of time-lags in disturbance responses in the evaluation of biodiversity monitoring data is difficult. Time-lags occur with respect to processes of both biodiversity loss (e.g. extinction following habitat fragmentation (Brooks et al, 1999)) and gain (e.g. regeneration (Liebsch et al, 2008)), which when superimposed on each other, and in the context of present-day impacts can be difficult to disentangle. Moreover, the level of inertia in biodiversity responses to past changes depends strongly on species-specific life-history traits, such as dispersal and reproductive rates. Nevertheless, research has demonstrated that the inclusion of historical disturbances in the analysis and interpretation of biodiversity survey data can be vital for contextualizing the importance of present day impacts. For example, Smyth et al (2002) found that the incorporation of fire and logging history variables into analyses of the abundance of hollow-nesting birds in Australia helped illustrate the limited importance of existing structural features (e.g. tree hollows) for the conservation of these species. In another study in the Atlantic forest of Brazil, Metzger et al (2009) found that differences in historical landscape dynamics can strongly affect the present distribution of species in fragmented landscapes.

Threshold effects and regime shifts

There is a growing amount of empirical evidence to suggest that human modifications may frequently induce nonlinear effects on the structure, composition and function of forest ecosystems with potentially irreversible consequences (Fischer and Lindenmayer, 2007). Support for this concern is given by three qualitatively very different examples regarding changes in the structure, composition and function of human-modified forest ecosystems. First, although difficulties in sampling across multiple landscapes mean that strong empirical support for a landscape deforestation threshold (where fragmentation effects act to compound the impact of habitat loss) is lacking, recent work indicates that severe deforestation can lead to marked changes in the distribution and flux of species across modified landscapes (Arroyo-Rodriguez et al, 2009; R. Pardini, *personal communication*). Second, in studying the positive feedback loop between vegetation structure and fire, Barlow and Peres (2008) recently reported that forest species composition in the central Amazon can almost completely turn over following recurrent fire events, leaving behind a suppressed and biotically impoverished early successional stand. Third, positive feedbacks between vegetation change and limiting nutrient resources can result in abrupt shifts in biogeochemical cycles, as shown by Lawrence et al (2007) who reported that soil phosphorus declined by 44 per cent after only three cultivation–fallow cycles in tropical Mexico.

Predicting the potential for such regimes shifts is extremely difficult, yet making explicit the potential for positive feedbacks in conceptual models of how human activities may impact biodiversity can go a long way towards pre-empting the risk of irreversible loss.

The importance of landscape context and regional deforestation

Although much of landscape ecology and biodiversity research has been conducted at the patch scale, differences in whole landscape mosaic properties, such as the amount and spatial configuration of native forest cover, are vital in understanding the value of modified forest landscapes for biodiversity conservation (Tscharntke et al, 2005; Bennett et al, 2006). Differences in topography and soil fertility ensure that the spatial extent and pattern of historical deforestation within and around individual landscapes is rarely random or consistent, and this can have a marked impact on levels of biodiversity retention (Kupfer et al, 2006). Arroyo-Rodriguez et al (2009) found that forest patch size was only important for explaining species density–area relationships of plants in landscapes where historical deforestation was highest – an effect that may be due to a landscape fragmentation threshold, and/or severe defaunation of primary seed dispersers. However, landscape context encompasses much more than simply differences in the amount of forest cover. Sampling across limited environmental and geographical gradients can generate unreliable extrapolations of biodiversity responses to land-use gradients elsewhere (Gillison and Liswanti, 2004). Moreover, when different landscapes are defined by similar environmental characteristics, small initial differences in disturbance regimes and human impacts can precipitate marked and cumulative divergences in species composition and ecosystem functioning through time (Laurance et al, 2007).

The dominant effect of regional deforestation on biodiversity responses to local human activities and conservation management efforts must not be underestimated yet addressing this problem presents a significant logistical and technical challenge as it requires monitoring across multiple sites, as well as an explicit accounting of the landscape features (remnant forest, dominant land uses) and human activities that lie outside the specific area(s) of study. To help deal with the problem of understanding landscape and regional context effects and multi-scale interactions on local biodiversity dynamics, DeFries et al (2009) suggest the use of a Zone of Interaction (ZOI) around individual study sites (Figure 13.6). The ZOI provides a way of explicitly accounting for key socio-economic, hydrological and landscape features that characterize adjacent areas and could have a strong influence on local biodiversity patterns. This same approach also helps conceptualize the importance of continental or global-scale influences of environmental and climate change. Even where landscape-scale factors cannot be included in analyses, their incorporation in our conceptual understanding of biodiversity processes can be very helpful in interpreting sample data.

Natural variability, stochasticity and surprise events

One of the most valuable pieces of advice in conceptualizing and interpreting the

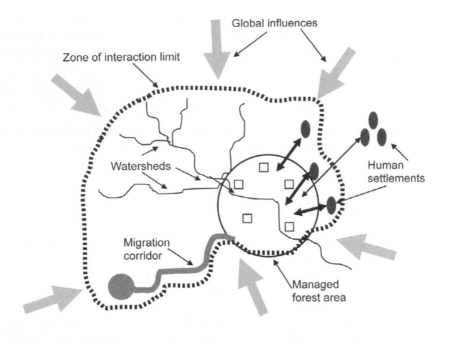

Figure 13.6 Hypothetical zone of local and regional interaction (ZOI) (thick dashed line) around biodiversity measurement plots (squares) within a managed forest area (solid black circle)

Note: The ZOI encompasses the upper watershed, migration corridors, and human settlements. Strong human interactions (thick arrows) occur between the protected area and nearby settlements; and weaker interactions (thin arrows) occur with more distant settlements. The ZOI is embedded within global influences, such as climate change and nutrient deposition
Source: Redrawn with permission from DeFries et al (2009)

response of forest biodiversity to human activity is to keep an open and humble mind regarding the possibility of unexpected and seemingly inexplicable changes. In relative terms, we have a very poor understanding of how forest ecosystems function, and even less understanding of how human impacts may be manifest on natural processes across different scales and levels of ecological organization.

One particular challenge lies in defining the extent of natural variability that exists within a baseline (see Chapters 6 and 9). While there are very few long-term monitoring data sets that have tracked changes in forest biodiversity, those that do exist often indicate surprisingly high levels of inter-annual variability. This variation can significantly confound attempts to isolate the contribution of a particular management impact to observed changes, or identify the importance of a particular forest attribute in determining the value of different areas of forest for conservation. For example, Pearce and Venier (2005) found that very high levels of inter-annual variability in small mammal populations in Ontario, Canada, made it impossible to quantify the impact of selective logging or identify reliable habitat relationships between study species and key structural attributes of the forest. Other work has shown evidence for general temporal trends in abundance across multiple species and study sites that are not demonstrably linked to particular human impacts. For example, Lindenmayer et al (2008b) found that patterns of occupancy of various bird species inhabiting a large plantation landscape in south-east Australia increased across the entire landscape during a ten-year survey period in a way that was not linked to spatial patterns of landscape transformation (Figure 13.7).

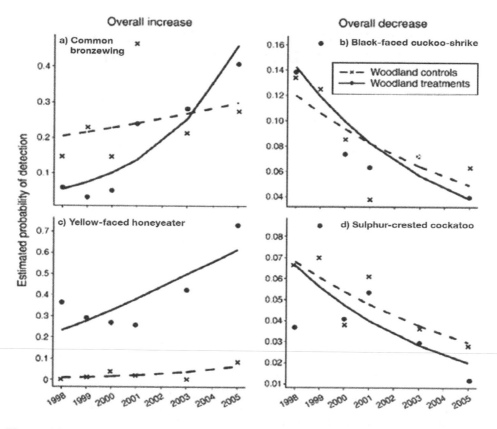

Figure 13.7 Linear trends and predicted detection rates for a selection of individual bird species that showed temporal changes across an entire multiple-use forest landscape (including both managed and control sites) in south-east Australia

Source: Redrawn with permission from Lindenmayer et al (2008b)

It is understandably very difficult to accommodate background levels of variability and seemingly stochastic fluctuations in species abundance and occupancy patterns in the design and analysis of biodiversity monitoring programmes. Surprising events are, by definition, unpredictable. Collecting and analyzing biodiversity data across multiple spatial and temporal scales can often be very valuable in discerning where real opportunities for conservation lie. There is no substitute for implementing long-term programmes of monitoring and research. Yet, perhaps the most important lesson is to recognize that there are serious limits to human knowledge and our ability to understand complex ecological processes. Accepting this uncertainty emphasizes the need for management to be adaptive as new information is gathered, and theories about biodiversity persistence in human-modified landscapes revised.

A SUMMARY OF THE ROLE OF CONCEPTUAL MODELS IN BIODIVERSITY MONITORING

It is hopefully clear from the preceding discussion that conceptual models, and the process of building them, can be exceptionally useful tools in the design, execution and dissemination of a biodiversity monitoring programme. In helping to clarify and make explicit our understanding of how a given study system works, conceptual models can be vital in helping to formalize alternative hypotheses about the ways in which human activities affect biodiversity, and thus how management strategies can be improved for conservation. Building a conceptual model is essentially an exercise in visually mapping out the thought processes that lie behind the design of a monitoring programme. A central part of this exercise is the process of synthesizing what we know and understand about the mechanisms that determine the persistence of biodiversity in human-modified ecosystems. Systematically evaluating the potential importance of key factors that may influence the relationship between human activities and biodiversity can help to provide important clarity, and enhance the reliability of any recommendations that are made for changes to management practice.

In well-studied systems, it is possible to create highly detailed conceptual models that depict multiple relationships and interactions among all hierarchical levels of a causal framework from drivers through changes in system state to changes in biodiversity (e.g. Barrows et al, 2005). Nevertheless, in places where less ecological knowledge is available, even simple models that make explicit what it is we do (and equally importantly what we *do not*) understand about the working of the system, can be powerful aids to the development of a monitoring programme. Yet, because conceptual models attempt to simplify very complex and poorly understood problems, they need to be used judiciously and with awareness of their limitations (Box 13.2). It is important to recognize that there are also a large number of constraints on understanding what can be imposed by inherent biases and uncertainty in the use of different sampling methods and approaches to data analysis. A more detailed discussion of these factors in the context of biodiversity monitoring is given in Chapters 14 and 15.

A final point of emphasis in the discussion about conceptual models is that, in order for the validation monitoring process to make a genuine contribution towards affecting real change in the management of forest ecosystems, it needs to be nested within our understanding of the wider management system (see Chapter 8). In particular, careful consideration needs to be given to the way in which any

Box 13.2 Issues to consider when using conceptual models for biodiversity monitoring and evaluation

- **No conceptual model is perfect** It is impossible to represent everything that influences a study site or research objective in a conceptual model.
- *Conceptual models may represent a biased or incomplete 'worldview'* Conceptual models are completely dependent on who develops them and what knowledge or information participants. Thus, models may not provide an accurate representation of reality.
- **The best conceptual models result from a team effort** Conceptual models require in-depth knowledge of a study area. Ideally, they encapsulate a detailed under-standing of context-dependent factors that influence patterns of biodiversity at the site. It is essential, therefore, to include in the development of a model people who are knowledgeable about the management system itself and the characteristics of the study region.
- **Conceptual models are dynamic** Conceptual models can and should change over time as managers and researchers learn more about their study system.
- *Getting the right level of information is an art and a science:* Model builders should strive to make sure their conceptual models provide enough information to reasonably portray reality but not so much as to confuse it. Conceptual models that include too many factors and show too many relationships can become bewildering, reducing their value as a communications, analytical, and evaluation tool.
- **Building a conceptual model requires time and commitment** Building an effective conceptual model as part of a monitoring programme can require a significant investment of time. Investing the required time, however, will allow those responsible for the programme to build a model that more accurately captures the key underlying assumptions that help to articulate cause–effect relationships. Ultimately, this will lead to a more meaningful evaluation of management performance.
- **Conceptual models will not effectively reach those readers who are not visual learners** People who do not learn visually will not experience or appreciate most of the benefits associated with conceptual models. For this reason, it is important that conceptual models are accompanied by adequate text descriptions.

Source: Adapted from Margoluis et al (2009)

conclusions regarding the impact of human activities on biodiversity are fed back to management, and how diverse stakeholders, with different positions concerning the value of biodiversity, influence and are incorporated into this process. In addition, it is important to work back from a focus on individual human impacts on a study system, and attempt to understand and identify the higher order drivers responsible for such impacts (whether they be economic, cultural or political) (see also Chapters 15 and 16). The process of making explicit many of the assumptions and contributing factors that characterize the wider forest management and social system shares many similarities with the process of understanding ecological cause–effect relationships. Indeed, visual conceptual models have been used to very powerful effect in evaluating conservation projects and understanding the complex web of direct and indirect ecological, social and economic factors that ultimately determine opportunities for implementing real improvements in biodiversity conservation (see Margoluis et al, 2009). The combination of conceptual models that are focused on both ecological and social processes respectively can be a highly effective and complementary approach to integrating monitoring and management.

CHAPTER 14

Sampling Design and Data Collection in Biodiversity Monitoring

Synopsis

While many decisions regarding sample design and data collection are context dependent, there are a number of general guidelines and principles that are common to all biodiversity studies. These guidelines emphasize the importance of giving careful attention to:

- The clear definition of research objectives in order to identify the types of management practice and levels of impact that arc the focus of evaluation, as well as the different types of indicators that need to be sampled and measured.
- Questions of experimental design, including the choice of reference sites, the distribution of existing management activities and disturbance regimes, and the definition of independent samples. Questions about stratification versus random or uniform designs require a careful assessment of potentially confounding factors, and the need to generate information that is useful for management while also accounting for unknown patterns of environmental variability in space and time.
- The choice of appropriate variables for measuring and explaining observed changes in biodiversity. Sampled variables need to provide a reliable and understandable description of biodiversity patterns while not being prohibitively expensive or time-consuming to collect.
- The choice of sampling methods and appropriate level of replication both between and within samples. Limited resources mean that all biodiversity monitoring programmes face a tension between securing samples of sufficient quantity (to permit extrapolation) and quality (to ensure reliability of individual estimates).
- Whether sufficient time, resources and expertise are available to ensure that a proposed monitoring design is feasible.
- The long-term viability and relevance of the monitoring programme, including the importance of preserving data integrity through careful recording and storage, and the need to be adaptive in the face of changing priorities and opportunities.

Once a clear idea of the *why* of biodiversity monitoring has been established (see Chapter 8), it is possible to move onto considering the *how*. Without careful prior planning to ensure that the right measurements are made it is all too easy to waste precious time and resources. Many of the particulars of a given sampling strategy will be largely determined by the type of objectives that are being tested, the type of indicators or target species under investigation, and the spatial scale and environmental characteristics that are particular to each forest landscape. However, despite this inevitable context dependency, there are a number of general guidelines and principles regarding sampling design, data quality and reliability that are common to all biodiversity field studies.

The general principles of sampling and experimental design have been well discussed and are largely beyond the remit of this book (see Table 14.1 for an annotated list of some tried and tested textbooks on sampling design). Instead, the purpose of this chapter is to provide a general overview of some of the key considerations and potential pitfalls in designing a biodiversity sampling protocol, while also acting as an entry point to the primary literature on some of the more technical issues. One limitation of standard books on sampling design is that they are largely focused on the application of well-designed, manipulative experiments, the likes of which are rarely possible (or necessarily desirable) in the real world. In discussing each of the main principles of sampling design, I attempt to maintain a realistic perspective regarding the extent to which theoretical ideals can be applied to real landscapes and resource-limited field projects while also emphasizing the importance of a rigorous approach to

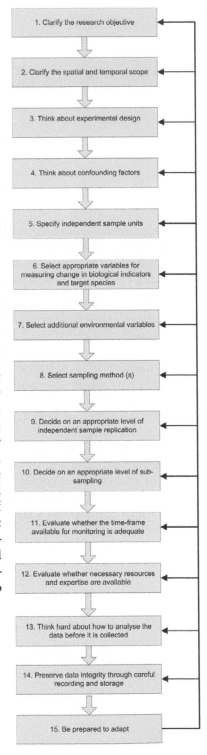

Figure 14.1 Fifteen-step process for developing a robust and cost-effective sampling and data collection strategy

Note: The process is not strictly linear and some steps may be evaluated together. It is important that the sampling design of any monitoring programme remains open to adaptation as initial data are collected, and new constraints and opportunities are encountered

Source: These steps are modified from a similar list presented by Feinsinger (2001)

Table 14.1 Some of the more commonly used textbooks for thinking about issues relating to the sampling design of biodiversity monitoring programmes

Book	Description
Designing Field Studies for Biodiversity Conservation Feinsinger (2001)	A must have. Perhaps one of the most accessible, engaging volumes available on devising and implementing biodiversity studies in the real world. The author's considerable practical experience is brought to bear on distinguishing the elements of the scientific process that really matter in the search for practical and reliable knowledge. Well grounded in good advice on the importance of natural history for biodiversity research. Consideration of data analysis is limited to traditional, frequentist statistics
Occupancy Estimation and Modelling MacKenzie et al. (2006)	A detailed synthesis of methods for analysing presence-absence data. Demonstrates how to calculate occupancy-related parameters while allowing for imperfect detectabilty. Includes an excellent chapter on the fundamental principles of statistical inference.
Experiments in Ecology: Their Logical Design and Interpretation Using Analysis of Variance Underwood (1997)	A now classic text on experimental design in ecology. Very clearly written and easy to read. Strong treatment of statistical power. On the down side, it is limited to balanced parametric analysis of variance that is simply not feasible for many natural experiments that are by nature 'messy' in design
Design and Analysis of Ecological Experiments Scheiner & Gurevitch (2001)	An edited, relatively technical volume covering an almost uniquely wide range of statistical techniques that are used in experimental ecology, including approaches for un-replicated and time series data, Bayesian statistics and meta-analysis. Limited discussion of basic sampling design
Experimental Design and Data Analysis for Biologists Quinn & Keough (2002)	A relatively advanced but otherwise very accessible volume on frequentist statistical methods and analysis, with a good coverage of experimental design, the scientific method and key topics like power analysis. A valuable last chapter on the presentation of results.
Analysis and Management of Wildlife Populations – Modelling, Estimating and Decision-making Williams et al. (2001)	A comprehensive and accessible tome that effectively links principles of study design, data analysis, modelling and population management in one coherent volume. Excellent opening sections on the scientific process and sampling design. Relatively unique in encompassing a broad range of statistical and modelling approaches under one conceptual framework. Limited in focus to the study and management of individual species.
Modern Statistics for the Life sciences Grafen & Hails (2002)	A novel introduction to learning statistics that employs the General Linear Model (GLM) as a powerful tool to analyse data that encompasses a great many traditional methods. Easy to read and freshingly thought-provoking. A good chapter on the principles of experimental design (randomisation, replication and stratification)

data collection. These principles are as relevant to a short-term research project that lasts only a few months as to a monitoring programme that lasts many years.

The chapter is structured around a series of 15 steps that define the process of designing and executing a sampling protocol for biodiversity monitoring (Figure 14.1). These steps are modified and restructured from a similar list proposed by Feinsinger (2001). For each step, I sketch out some of the important design considerations, and highlight examples of where things have gone wrong in the past, or where new approaches and tools can help overcome persistent difficulties. Because the process of data collection is superimposed on the foregoing stages of the overall programme, I frequently refer back to issues raised in earlier chapters. As for other chapters in Part III of the book, the focus of discussion is on validation monitoring and the process of attempting to link human impacts and management interventions to changes in biodiversity.

Step 1: Clarify the research objective

Paramount to the success of any monitoring programme is an early and clear definition of the research objective. Key components of a research objective that are relevant to sampling design issues include:

● The human activity or management intervention that provides the focus of research and monitoring interest (i.e. the experimental 'treatment' or 'design' factor). This can be thought of as the leverage point in the management system that could potentially be adjusted (or even halted) to generate benefits for conservation (e.g. the design of riparian corridors, level of logging intensity etc.; see Chapter 10, and Table 10.1 for examples). Consideration is also needed as to whether it is appropriate and possible to evaluate multiple types of human impact. This may be advisable if there is strong evidence for interaction or synergistic effects, such that the management of a single, isolated impact would be an inefficient or even futile mechanism for achieving progress towards conservation goals (e.g. management of selective logging without controlling the spread of wildfire) (see Chapter 13 for a more detailed discussion of interaction effects).
● The specific type and number of levels of the experimental design factor. If riparian corridors are selected as the management focus for research, what specific management options for corridor design (e.g. corridor length, width, age, etc.) will provide the focus of evaluation? Once the type of factor is specified, what range of levels will determine the extent of the study? Inevitably this choice will represent a balance between what is possible to ask in reality (e.g. the variability in corridor length that exists within the study landscape, as well as resources available for monitoring), what variation is likely to be ecologically meaningful in the context of other threats to biodiversity, and what variability poses a realistic lever for change in the management system (e.g. how logistically and economically feasible would it be to implement changes to the chosen management control variable?). It is also necessary to clarify whether the design factor has continuous or categorical levels. There are advantages to both and the choice depends on: the nature of the conservation concern (e.g. to prevent selective logging altogether or identify an ecologically responsible threshold of harvest intensity); the nature of the variability that is available for study in the landscape; and the complexity of the objective (e.g. if two factors are being evaluated, a categorical design can be easier to analyse for interaction effects) (Feinsinger, 2001).

- The management process and performance indicators that are associated with the impact being studied. These indicators are central to the validation monitoring process and provide the mechanism by which research findings can be translated into improved management standards (see Chapter 8 for discussion of full monitoring-management framework). As discussed in Chapter 11, there are powerful arguments for choosing structural indicators and indicators of disturbance processes (e.g. fire return-intervals and livestock grazing) as measurements of performance rather than species-based indicators.

- The biological indicators and/or target species that provide the focus of evaluation against long-term conservation goals (see Chapter 12). Single species monitoring may often require specific design considerations that relate to species-specific life-history strategies.

- The *a priori* research hypotheses that help predict ways in which a particular management practice or human activity may drive changes in the biodiversity of interest, how this relationship changes across different levels of impact, and whether any interaction effects may exist. A good demonstration of the value of articulating alternative hypotheses to clarify research objectives and help guide the design of a field study is given in ongoing and unpublished work of Renata Pardini, Jean-Paul Metzger and colleagues in a long-term project evaluating the influence of forest fragmentation on biodiversity in the Atlantic forest of Brazil (Box 14.1).

Box 14.1 The value of formulating *a priori* research hypotheses: Discovering context-dependent fragmentation effects in the Atlantic forest of Brazil

The Atlantic forest, the second largest tropical forest in South America and one of the most imperiled tropical forest ecosystems in the world, is characterized by both an extremely rich and endemic biota while having suffered an intense and long-standing human occupation (Metzger, 2009). The forest has been reduced to less than 12 per cent of its original extent, mostly distributed in patches smaller than 50ha and closer than 250m to the nearest edge (Ribeiro et al, 2009). Although the effects of forest fragmentation are generally quite predictable at the scale of an individual landscape (smaller fragments generally hold fewer species), very few studies have evaluated how biodiversity responses to fragment size may vary among landscapes with different overall levels of forest cover. Understanding the importance of landscape context on species–area relationships is vital for guiding effective forest conservation strategies in a region that is characterized by highly variable levels of deforestation.

To examine the potential for a landscape-dependent effect of fragment size on forest biodiversity, Renata Pardini and colleagues developed a series of alterative *a priori* research hypotheses to be tested against field data collected from three study landscapes. The study is currently ongoing and is sampling a number of different species groups across varyingly sized forest patches in three different landscapes – one with relatively high forest cover (50 per cent), one with 30 per cent forest cover and one highly deforested (10 per cent). Adding more landscapes would, of course, provide greater inferential strength to the study but very few projects have yet achieved this – giving testimony to the difficulties of implementing landscape-scale field research (Bennett et al, 2006). By developing alternative hypotheses, associated with different theoretical predictions regarding biodiversity responses to fragmentation, the researchers were able to pose clear research objectives. Moreover, by clarifying their objectives in this way, Pardini and colleagues identified key *a priori* considerations of how the data should be collected (specification of design factors and levels), as well as an analytical framework for identifying the most likely explanations for any observed patterns.

The set of six alternative hypotheses is illustrated in Figure 14.2. The sketches represent the expected relationship between abundance or richness (y axis) and fragment size (x axis) in landscapes with different amounts of remaining forest (different lines). The null hypothesis is of no relationship with fragment size irrespective of forest cover – a situation that may be predicted if the species assemblage is dominated by habitat generalists (A). The species–area relationship from the theory Island biogeography predicts a relationship between richness and abundance of forest biota irrespective of forest cover at the landscape scale (B). Alternatively, the amount of forest at the landscape scale may itself determine patterns of species abundance and richness regardless of (C) or in combination with (D) fragment size area. Finally, forest species may respond to the effect of fragment size only above a threshold level of deforestation (E) or only at intermediate levels of forest loss if highly deforested landscapes cannot support viable populations of forest species and maintain the flow of species between fragments.

	A	B	C	D	E	F
Patch-area effect	NO	YES	NO	YES	Above a given degree of habitat loss	At intermediate degree of habitat loss
Landscape-context effect	NO	NO	YES	YES	YES	YES
Landscape context-dependent patch area effect	NO	NO	NO	NO	YES	YES
Underlying explanation or theory	Null hypothesis	Island biogeography theory	Habitat loss only at the landscape scale	Habitat loss at both landscape and patch scale	Fragmentation threshold, Andrén, 1994	Intermediate landscape effect
Sketch	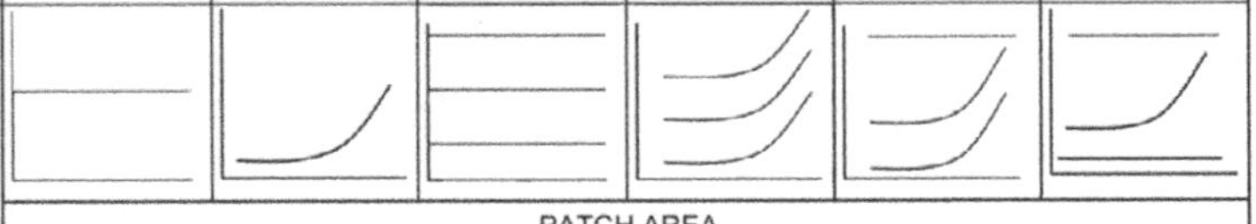					

PATCH AREA

Figure 14.2 Alternative hypotheses for explaining the response of forest species to fragmentation in landscapes characterized by different levels of historical deforestation

Source: Based on an ongoing and unpublished study by Renata Pardini, Jean-Paul Metzger and colleagues at the University of São Paulo, Brazil (figure courtesy of Renata Pardini)

The process of planning a monitoring programme should not be thought of as a strictly linear, step-by-step exercise. Instead, many components require simultaneous consideration and the whole process involves repeated fine-tuning and modification of earlier design choices in response to new insights, or the receipt of new information. This includes definition of the research objective itself, which may often be modified in response to constraints imposed by design criteria during the planning process (Feinsinger, 2001; and see below).

Step 2: Clarify the spatial and temporal scope

To understand the appropriate context for interpreting any research findings, and the limits within which extrapolations may be justified, it is necessary to clearly define the spatial and temporal scope of a monitoring programme. Biodiversity monitoring programmes for forest management need to support inferences that are relevant to entire landscapes, yet measurements are commonly restricted to the scale of individual stands (Kneeshaw et al, 2000). As a consequence, there is often a mismatch between the scale of scientific investigation and the scale at which management is conducted (Hobbs, 2003).

If a research objective is concerned with studying the possible effects of a particular management activity (e.g. selective logging) in a given landscape it is necessary to first check whether sampling across the entire landscape is indeed possible. If it is not, then it is important to establish whether there are any obvious differences (e.g. soil type, elevation, distance to water) between the areas that will be monitored and those that will not, and then qualify the research objective accordingly to reflect any possible constraints on inference. This problem is actually quite common because native forest fragments and areas of less disturbed forest are often restricted to steep slopes, river edges, flooded and unproductive areas (e.g. Figure 14.3).

Figure 14.3 Deforestation of flat, low-lying areas to make way for cattle ranches in São Félix do Xingu, Brazilian Amazon

Source: Photograph by the author

It is also important to clarify the temporal scope of a monitoring programme, and particularly whether it will be possible to collect data in different seasons. For species groups that exhibit highly seasonal activity patterns the effects of a particular type of forest management or human land use may vary significantly depending on the time of year. An example of this was found by Barlow et al (2007d), who compared patterns of butterfly diversity between primary and second-growth forests in the Brazilian Amazon, and found that their relative conservation value can depend entirely on the season when the sampling was conducted (Figure 14.4). If such seasonal interaction effects are expected to be important *a priori* then they should ideally be included as part of the sampling design. If they cannot then the research objective should be further refined, and the generality of any conclusions constrained appropriately.

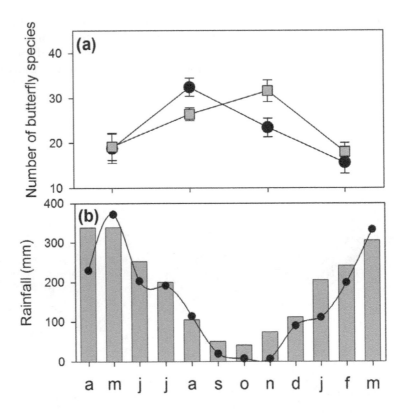

Figure 14.4 (a) Patterns of butterfly species richness in primary (circles) and secondary (squares) forest in Jari, Brazilian Amazonia, sampled in four quarters of one year. Lower panel (b) shows monthly rainfall variation, both for the year of study (bars) and across a 38-year average (solid line). There are clear differences in sampled species richness of different forests depending on the season of sampling.

Source: Redrawn with permission from Barlow et al (2007d)

Step 3: Think about experimental design

The design of any validation monitoring programme has two basic field components: a descriptive component that establishes a baseline condition within reference areas, and an experimental component that evaluates the effects of specific management interventions and human impacts ('treatments'). True manipulative experiments are rarely possible (or even desirable) in large-scale and complex ecological systems and instead it is necessary to rely on so-called 'natural experiments' where existing management and disturbance regimes provide the treatments for evaluation (see Chapter 10). By adopting a more systematic approach to evaluating the opportunities available for natural experiments in managed forest landscapes, it may often be possible to develop monitoring programmes that have a rigorous study design, but also – and equally importantly – capture realistic patterns of human disturbance at spatial scales relevant to biodiversity conservation (Chapter 10). A good example of this is the Tumut Fragmentation Experiment in south-east Australia that was established in the 1990s by David Lindenmayer and colleagues (Box 14.2).

In a manipulative experiment, control samples would ideally be taken before and after any impact or change in management practice occurred (replicated across different sites and over time to minimize confounding factors, see below) (e.g. Azevedo et al, 2006). However, this is very rarely possible for many real-world situations, particularly when studying long-term biodiversity responses to disturbance. Instead, it is normally necessary to 'substitute time for space' and evaluate any impacts on biodiversity against spatially discrete control sites or a 'reference condition'. The reference condition provides the gauge against which management impacts are measured in the form of deviations from a more desirable state. The accurate selection and measurement of a reference condition as a benchmark for judging management performance has many challenges (see Chapters 6 and 9). However, progress is possible by stepping back from the somewhat philosophical discussions about 'what is natural' and identifying sites that provide the most appropriate controls based on what we know about the history of human disturbance in the region (see Chapter 9).

Accurate definition of appropriate reference conditions for biodiversity monitoring and research is an active area of scientific inquiry (Nielsen et al, 2007; and see Chapter 9) and one that is hugely important to the development of meaningful guidelines for conservation management. Closer integration of research in aquatic and terrestrial systems is long overdue in this area and is likely to deliver substantial benefits for work in terrestrial systems.

Step 4: Think about confounding factors

Confounding factors are the things that confound our understanding of the link between a particular management impact and changes to biodiversity. They create the variability that needs to be accounted for by repeat sampling (i.e. we can very rarely depend upon single samples to draw reliable inferences).

Box 14.2 The Tumut Fragmentation Experiment

The Tumut Fragmentation Experiment is a long-term research study concerned with understanding the persistence of biodiversity in human-modified forest landscapes. Situated in New South Wales, Australia, the site encompasses a 50,000ha plantation of radiata pine (*Pinus radiata*), an exotic softwood species established on areas that formerly supported native eucalypt forest. From 1932 until 1985 native forest was cleared to make way for pine plantations, leaving isolated fragments of native habitat ranging in size from half a hectare to 200ha (Figure 14.5). Beyond the boundaries of the plantation, there remain large continuous areas of native eucalypt forest.

Figure 14.5 The natural experimental landscape at Tumut, south-east Australia

Preliminary investigations led to the Tumut landscape being selected for study because it permits a particularly strong experimental design. Most notably the fragments of native eucalypt forest that remain within the plantation areas are not all limited to economically undesirable parts of the landscape (rocky hills and areas next to rivers and streams), but instead are largely representative of the types of forest found in neighbouring reserve areas. By superimposing a monitoring programme on top of this natural experiment, David Lindenmayer and colleagues at The Australian National University have been able to address a host of novel and important questions about the relationships between landscape change, habitat fragmentation and biodiversity conservation. Of course, it is not always possible to superimpose a monitoring project on such an ideal landscape; instead you need to adapt your study design as best you can to account for reality on the ground. However, flagship demonstration sites such as Tumut can be extremely useful for asking questions that cannot be easily addressed elsewhere.

Note: Further information on the Tumut Fragmentation Experiment can be found at www.fennerschool-research.anu.edu.au/cle/tumutstudy/ and in a synthesis by Lindenmayer (2009)

The process of building a conceptual model of the study system can greatly facilitate the identification of potentially important confounding factors (see Chapter 13). Confounding factors can be both obvious (like forest type or altitude) and quite subtle. For example, differences in soil type are not always clear without conducting explicit soil analyses, yet despite the fact that edaphic constraints can have a very strong influence on patterns of biodiversity, few biodiversity research projects include a serious soil sampling component. Cryptic environmental heterogeneity may also often be present with respect to differences in distance to surface or ground water. Confounding factors can also be temporal, such as the example given earlier in the chapter regarding seasonality and butterfly responses to land-use change (Barlow et al, 2007c).

Once identified, there are a number of options for dealing with confounding factors (adapted from Feinsinger, 2001):

- Randomly distribute sample replicates for different levels of a treatment factor (e.g. logging intensity) with respect to potential confounding factors (not all of which may be known). The hope is then that the sample size is sufficient, or the effect size of interest sufficiently large in comparison to the influence of any confounding factor, that it remains possible to pick out any relationship between human impacts and biodiversity, should one exist. While this approach has many advantages (including the avoidance of subjective *a priori* assumptions regarding ecologically important heterogeneity) the danger is that a huge number of samples may be needed to swamp any confounding variation or account for a large number of covariates in data analyses.
- Stratify samples with respect to the most obvious potential confounding factors and their relative distribution in the landscape. This is often seen as an attractive option because a more balanced sampling design greatly facilitates the ability to minimize the influence of confounding variation through statistical analyses. Moreover, if the confounding factor is responsible for driving important interaction or synergistic effects, then understanding its role (rather than just eliminating its contribution to sampling variance) is an important research objective in its own right.
- Control the most obvious of the potential confounding factors by restricting sampling. This obviously limits the level of extrapolation that can be made from the data but may be essential in some cases (e.g. with respect to gross differences in altitude or major dispersal barriers such as rivers).
- Perform a manipulative experiment that accounts for any confounding factors by randomly allocating treatments to samples. While this is very powerful, it is rarely possible in field studies and may not accurately reflect the real-world management situation that is in question.
- Accept the known potential confounding factors and reword the research objective so that they are no longer confounded. This approach inevitably makes the research question less interesting from a management perspective as the focus is no longer on disentangling the effect of a particular impact. Nevertheless, prior ecological and natural history knowledge can often go a long way towards understanding differences for observed patterns even if they were not accounted for in the sampling design and/or data analysis.

The decision as to whether to randomize, uniformly distribute or stratify samples with respect to potentially confounding factors is often a source of enormous controversy and contention among ecologists. Some researchers take the view that if you know enough to stratify sensibly, you probably know so much about the system that you don't need to do the study in the first place (or at least there are probably many more pressing studies that need to be done) (William Magnusson, *personal communication*). I have sympathy with this position but feel that a compromise solution may often be the best approach. It is rarely the case that researchers are confronted with only two options – 'to stratify or not' – but rather there is usually a range of choice with some elements of stratification being useful in some situations but not in others.

Stratification is essential if there is research interest in the interaction effects between management impacts and known environmental heterogeneity. Stratification is also a sensible, if not necessary, approach when confounding factors are known *a priori* to be important yet cannot be eliminated from the study design without compromising the opportunity to sample across a wide-range of treatment levels (e.g. when certain levels of logging intensity or sizes of forest fragment may only be found in a particular part of the landscape or on a particular soil type). However, stratification becomes difficult when there is large number of confounding factors that require huge sample sizes to be adequately accounted for (as is invariably the case in complex multiple-use forestry landscapes). An additional difficulty is that some sources of environmental heterogeneity are very poorly understood or even completely unknown. In such cases, a random or uniform sampling approach that intersects the landscape at multiple points may be more appropriate and help ensure that important spatial patterns are not overlooked. An example of such an approach is the Brazilian Government Research Programme in Biodiversity (PPbio) (Figure 14.6 and Magnusson et al, 2005).

The major disadvantage of a uniform-grid approach is that it demands a high level of sampling effort, and thus may not be feasible when resources for monitoring are limited and representative data is required from across a large area and/or a number of different types of management. To account for this a compromise stratified-random or stratified-uniform approach may often provide the most appropriate design. To generate information that is relevant to local management, the basic sampling design should be based on management-orientated objectives that are linked to our current understanding of the study system and priorities for conservation action (Williams et al, 2001; Nichols and Williams, 2006; and see discussion in Chapter 10). This encourages a design that attempts to provide maximal discrimination between samples of different management treatments and treatment levels (determined by their availability within the landscape), while also attempting to minimize the influence of any systematic bias due to the presence of major confounding factors. Within the different sampling strata, a randomized or uniform sampling approach can then be adopted to take account of unknown sources of environmental heterogeneity. In cases where we have a poor understanding of environmental heterogeneity and its interaction with management, yet do not have the resources to distribute uniform or random samples across the entire study landscape, one option is to limit any *a priori* stratification to only very coarse land-cover classes (e.g. forest and non-forest), thereby ensuring that the major landscape patterns are represented with minimum assumptions (see Figure 14.7).

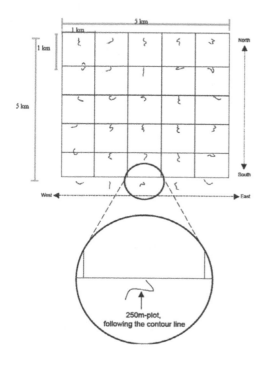

Figure 14.6 The standardized sampling grid used by the Brazilian Government Research Programme in Biodiversity (PPBio)

Note: The grid is 5km on its side with a minimum distance of 1km between permanent terrestrial sampling plots, each of which are comprised of a 250m transect (that follows the altitudinal contour line to minimize edaphic variation within plots). Medium and large mammals are surveyed along the full 5km transects, while plants and all other terrestrial faunal groups are surveyed in the 250m plots. This standardized 25km² grid is considered adequate for population studies of most organisms, as well as studies of watershed chemistry, erosion, distributions of introduced organisms, biomass change and other landscape processes. A further description can be found at the PPBio website:
www./ppbio.inpa.gov.br/Eng/sobreppbio and in Magnusson et al (2005)
Source: Figure courtesy of the PPBio programme

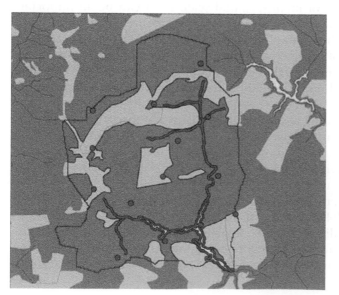

Figure 14.7 A hybrid stratified-random sampling design of a micro-watershed where a predetermined density of sampling plots is distributed between major strata (depicted here simply as forest and non-forest, in dark and light grey respectively) in proportion to their spatial extent in the landscape

Note: Within each land-use stratum sampling points are distributed randomly (while also guaranteeing a minimum separation distance among points) to ensure a roughly even coverage of underlying spatial environmental heterogeneities

The difficulties of stratification when planning for the long term or multi-site comparisons

It is, of course, hoped that a monitoring programme, once established, will survive long into the future. However, many human-modified landscapes are highly dynamic and even areas of currently continuous forest may undergo substantial changes due to new human activities or large-scale threats from climate and environmental change. For a one-off evaluation study of a specific management problem, a highly stratified design can make a lot of sense. However, as the size and distribution of the original sampling strata (i.e. treatments and associated confounding factors) change over time, the value of a highly tailored design may become severely compromised. Thus, the value of stratification depends on both the expected time-frame of the programme and the dynamics of the landscape being studied.

Forest researchers and managers concerned about biodiversity in different places will inevitably have different priorities and interests regarding areas of the management system that need investigating (see Chapter 10). Sampling designs need to accommodate such differences and ensure that the data generated are reliable for tackling local problems. However, sampling designs that are highly stratified with respect to local patterns of management and local landscape features may be limited in their ability to compare results across different places.

Achieving a satisfactory balance between short-term, local research priorities and the ability to assess long-term trends and large-scale patterns can be difficult. To safeguard the potential for tracking long-term and large-scale changes it is worthwhile including a minimum number of evenly or randomly distributed samples across the study landscape wherever possible. These samples can then be complementary to the more targeted samples that are needed to tackle specific, local-scale research objectives.

The importance of spatial scale as a confounding factor in sampling designs

In addition to environmental variability, spatial scale itself can be an important confounding factor in comparative studies of forest disturbance on biodiversity. Limitations on the scale at which sampling is conducted in treatment and control areas have been shown to often have an important confounding effect on our perception of management impacts (Hamer and Hill, 2000; Hill and Hamer, 2004; Drumbell et al, 2008). Dumbrell et al (2008) found that this scale dependence occurs because alpha (point) and beta (between site) diversity invariably increase with spatial scale at a significantly faster rate in undisturbed forest compared with disturbed forest. This pattern is most likely due to the often negative effect of disturbance on habitat heterogeneity and associated reduction in the spatial turnover in species composition. While it is impossible to know *a priori* the most suitable scale at which a given taxon should be sampled in relation to a particular type of management impact, the most effective approach is often to sample across multiple scales (Hobbs, 2003; Dumbrell et al, 2008). Sampling across multiple scales can help partition the variability that can be explained in the abundance and distribution of the biodiversity of interest by processes operating at different scales – for example, through variance partitioning analysis on spatially nested data (Cushman and McGarigal, 2002).

There are a number of sampling designs that can facilitate data collection across multiple spatial scales. The simplest is to overlay a uniform grid on the study landscape and collect samples at the intersections of each grid cell. An alternative method that is designed to capture variability across a range of spatial scales is to use a fractal-based design, which maximizes the range of spatial scales that can be obtained for a given minimum sampling effort (Figure 14.8; Robert Ewers, *personal communication*). In addition to capturing unknown patterns of variation, standardized sampling designs, such as grids or fractals allow direct comparison of data collected by different studies and in different landscapes (Magnusson et al, 2008).

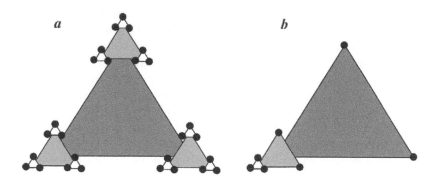

Figure 14.8 Fractal geometry of a network of sampling sites designed to investigate patterns of beta diversity across multiple, nested spatial scales (a), and (b) an optimal method for sub-sampling that greatly reduces the sampling effort required while retaining estimates of beta-diversity for all spatial scales

Note: Points located on the vertices of the white triangles represent sampling locations, and triangles of progressively darker shades indicate a progression from the first to third order of the fractal pattern respectively
Source: Figure courtesy of Robert Ewers

Step 5: Specify independent sample units

This is the step where controversial decisions are frequently made regarding the definition of a 'true' or independent replicate, and represents a long-established problem in ecological research (Hulbert, 1984). True or independent samples are needed to take account of spatial variation in the treatment or design factor, while sub-samples are required from within individual samples to ensure adequate levels of representation are achieved (see also Steps 9 and 10). Because sub-samples represent repeated measurements of the same sample population (e.g. the total pool of individuals of a study species that inhabit a given area) they do not represent independent pieces of information and are therefore often termed 'pseudo-replicates'.

There is no easy solution for identifying what constitutes an independent sample. First, the choice depends on the objective of the research – samples that are independent for addressing one objective may not be for another. For example, a set of replicate samples that are limited to within one small and one large fragment are appropriate for asking about differences between these two fragments, but are inadequate for evaluating the importance of fragment size in general. Sample independence also depends on the distribution of confounding factors. Independent samples should ideally be distributed randomly with respect to potentially confounding variation so as not to obscure the effect of the factor of interest. Finally, what defines an independent sample depends on the scale of the management impact being studied combined with the scale at which the sampled biodiversity perceives the landscape. For example, independent samples of large mammal assemblages need to be much more widely spaced than samples of earthworms. For many questions that relate to the composition and configuration of remnant forest patches, the appropriate independent unit of study is the landscape itself, although for logistical reasons very few studies are successful in achieving this (Bennett et al, 2006). If species' ranging patterns differ between different land uses, the definition of an independent sample for each land-use may differ accordingly. For example, many species are known to range further in sub-optimal habitat, thus requiring a wider dispersion of sampling sites than in good quality areas to ensure comparability.

It is important to not confuse pseudoreplication *sensu stricto* (i.e. the inappropriate interpretation of non-independent samples as independent replicates in an analysis), with non-independent samples that can be accounted for through statistical analysis (Scheiner, 2001). For example, it is often appropriate (and indeed desirable) to have a hierarchical sampling design where samples are nested within sites that are then distributed among treatments. In this case, 'site' can be included as a random factor in analysis as one way to account for the non-independence of samples. Accounting for random factors in analysis (so called 'mixed-effect' models), can be difficult and often benefits from consultation with a statistician. Pinheiro and Bates (2000) is the standard statistical textbook on the subject, while Bolker et al (2009) provide a practical review of some of the main issues when dealing with mixed models in ecological data. More subtle and powerful techniques exist to explicitly incorporate the spatial structure of sampling locations into statistical models and help identify underlying mechanisms – for example, Principal Coordinates of Neighbourhood Matrices (Borcard et al, 2004).

Step 6: Select appropriate variables for measuring change in biological indicators and target species

Once the main elements of the sampling design have been decided it is necessary to identify what it is that will be measured and how. Deciding on measurements is a prerequisite to thinking about sampling methods and replication as it determines the cost and time commitments associated with field and laboratory work.

The process of selecting biological indicators and/or target species identifies the *focus* of monitoring (see Chapter 12) but further decisions are necessary to identify the most appropriate variables for measuring and explaining observed changes in the biodiversity of interest. The choice of variables needs to be sufficient to ensure that sampling provides a reliable description of important changes in the selected species and populations, while not so laborious to measure that it is prohibitively expensive or time-consuming. There are a number of questions that can be asked when selecting important variables for study, including:

- Is it necessary to count individuals or captures of each species or are occupancy data sufficient?
- Are there options for directly sampling measures of reproductive success?
- What species-specific traits can be measured to help understand the functional significance of biodiversity loss?
- What species-specific traits can be measured to help understand the causes of biodiversity loss?
- Do we need to sample all species from a target assemblage?
- To what extent is it necessary to identify entire biodiversity samples to the level of species?
- Are taxonomic identifications the only way of measuring biodiversity?

Is it necessary to count individuals or captures of each species or are occupancy data sufficient?

The vast majority of studies on the distribution of species across modified landscapes thus far have relied on presence–absence or occupancy data (Scott et al, 2002), and a focus on occupancy data has been recommended as a cost-effective strategy for monitoring (Manley et al, 2004). Yet, while occupancy data may be adequate under some circumstances, their ability to capture underlying population changes in indicator or target species depends critically on the relationship between occupancy and abundance (Holt et al, 2002; VanDerWal et al, 2009). Unfortunately, there are very few studies that have compared disturbance–response patterns for both abundance and occupancy data (e.g. Nielsen et al, 2005). A major problem is that it is often difficult to understand what occupancy data mean in biological terms. If a given area is observed for a long enough period of time then it will be 'occupied' (if only briefly) at some stage by individuals of many different species. As such, presence–absence data can often represent as much a sampling artefact as a genuine reflection of differences in habitat viability for a particular species. Robust analytical procedures exist to account for vagaries of sampling effort and species detectabilities in determining occupancy data (see Step 10), although few studies of entire species assemblages have access to enough data to implement them (e.g. Ferraz et al, 2007). In multi-species analyses, it is also often difficult to distinguish the detectability of individuals from the detectability of species, as individuals can be difficult to detect either because they are cryptic (with respect to a given sampling method) or because they are scarce.

Given these problems with occupancy data, collecting some form of abundance estimate (that can, if data allow, be corrected for differences in detectability; e.g. Schmidt, 2003) is always preferable where viable, and helps provide a richer basis for attempting to understand species responses to habitat change (e.g. McGeoch et al, 2002). Caution is still needed in interpreting abundance data, however, as source-sink dynamics can result in even low quality areas of habitat exhibiting relatively high population densities (Dunning et al, 1992). In the case of studies on eusocial insects (including most species of ant, termite and bee), occupancy (or incidence) data should usually be used, as multiple captures of the same species at the same site cannot be considered independent.

Are there options for directly sampling measures of reproductive success?

Data on both species occupancy and abundance patterns represent uncertain prox-ies of actual population viabilities. Where resources allow, it is often desirable to collect direct evidence of productive success, or measures of population dynamics, more broadly. For example, Sekercioglu et al (2007) used radiotelemetry to detail the habitat use, movement, foraging and nesting patterns of three bird species in a human-modified forest landscape in Costa Rica. This kind of work is necessarily labour and resource intensive. However, such data provide an important validation of biodiversity sampling data and it may often be worthwhile implementing such studies as complementary research projects (e.g. by masters or doctoral students) to the main monitoring programme.

What species-specific traits can be measured to help understand the functional significance of biodiversity loss?

For studies that are concerned with inferring differences in ecosystem function, and the extent to which functional processes are linked to biodiversity, it may often be worthwhile recording trait information in addition to just counts of individuals. There are a vast number of species life-history traits that could be measured but one that is easy to measure and often of considerable functional importance is body mass. Recording differences in body mass is particularly important for invertebrates where individual species from the same taxonomic group can vary in body mass over a number of orders of magnitude. Depending on whether abundance or biomass data are used, it is possible to reach quite different conclusions regarding patterns of community structure among sites (Gardner et al, 2008b). Moreover, when species are known to perform important biomass-related functions (e.g. dung burial by dung beetles), then biomass information can be important for under-standing patterns of functional impairment following disturbance (Nichols et al, 2008).

What species-specific traits can be measured to help understand the causes of biodiversity loss?

To understand the ecological mechanisms that drive biodiversity responses to forest loss and modification, we need to identify those species traits or life-history char-acteristics that confer vulnerability (or conversely, resilience) to land-use change. Henle et al (2004) reviewed the scientific literature on traits that confer vulnerability to habitat fragmentation and found good empirical support for a number of traits, namely: population size and population fluctuation; traits associated with competitive ability and disturbance sensitivity in plants; micro-habitat specialization and matrix use; rarity; and relative biogeographic position. An important conclusion of this and other work is the need to consider interactions between species traits and environmental heterogeneity if we are really to improve our understanding of observed changes in biodiversity. Some traits can be responsible for conferring both functional importance and vulnerability to disturbance. A good example of this, once again, is body mass in dung beetles where larger beetles are more functionally significant (at least for certain processes), as well as being more extinction prone (Larsen et al, 2005; Nichols et al, 2009). Trait information can either be used directly as a response variable for analysis (i.e. what is the impact of selective logging on dung beetle biomass?), or as additional explanatory variables to help understand observed changes in species abundance and distribution.

Do we need to sample all species from a target assemblage?

In some instances, it may be possible to select individual indicator species whose presence or absence can provide a reliable indication of patterns of diversity or abundance of an entire species group. Empirical support of this idea has been shown for data on birds and butterflies in the Great Basin of the US (Mac Nally and Fleishman, 2002; Mac Nally and Fleishman, 2004). However, a large amount of field research would be needed to establish and validate such an approach for many areas of the world, and it is only likely to work if individual indicator species are especially easy to recognize and sample.

It is often the case that many of the rarer sampled species contribute little towards our understanding of biodiversity patterns due to the uncertain nature of such data. There may, therefore, sometimes be an argument for focusing sampling efforts on more abundant species, although such an approach should be adopted with caution as it is likely to ignore the very taxa that are of greatest conservation concern.

To what extent is it necessary to identify entire biodiversity samples to the level of species?

Sorting and identification of biodiversity field samples, especially invertebrates, can be extremely laborious. To get around this, it has been proposed that individuals that are only identified to higher taxonomic (e.g. family) levels can provide adequate and highly cost-efficient surrogates of species diversity for some conservation applications (e.g. large-scale conservation planning; see Balmford et al, 1996). However, the individualistic nature of disturbance response trajectories of many, if not most, species to gradients of human impact (Fischer and Lindenmayer, 2006) suggests that for monitoring purposes it is invariably necessary to identify indicators to the level of species or morphospecies (see Basset et al, 2004a). Nevertheless, one potentially promising compromise is to re-sample field collections in the laboratory and only identify a proportion of the total from each sample locality. Few studies have evaluated this approach but recent work on oribatid mites from the Amazon suggest that it can greatly increase the cost-efficiency of biodiversity monitoring for some groups (Santos et al, 2008).

Are taxonomic identifications the only way of measuring biodiversity?

This question may seem paradoxical but a variety of classification approaches have been suggested that can provide direct measurements of biodiversity that do not rely on taxonomy. These methods mostly relate to plants. For example, Slik et al (2008) presented an approach based on measuring patterns of wood density to quantify patterns of forest disturbance. Gillison (2002) generated a computer-aided method for rapid vegetation surveys based on variation in physical plant morphology. Also, a growing number of studies have demonstrated the potential for satellite technology to classify variability in tree diversity in the forest canopy (Chambers et al, 2007). The advantage of all these methods is that they greatly reduce the reliance on a small number of experts for species identification and the amount of time needed to process samples. With continued research and field trials, it is possible that such indirect classification techniques could make a significant contribution to the development of cost-effective biodiversity monitoring programmes.

Step 7: Select additional environmental variables

To improve our understanding of the cause–effect relationships between management impacts and biodiversity, it is necessary to collect additional environmental information beyond the main treatment factors (and associated management performance indicators). This information can then be used to help explain the mechanisms that lie behind observed patterns, thereby improving our confidence in the choice of recommended performance indicators and helping to further refine research hypotheses for ongoing validation monitoring work. Where such variables represent large-scale spatial heterogeneities, they can be incorporated into a stratified sampling design (see Step 5). Additional measurements can also be recorded at the scale of individual sampling sites and used as covariates in data analyses. Selection of such variables should be made in accordance with our understanding of natural disturbance dynamics, habitat and resource requirements of the biological indicators and/or target species that form the focus of evaluation. Caution is needed not to adopt a potentially wasteful 'shot-gun' approach of collecting lots of variables in the hope that some of them will be useful when it comes to analysing data.

Step 8: Select sampling method(s)

Depending on the choice of indicator group or target species, you counselled a huge number of field sampling methods. There are many books and field-manuals available that detail how to employ different methods, as well as their various advantages and disadvantages (see Table 14.2). Most of these publications encompass a variety of ecosystems and not just forests.

Table 14.2 A selection of popular books and manuals on biodiversity survey methods

Book	Description
Measuring and Monitoring Biological Diversity: Standard Methods for Amphibians Heyer et al. (1994)	A tried and tested volume covering a wide range of field techniques for sampling amphibians, as well as protocols for specimen collection and processing sample data. Also includes a valuable introduction to research design and keys to developing a successful field project
Measuring and Monitoring Biological Diversity: Standard Methods for Mammals Wilson et al. (1996)	Similar to the amphibian volume in the same series but focused on mammals
Bird Ecology and Conservation: A Handbook of Techniques Sutherland et al. (2004)	An authoritative volume covering bird census and survey techniques, but also a host of other chapters on techniques relevant to studies of bird ecology and conservation, including such topics as breeding, foraging, migration and survival, and population and habitat management
Invertebrate Surveys for Conservation, New (1998)	A very readable and comprehensive summary of field techniques for surveying both terrestrial and aquatic invertebrates. It also includes very useful chapters on sampling design, processing of invertebrate sample data, and the application of this data to problems in conservation and management. An indispensable book for any invertebrate survey work
A Handbook of Tropical Soil Biology, Moreira et al. (2008)	Soil biodiversity is almost always ignored in biodiversity monitoring and this book should help address this bias. Chapters include techniques for macrofauna and soil microbes, including bacteria, mycorrhiza and free-living fungi. There is also a useful chapter on sampling design
Ecological Census Techniques: A Handbook, Sutherland (1996)	A now classic book, outlining methods for surveying plants, vertebrates and invertebrates (including aquatic organisms). Additional useful chapters on measurement of environmental variables and designing field surveys
Handbook of Biodiversity Methods: Survey, Evaluation and Monitoring Hill et al. (2005)	A large and comprehensive volume covering a diverse range of topics relating to biodiversity surveys, from planning of a research or monitoring project, habitat assessments and a very comprehensive treatment of techniques for surveying and evaluating different taxonomic groups. Biased towards temperate surveys (with a particular focus on the UK)
Forest ecology and conservation: a handbook of techniques, Newton (2008)	An authoritative volume on a wide range of techniques and research methods relevant to forest ecology, with a particular focus on forest management. Chapters encompass a range of topics, including plant surveys, forest dynamics, landscape structure, genetics and modelling

Care is needed in choosing a method that is appropriate for the task in hand and capable of generating sample data that reliably reflects the underlying spatial and temporal variability in target measurements (be they relative abundances of a set of indicator taxa or population estimates of an endangered species). The most important fact to recognize in selecting a sampling method is that *none are perfect*. More specifically, all survey methods are subject to what is commonly termed *detection error* (although this is much less of a problem for plants and other sedentary organisms). Detection error comes about because any given sampling method inevitably fails to detect all individuals, or even all species, that are present in a sampling site (Yoccoz et al, 2001). In technical terms, this is referred to as a 'false negative' problem as the species-sample occupancy matrix has more zeros than would be the case if the method was 100 per cent reliable (Tyre et al, 2003).

Detection error is likely to vary among forest types and species as well as among observers or data recorders. The consequence of such imperfect and variable detection rates is that differences can be generated between treatments (e.g. different types of management intervention) that are due to biases in the sampling methods used rather than actual treatment effects. This problem is often encountered when a given sampling method is more effective at sampling an assemblage of study species in certain habitats compared to others.

For example, mist nets are known to be highly effective at catching understorey species of bird, yet unsurprisingly fail to catch the canopy species. This limitation can introduce substantial bias when comparing dense low-canopy disturbed habitats with high canopy open-understorey primary forest, as demonstrated by Barlow et al (2007) in the Brazilian Amazon (Figure 14.9). This kind of general bias is relatively easy to overcome by applying complementary methods to ensure that each habitat is fully sampled (e.g. combining mist nets with point counts). A more difficult problem is to account for variability in detection probabilities for species that can reasonably be expected to be detected by a given sampling technique (see Stage 10).

In addition to biases in species detection, another important consideration in deciding on a sampling methodology is that all sampling techniques cost time and money to implement. Although this seems a very obvious point, extremely few studies have actually evaluated the cost-effectiveness of different techniques for sampling a given species group. Where this has been done, there is strong evidence to suggest that significant cost savings can be made following a careful evaluation of alternatives (Franco et al, 2007). Assessments of cost-effectiveness do not need to be complicated in order to be useful, and may just entail a basic comparison of numbers of species or rare species sampled per unit of investment (e.g. Garden et al, 2007).

Finally, it may sometimes be worthwhile to select a sampling methodology that matches that used previously by researchers working at the same site. For much of the world, we are desperately lacking the longitudinal data series that are so necessary to understand the importance of historical disturbances. As such, there is a lot of value in ensuring that new data are comparable with historical datasets, even if the original sampling methods employed are sub-optimal. Quintero and Roslin (2005) provide a good example of this approach by demonstrating the recovery of dung beetle communities in forest fragments in the Brazilian Amazon 15 years after initial data were collected. This was only possible by guaranteeing that the exact same methods as the original ecological study (Klein, 1989) were employed .

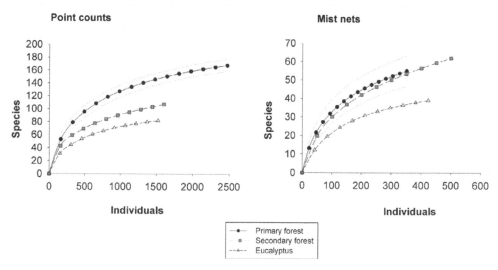

Figure 14.9 Species rarefaction curves comparing bird communities sampled in primary, secondary and plantation forests in Brazilian Amazonia, using both point counts and mist nets

Note: Overlapping confidence intervals demonstrate that mist nets fail to distinguish differences in richness between primary and secondary forest while the pattern is clear when using point counts
Source: Redrawn with permission from Barlow et al (2007)

Step 9: Decide on an appropriate level of independent sample replication

Independent samples are absolutely essential to ensuring that any conclusions regarding management can be generalized beyond the limited distribution of sampling locations. Understanding the number of replicates that are necessary to confidently detect an effect size of interest from background ecological and sampling variability is achieved through the use of power analysis. Obtaining adequate statistical power helps to ensure that important effects (e.g. changes in biodiversity following management impacts) do not pass unnoticed, and is therefore central to the design of any monitoring programme (Di Stefano, 2001; Lindenmayer and Burgman, 2005; Legg and Nagy, 2006). Statistical power is defined as one minus the Type II error rate. Understanding the difference between Type I and Type II error rates is essential to understanding the concept of statistical power. A Type I error results in falsely ascribing importance to an impact that is in fact not important. Statisticians have traditionally used an arbitrary 5 per cent significance level in hypothesis testing to minimize the risk. By contrast, a Type II error represents a failure to identify an important effect where one exists.

To calculate the statistical power of a study requires information on the sample size, the effect size to be detected, the Type I error rate and the variability inherent in the data. By defining *a priori* the desired level of statistical power, it is possible to calculate the number of samples necessary to achieve an effective sampling design.

If the data are highly variable, and/or the effect size of interest is quite small, then very large sample sizes may be required to guarantee a minimum level of power. This may be unfeasible for many monitoring programmes that only have access to limited resources. To tackle this problem, an alternative approach to increasing the sample size (or reducing sample variability) is to relax the significance level to above the conventional 5 per cent mark. This approach has rarely been adopted but may be essential if we are to develop monitoring programmes that are both powerful and cost-effective (Field et al, 2005, 2007).

Despite their theoretical appeal power analyses are not without their problems. Two common difficulties are the lack of *a priori* information on sample variability, and the difficulty of deciding what represents an effect size of conservation concern (or conversely, what constitutes a tolerable amount of loss) (Walshe et al, 2007; and see Chapter 15). Nevertheless, even while estimates may be quite crude, power analysis can still provide an invaluable guide to deciding the number of samples necessary to detect important effects of human activity on biodiversity.

Getting the balance right between sample quantity and quality

Limited resources mean that the design of any biodiversity monitoring programme face, a tension between securing samples that are of both sufficient quantity (to ensure that results can be extrapolated across scales that are relevant to management) and quality (to ensure they reliably capture the underlying variability in the measurement of interest) for addressing the research objective at hand (see also Step 10). More representative samples will confer greater statistical power by reducing sampling variance. In turn, this reduces the number of samples that are needed to detect a given effect. However, continued investment in sub-sampling to improve representation will ultimately result in less powerful sampling designs if a lack of resources compromises the number of true replicates that can be distributed across the study landscape.

Getting this balance right is obviously critical and requires careful consideration as to the nature of the research objective being addressed, the characteristics of the study landscape (scale and complexity of confounding factors) and the ecology of the study species. For example, if the focus of research is on understanding temporal trends, it is preferable to employ the minimum number of individual sample sites necessary to be representative of that treatment class, while trying to maximize the number of time periods over which data are collected. By contrast, if the interest is more in spatial questions (or in substituting space for time to evaluate long-term responses to disturbance (e.g. Murilo et al, 2009)), then it is preferable to use a minimum number of temporal replicates to guarantee adequate sample quality and maximize the amount of sampling effort that can be distributed across different localities.

Some response variables are much more sensitive to differences in sampling effort than others (Magurran, 2004). Species richness is known to be particularly sensitive and for a given level of sampling effort greater levels of statistical power can often be achieved by using more robust response variables, including some indices of diversity (e.g. Simpson index; Lande et al, 2000) or measures of community structure and composition (e.g. community similarity indices; Barlow et al, 2007a). It is also often possible to conduct analyses of management impacts using estimates of standardized effect sizes (rather than absolute values) as the response variable; for example, the reduction in the number of species of bird in logged compared to undisturbed forest. Because survey errors are shared across all sample sites, standardized effect sizes can be highly robust to changes in sampling effort which means that a larger number of samples can be collected without sacrificing reliability (Figure 14.10)

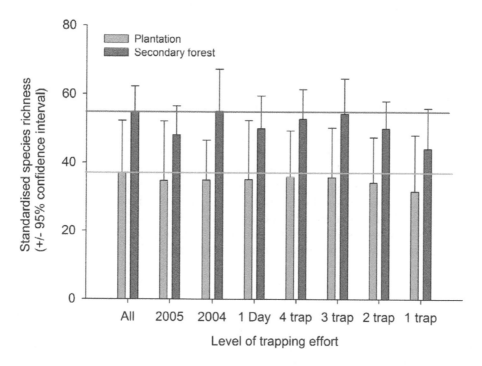

Figure 14.10 Robustness of standardised effect sizes to differences in sampling effort

Note: Plot shows the reduction in the number of dung beetle species from primary forest to plantations (light shading) and primary to secondary forest (dark shading) in Jari, Brazilian Amazonia. Different bars represent the same measurements using subsets of the data that draw on different levels of sampling effort, including only using data from one year (2004 or 2005), only using one temporal sub-sample from each site visit (rather than two), and only using data from a decreasing number of spatial sub-samples (one, two, three or four traps rather than five)

Source: T. A. Gardner, unpublished data

Re-sampling analyses can be a highly effective way of identifying more cost-effective sampling designs for biodiversity monitoring. For example, Mac Nally (1997) successfully employed a re-sampling approach to evaluate alternative bird sampling designs and guide improvements in environmental impact monitoring in south-east Australia. More research in this area would be of considerable practical benefit to forest managers worldwide.

Step 10: Decide on an appropriate level of sub-sampling

Once a sampling method or set of sampling methods is selected, and a minimum number of sample replicates identified, the next step is to decide on an appropriate level of sub-sampling at each of the individual sampling sites. Irrespective of whether the sample data are being used to estimate demographic parameters of a target species, or indices of species richness and diversity, there is always a level of natural variability among sub-samples that can bias estimates and result in unrepresentative samples. If the level of sub-sampling effort (defined as the combination of repeat visits together with the number of spatial sub-samples) invested in individual sites is low, this error can be very large, diluting confidence in the reliability of any between sample comparisons.

Decisions need to be made concerning the number of both temporal and spatial sub-samples that are necessary to achieve satisfactory sample estimates. How can this be done? First, it may be possible to draw upon past experience or recommendations from experts on the level of sampling effort (number of traps, number of survey hours, etc.) that is necessary to collect a representative sample of a given species assemblage, or determine patterns of occupancy of a target species. Such recommendations can be invaluable, but caution should be exercised in accepting them at face value, especially if they are based on work done in a different forest type or on a different assemblage of species. Where possible, it is preferable to employ an analytical approach (based on data from control or reference sites) to determining the level of effort that is necessary in order to achieve representative samples. This may be based on preliminary data sets, or if not available, built in as an adaptive component of the monitoring programme itself.

Deciding on the number of temporal sub-samples

To have confidence in a sampled estimate of a given biodiversity variable (e.g. population size, relative abundance, species richness, etc.) at a given location, sampling should continue through time until the point at which variability in estimate values stabilizes to within acceptable limits.

If the objective is to determine patterns of species occupancy and composition, then it is important to know the number of repeat samples necessary to guarantee a minimum probability of species detection at any given sample locality. For many situations, it is unclear the extent to which differences in management may interact with the effectiveness of different sampling methods, and bias the detection of individual species. Consequently, it has been argued that without some knowledge of variability in species-specific detection probabilities, it is not possible to draw strong inferences from biodiversity monitoring data (Thompson and Seber, 1996; Yoccoz et al, 2001; Schmidt, 2003). The problem of variability in detection rates has been studied in considerable detail and methods for estimating actual detection rates between sites and across time have been proposed (e.g. MacKenzie and Kendall, 2002; MacKenzie et al, 2006), and can be calculated using the freely available software programme PRESENCE (www.mbr-pwrc.usgs.gov/software/presence.html).

Few researchers would argue that any sampling method has an equal probability of detecting all individuals of all species, and that all species present in a given site will be detected. However, it is often logistically impossible or financially prohibitive to calculate detection probabilities for all study species. A compromise solution is to focus on calculating detection probabilities (e.g. through distance or mark-recapture sampling) for a subset of survey locations or particularly important indicator species in the first few years of monitoring. The results of this assessment can then be fed back into an AM programme, thereby allowing any necessary adjustments to be made to the sampling design (e.g. increased sampling effort in vegetation types where detection probabilities are lower). In cases where no *a priori* data are available, careful attention is needed to avoid obvious biases in selected sampling techniques, and wherever possible employ multiple and complementary sampling methods (e.g. Barlow et al, 2007; Ribeiro Jr et al, 2008).

When deciding on the number of sub-samples (temporal or spatial) for comparative studies, it is important to note that differences among sites in the density of individuals, as well as their patchiness in space, mean that a simple standardization of sub-sampling effort by the number of samples (e.g. trap nights, repeat transects, etc.) will likely result in erroneous conclusions about patterns of species richness. To account for site-level differences in the density of individuals (e.g. due to productivity), it is therefore necessary to recalibrate sampling effort from samples to numbers of individuals (see Gotelli and Cowell, 2001).

Finally, it is important to ensure that temporal sub-samples are as independent as possible. That is to say, the probability of detecting a species in one survey should not depend on the outcome (detection or non-detection) of another survey (MacKenzie et al, 2006). This can be achieved by leaving sufficient time between repeat samples, or marking individuals to identify (and exclude) re-captures.

Deciding on the number of spatial sub-samples

Once it can be determined that individual sub-samples provide a reasonable reflection of the underlying ecological variability (i.e. they are reliable estimates of reality and are not dominated by sampling error), it is possible to estimate the mean and spatial variability of the target population ('population' is used here in the statistical sense to mean the full set of measurements available for study, such as trap-level species richness or abundance). With this information, it is then possible to calculate how many spatial sub-samples should be collected within each independent sampling site.

Feinsinger (2001) presents two simple ways in which this can be achieved. The first is termed the 'relative precision' method, where the goal is to estimate the minimum number of samples that will lower the estimated error in the mean to some acceptable relative value (e.g. 5 per cent of the sample mean). Alternatively, an 'absolute precision' method can be used that calculates the sample size that will provide an estimate of the population mean with a desired degree of precision and confidence (see also Lindenmayer and Burgman (2005); and Step 9 for discussion of statistical power, the calculation of which is related to that for estimating sample sizes).

One problem that doesn't have an easy solution is the spatial scale over which sub-samples should be distributed. Here, the answer will depend on the scale of impact under investigation and the ecology of the study species. Sub-samples of species with large home ranges need to be distributed widely to provide a fair reflection of the study population in each independent sample.

Balancing the number of spatial and temporal sub-samples

Deciding on an appropriate balance of spatial and temporal sub-samples at a given sampling site is a tricky exercise, and one that has been given relatively little attention by ecologists. If you are sampling dung beetles, to what extent does it matter whether you employ 30 traps arranged 50m apart and sampled for one day, or five traps arranged 50m apart and sampled for six days? Options to help resolve this problem include drawing on experience from past research that has employed different designs, conducting a re-sampling analysis of available data or designing a dedicated study to evaluate the effectiveness of different approaches. Careful consideration is needed of the ecology of the study species and what we understand about the factors that determine variability in activity patterns and hence variation in detection probabilities (MacKenzie et al, 2006).

Step 11: Evaluate whether the time frame available for monitoring is adequate

Human-modified forest ecosystems are highly dynamic and subject to constant change, not only from ongoing management and disturbance events but also from the legacy of past human impacts and natural ecological processes. As discussed in Chapter 13, these complex dynamics can result in time-lags in the regeneration and recovery of affected species, and in the expression of interaction effects among different drivers of change, including cascade effects (e.g. through the disruption of trophic interactions or the loss of keystone species) in the biological system. While modified forest ecosystems will never stabilize, and there will always be uncertainty in management recommendations, it is important to reflect carefully on the extent to which the time-frame of the monitoring programme matches the time-frame of any likely disturbance responses in the ecological system.

Step 12: Evaluate whether the necessary resources and expertise are available

Once all of the above steps have been followed, a reality check is needed to ensure that the proposed sampling design is feasible. Is there enough time to allow the desired level of sample replication? Are there budgetary or logistical constraints on the number of sites that can be visited, and are there any limitations on the distances that can be travelled in the study landscape? Are sufficient skilled observers and data collectors available to conduct the work? By carefully mapping out the proposed design (e.g. on a spreadsheet) and comparing against these practical constraints, it is possible to conduct an iterative process of refinement where scale and level of replication of the study is adjusted until an effective, yet realistic design is found.

Specific considerations of cost and time are inherent in many of the individual steps outlined above, but it is important to bring all these choices together at the end of the planning process to help identify the most acceptable areas for compromise. In summary, any biodiversity monitoring programme can be thought of as representing a trade-off in three dimensions – between sample detail (the number and resolution of species and environmental measurements that will be taken at each site), sample quality (the number of sub-samples that will be collected to achieve satisfactory representation) and sample quantity (replication across space) (Figure 14.11).

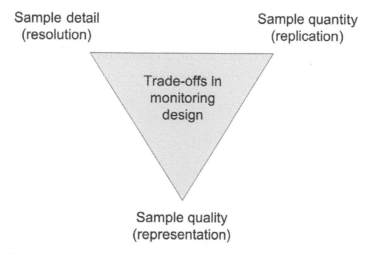

Figure 14.11 The inherent trade-offs that exist in the design of any biodiversity monitoring programme between sample quality, quantity and detail

Note: Each of these components requires its own investment of time, money and expertise

By systematically evaluating each step of the monitoring planning process, it is almost always possible to find significant opportunities for improving cost-effectiveness. Key steps include defining the scope of the study (including goals and objectives), the choice of indicators, sampling methods and variables used to measure change in selected indicators, and the level of sample replication and representation that is necessary to address the stated objective. Care is needed to identify possible economies of scale, such as the opportunity to address an additional and complementary research objective, or include an additional set of indicators, for only a relatively small marginal cost increase. For example, in an analysis by Gardner et al (2008[!Q92!]) on the cost-effectiveness of biodiversity surveys in tropical forests, a saving of 16 per cent of the total budget was made possible by surveying different taxonomic groups that are amenable to the same sampling methodology with little additional modification (e.g. leaf-litter amphibians, reptiles and small mammals), and by surveying different taxa that can be simultaneously sampled by the same field workers.

There is an additional trade-off that should be considered that goes beyond concerns about sampling design, power and cost-effectiveness of monitoring, and exists between the overall investment in monitoring and the need to make expedient decisions for management (Joseph et al, in press). Monitoring can cost a lot of money and, in some circumstances, fixed budgets may mean that allocating conservation funds to monitoring will result in less funding being available for biodiversity management. Figure 14.12 illustrates this optimal trade-off with a schematic diagram. As monitoring effort increases, we can assume that the effectiveness of management increases as a result of learning, but at the same time management effort decreases due to an increased allocation of limited resources to monitoring. When the relative gains from monitoring cease to compensate for the reduced management effort, it is time to stop investing in monitoring. Very few monitoring programmes have performed such a critical evaluation.

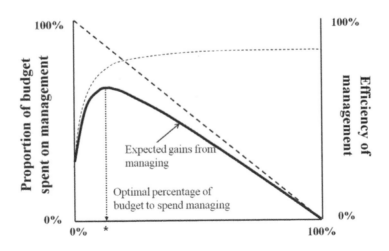

Figure 14.12 The optimal allocation of resources to monitoring and management

Note: This is a trade-off between learning and doing. The dashed line (– – –) represents the conservation outcome (management efficiency) achieved with 100% accuracy (i.e. no error due to lack of understanding, and therefore no need for validation monitoring) as an increasing proportion of the budget is spent on monitoring. The dotted line (- - - -) shows the knowledge gained from monitoring, and associated improvement in management efficiency with increased investment where there is uncertainty in the management system. The solid line (——) represents the monitoring–management trade-off where learning improves the efficiency of management but only until a point, beyond which continued investment in monitoring reduces conservation outcomes because less of the budget is available for management

Source: Figure redrawn with permission from Joseph et al (in press)

Not all constraints on monitoring are related to cost or time. Access to appropriate levels of expertise for sampling and identifying different groups can often be a major challenge and almost always limits the number and type of species groups that can be studied in a given place. A particular difficulty is in groups such as large mammals, birds and plants that require expert field identification skills (in contrast to most invertebrates where specimens can be transported for identification elsewhere). If long-term biodiversity monitoring programmes are to be implemented in areas of the world where they are most needed, much more investment is required for the training and capacity building of taxonomists, students, technicians and local people, all of whom have vital roles to play in the monitoring process (see Chapter 16).

Step 13: Think hard about how to analyse the data before it is collected

How biodiversity data are collected and the limitations of a particular sampling design greatly determine what analyses are possible. It is therefore essential to think ahead and consider how monitoring data could be analysed, and the design requirements that are necessary to improve the strength of any inferences regarding management impacts. Careful reflection of the data analysis process can help ensure that important measurements are not forgotten when in the field – and, equally importantly, that precious time and resources are not wasted collecting information that will never be used. The process of building a conceptual model and identifying alternative research hypotheses that articulate our current understanding of cause–effect relationships between forest management and biodiversity can be very important here as it provides a framework for thinking about both the design and implementation of a monitoring programme (see Chapter 13). Chapter 15 reviews some important considerations in the analysis and interpretation of biodiversity monitoring data.

Step 14: Preserve data integrity through careful recording and storage

Fundamental to the longevity of any monitoring programme is strict adherence to a transparent and comprehensive approach for recording and storing field data. All too frequently, the integrity of monitoring data is lost because particular sites or years are missing from files. It is also common for raw data from early years of monitoring to be discarded or misplaced once summary findings have been published. This can be a tragic loss because as we gain new theoretical insights, and develop novel and more powerful analytical approaches, historical data become increasingly valuable over time. Magnusson and Mourão (in press) propose the following series of practical recommendations to help ensure that data integrity is maintained, and relatively little pre-planning show how with significant gains in scientific output can be achieved:

- Careful recording of primary metadata. Raw data are useless without information that describes the context in which they were collected and the meaning of individual variable names. Magnusson and Mourão (in press) term this information primary metadata, and no raw data should be stored or presented in its absence. Anybody who has ever tried to conduct a meta-analysis will understand the frustration of trying to tease out even basic metadata on study design and location from poorly prepared scientific papers and reports. Standardized approaches for recording metadata exist and can provide useful models for new programmes – such as that used by the Long Term Ecological Research Network (www.lternet.edu/).

- Careful recording of raw data. Many research projects summarize data into tables and graphs for annual reports. Over time, the underlying raw data can frequently be lost, greatly limiting the opportunities for data analyses of the full time series. Raw data should be tabulated and databased in a format that can be readily queried by researchers interested in a particular question. The basic structure of a data table is to have objects (the minimum unit of interest in a study) as rows and attributes or variables (the things that define characteristics of the objects) as columns. In the case of species data, table rows should represent actual individuals rather than pooling data by traps or samples. Clear location data (site, transect, trap, etc.) are vital to ensuring that different tables containing different information on species and environmental variables can be easily linked and compared.
- Conduct independent error checks wherever possible. It is very easy to make errors when entering raw data and mistakes can remain hidden for a long time, generating perplexing and misleading results and generally frustrating the analyst and wasting time.
- Wherever possible, data should be made available publicly. This is the surest way of ensuring the long-term integrity of costly monitoring data while also guaranteeing that the maximum scientific benefits will be derived. Many researchers are concerned about data being stolen and the loss of intellectual property. However, there are very few cases has been where primary data are taken and published by a third party without the knowledge and consent of the originator. Generally, the benefits of collaboration and sharing of ideas and research questions far outweigh any risk of thievery (see also Chapter 16 for discussion of the importance of collaborative research networks).

Step 15: Be prepared to adapt

The final step in planning the sampling design and data collection process of a monitoring programme is to recognize that any decisions that are made are not immutable. A lot of important insights about what is and is not possible can only be acquired when actually in the field. Moreover, as monitoring proceeds, initial data can be used to adjust and fine-tune the design in ways that will enhance our ability to draw reliable inferences about the study system and identify opportunities for continuous improvements to management. This is known as 'adaptive sampling' (*sensu* Thompson and Seber, 1996), and although it is a logical and intuitive approach, there have been very few applications in the field (Field et al, 2007), and none to my knowledge for multi-species monitoring programmes. Despite the importance of learning while doing, a strong note of caution is warranted when considering adaptations to an ongoing monitoring programme, as modifications to address short-term problems can often compromise the value and reliability of data to address long-term trends. For example, there is often a temptation to adjust sampling regimes to reduce the number of zeros in sample data, yet understanding a species resource requirements can depend on our understanding where (and when) it is not found as much as where it is.

CHAPTER 15

Analysis and Interpretation
of Biodiversity Data

Synopsis

- The analysis of biodiversity data is a highly context-dependent process that is defined by the underlying goals and objectives of monitoring, the limitations on any potential inferences that are imposed by the choice of indicators and sampling design, and the availability of relevant theory and natural history knowledge to help interpret the ecological significance of any results.

- The four main types of data analysis that are relevant to monitoring relate to problems of: describing the biodiversity to be used in analyses; detecting change with respect to performance standards (i.e. as part of effectiveness monitoring); understanding the reasons behind observed changes (i.e. as part of validation monitoring); and predicting patterns of biodiversity across space and time.

- Decisions about how biodiversity data are summarized and manipulated prior to any modelling work can have a significant influence on our perception of management impacts. Important considerations for improving the quality of sample data include: analysing the response patterns of individual species, as well as entire assemblages; employing caution when using estimates of species richness and diversity indices; testing the sensitivity of observed patterns to differences in spatial scale; understanding species-mediated ecological functions; assessing species-specific conservation priorities; and using biotic integrity indices to improve dissemination of monitoring results.

- Confidence intervals provide a more defensible and practical approach to assessing management compliance against performance standards than significance tests of a null hypothesis of no change in indicator values.

- Statistical modelling provides a means of formalizing our understanding of the environmental impact of human activities and identifying the extent to which our predictions are supported by sample data.

- Information-theoretic and Bayesian approaches to the analysis of monitoring data have distinct advantages over more traditional methods that are based on null hypothesis testing (NHT) because they are explicit about uncertainty and allow the evaluation of a range of alternatives.

- While statistics provide an essential and powerful tool for biodiversity monitoring, they are only one piece of the puzzle and their usefulness can be readily compromised by uncertainty in sample data or a lack of relevant ecological knowledge about the study system.
- To be most useful the analysis of biodiversity data needs to be conducted in the context of trade-offs between multiple management objectives.

Once sample biodiversity data have been collected, the next obvious step in the monitoring process is to figure out what they mean, and more specifically, how their interpretation can shed light on the underlying programme objectives. Far from being an isolated exercise, the analysis of biodiversity data is a highly context-dependent process that is defined by: the underlying goals and objectives that define the overall purpose of monitoring and the relevance and application of any research findings to the wider management system; the limitations on any potential infer-ences that can be drawn from monitoring data as imposed by earlier decisions regarding indicator choice, sampling design and data collection; and the availability of relevant theory and natural history knowledge to help interpret the ecological significance of any results. If insufficient attention is given to this wider context, a lot can go wrong, with the implications of any mistakes being carried forward to result in potentially misleading or inappropriate recommendations for management.

The subject of ecological data analysis is a vast and rapidly growing area of research in its own right, the majority of which is beyond the scope of this book. The purpose of this chapter is to provide an entry point to the existing literature, as well as a broad overview of some of the important (yet, often poorly appreciated) principles, considerations and potential pitfalls that need to be accounted for when analysing biodiversity data. Although there is rarely any 'right' answers to the question of how to analyse biodiversity data, there are frequently a lot of wrong ones.

There are at least four types of data analysis relevant to biodiversity monitoring, relating to the problems of:

1   describing the biodiversity to be analysed;
2   detecting change;
3   understanding the reasons behind observed changes; and
4   predicting patterns of biodiversity across space and time.

Different types of analyses serve different functions in the overall monitoring frame-work (see Chapter 8). Testing for changes to forest attributes is central for assessing compliance against performance standards (i.e. effectiveness monitoring). By contrast, efforts to understand the cause–effect relationships that lie behind observed changes in biodiversity is essential for validating performance standards and identifying ways to enhance progress towards conservation goals (i.e. valida-tion monitoring). As discussed in earlier chapters, the process of validation monitoring is the most challenging component of the wider monitoring process and requires, of the most careful analysis. At its heart, lies the problem of balancing evidence in support of alternative explanations for observed biodiversity patterns, and understanding the extent to which any favoured explanation can be reliably extended to other areas and into the future. Predicting patterns of biodiversity beyond the sample data is a logical next step once robust species–environment relationships have been developed, and provides the basis for conservation planning and scenario analyses.

This chapter is split into four sections to cover the different aspects of data analysis. In the first section, I cover the problem of describing biodiversity, highlighting some of the factors that need to be considered when selecting units for analysis. Second, I deal briefly with the challenge of detecting change in performance indicators, building upon the problems discussed in Chapter 6. The third section covers some of the issues surrounding the modelling of biodiversity data and the ongoing process of developing robust species–environment relationships. I emphasize the relative advantages of different approaches to statistical inference, while also stressing the need to 'go beyond the numbers' and consider the many factors that can help or hinder ecologically meaningful interpretations of modelling results.

In the final section of the chapter, I zoom back out to the broader context of responsible forest management and touch on the importance of considering multiple management objectives and trade-offs among competing objectives when analysing biodiversity data.

DESCRIBING BIODIVERSITY

The process of describing biodiversity within any given monitoring programme has three main stages. First biological indicators and/or target species are selected to provide a focus for monitoring efforts (see Chapter 12). Second the attributes of species (e.g. occupancy, abundance, life-history traits and measures of reproductive success) and/or species groups (proportion of species sampled, taxonomic resolution) that will be measured in the field (see Chapter 14) are selected. The final stage describes the ways in which sampled biodiversity data are summarized and manipulated prior to any subsequent hypothesis testing or modelling work. This defines the type of response variables that are ultimately used to make inferences about the impacts of human activity on biodiversity and guide recommendations for management. There is a huge literature on the subject of describing biodiversity and the multitude of ways for measuring and estimating patterns within and between different spatial scales. A comprehensive and insightful book on this topic is Magurran (2004), and is recommended reading for anybody who is serious about working with biodiversity data. There is also an increasing number of software programmes, some of them free, which provide access to a powerful array of analyses for summarizing and describing biodiversity data (Table 15.1).

Table 15.1 Commonly used software packages for the analysis of biodiversity data

Program	Website	Description
EstimateS*	http://viceroy.eeb.uconn.edu/estimates	Enormously popular application that computes a variety of biodiversity functions, estimators, and indices based on biotic sampling data. Regularly updated to include new advances from theoretical work
Ecosim*	http://www.garyentsminger.com/ecosim/index.htm	Focused on application of null models in ecology. Calculates rarefaction and some diversity indices
Species diversity and richness	http://www.pisces-conservation.com/indexsoftdiversity.html	Comprehensive software package for a large range of alpha and beta diversity indices. User friendly
Primer-E	http://www.primer-e.com/	Standard and proven software package for analyses of multivariate biodiversity data. Focused on non-metric methods and permutations that make few assumptions about sample data. Supported by very good guidance manual
PcORD	http://home.centurytel.net/~mjm/pcordwin.htm	Standard and proven software package for analyses of multivariate biodiversity data. Offers a huge range of non-parametric and parametric approaches and a vast array of data processing and interpretation options. Very good graphics.
Vegan*	http://cc.oulu.fi/~jarioksa/softhelp/vegan.html	A package for the R open-source statistical environment. Extremely comprehensive coverage of diversity analyses, ordination and dissimilarity indices. Supported by a clear online tutorial
PAST*	http://folk.uio.no/ohammer/past/	A software package originally for palaeontology but includes many functions relevant to biodiversity studies in general, including richness and diversity indices, and many different multivariate analyses
CANOCO	http://www.pri.wur.nl/uk/products/canoco/	Standard and proven software package ordination analyses of species community data and specie–environment relationships. Focus is on metric, linear and unimodal methods

Note: Packages marked with an * are freely available

Because there are so many options available for describing biodiversity data, with more appearing in scientific papers all the time, making appropriate choices can often be difficult. In the following sections, I briefly flag some of the most important considerations for thinking about biodiversity data and deciding which approach to use. Space limitations mean this treatment is far from comprehensive and I encourage the reader to look to the primary literature for more detailed guidance.

Data on individual species patterns should not be discarded

Although the focus of this book is on monitoring entire groups of species (i.e. biodiversity, rather than single-species monitoring) this does not mean that information about the ecology and responses to disturbance of individual taxa should be lost during data analysis.

The complexity and high species diversity of many forest ecosystems means there is a common tendency to collapse multi-species sample data immediately into summary analyses of richness, and indices of diversity and community structure and composition. However, because different species often exhibit different responses to disturbance, analyses of species-specific patterns can be useful to understand the impacts of management on forest biodiversity. As discussed in Chapter 12, there are various analytical methods for distinguishing individual indicator species that can be used to help evaluate particular forms of disturbance or management; with the IndVal approach of Dufrêne and Legendre (1997) offers particular promise (see McGeoch et al, 2002; da Mata et al, 2008). Isolating the responses of individual species also facilitates the use of existing ecological and natural history information to help interpret underlying explanations for observed patterns. For example, Lees and Peres (2008a), who used life-history traits of birds, including independently derived estimates of disturbance sensitivity and primary forest dependence, body size, territory size, flock size, dispersal ability and foraging and feeding guilds to help understand species responses to habitat fragmentation, as well as possible implications of any changes for ecosystem function. Larsen et al (2005), employed data on species body-mass to predict the functional consequences of dung beetle extinction on dung burial rates.

Species richness is deceptively difficult to estimate reliably

Species richness is unquestionably the most popular measure of biodiversity in ecological and conservation research. Nevertheless, the estimation of species richness is deceptively difficult and frequently confounded by the vagaries of sampling effort and bias (Gotelli and Colwell, 2001; Colwell et al, 2004; Magurran, 2004; O'Hara, 2005). Moreover, it is not always clear what differences in the number of species signify in terms of conservation value. In thinking about numbers of species as a unit for analysing biodiversity data, the following considerations (and this is not an exhaustive list) should be born in mind:

● The detectability of a particular species to a given sampling technique can vary between sampling units and over time. This is especially true when comparing patterns of biodiversity between different vegetation types or seasons.

- The number of species sampled by a given technique is a function of the effectiveness of that technique and not necessarily the richness of a particular species group in a given area. Clear definitions of target species groups, such as 'beetles attracted to human dung-baited pitfall traps of X design', should be used to avoid ambiguity about what is actually being monitored.
- Raw counts of species numbers are rarely reliable for comparative work due to differences in total species abundances, species detection rates and species abundance distributions (Gotelli and Colwell, 2001). Some form of standardization by sampling effort is required.
- Because the number of species is both area, and time-dependent, the total area and time period over which samples are distributed are often the most important variables affecting the asymptote of a taxon-sampling curve for a given sampling. Standardized spatial arrangements of samples or random sampling within a given area are necessary to guarantee comparable estimates of species richness (or species density if area limited).
- Not all species are equally detectable by a given sampling technique. Almost all biodiversity data are dominated by species that are only captured a few times the infamous 'long tail' of species–abundance distributions (SADs, also called dominance-diversity, or Whittaker plots) (Magurran, 2004). It is possible some rare' species are not rare at all; rather they are just not susceptible to being detected by the technique that was used. Getting around this problem is not easy. Ideally, species-specific estimates of detection probabilities should be incorporated into comparative analyses (e.g. Ferraz et al, 2007; and see Chapter 14) but it is rare that there are sufficient data available to do this (especially at the start of a monitoring programme). Often the use of complementary sampling methods (e.g. mist nets and point counts for bird surveys) is the easiest way to improve the reliability of results. Another common approach is simply to discard data for species with very low capture rates – although this obviously comes with the risk of losing important information.
- Differences in SADs between sites can mean that even standardized sample sizes may not produce comparable, representative results (Cao et al, 2002). To get around this problem Cao et al (2002) proposed a method of standardization based not on sample size, but on auto-similarity between sequential samples (i.e. once two repeat samples from the same locality have a sufficiently similar species composition they can be considered as representative samples of that locality).
- Species accumulation curves can give different results depending on whether they are sample- or individual-based (Gotelli and Colwell, 2001). It is not always clear what is the most appropriate choice, although sample-based curves are preferred if the target species are patchily distributed. If interest is in species richness (not density), then taxon curves should be scaled by numbers of individuals rather than samples to account for differences in total abundance (see Figure 15.1).
- Accumulation and rarefaction curves are only useful if sample sizes are sufficient (see Chapter 14) because all curves converge at low abundances. Rarefaction analyses that compare across small fractions of the total community are meaningless.

- Statistical estimators of total species richness are notoriously unreliable, as they are strongly influenced by differences in total abundance, sampled species richness, abundance distributions, and sampling effort (O'Hara, 2005; Ulrich and Ollik, 2005). This is an active area of research but at present such estimators should be relied upon only to give upper and lower bounds on species richness.
- Even if it is possible to calculate reliable estimates of species richness or density, it is not often clear how relevant such values are for conservation purposes. This is because species numbers are often inflated in human-modified forests by an influx of generalist and open-area species that are generally of low conservation importance, resulting in underestimates of biodiversity loss. For example, Pearman (1997) showed that amphibian species richness – an often popularized indicator of environmental degradation – is a poor measure of human impacts in the Ecuadorian Amazon because the absence of sensitive *Eleutherodactylus* spp. frogs from disturbed sites was often masked by increases in disturbance-tolerant species (mostly *Hyla* spp.). Similar examples are commonplace in the ecological literature (Hobbs and Huenneke, 1992).

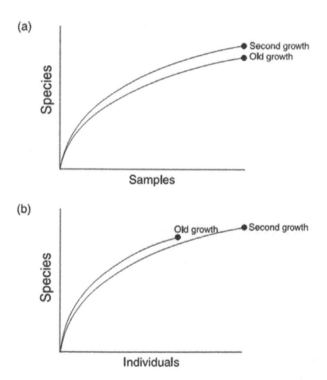

Figure 15.1 The effect on species richness of rescaling the x-axis of sample-based rarefaction curves from samples to individuals when individual densities vary

Note: In this example, species richness appears to be higher in secondary than primary forest when plotted using the accumulative number of samples. Yet, because the density of individuals is much higher in secondary forest, the pattern is reversed when the curves are re-scaled based on the number of individuals

Source: Redrawn with permission from Gotelli and Colwell (2001)

Despite all these difficulties, species richness is likely to remain at the heart of conservation priorities and decision-making, if only due to its popularity. However, it is vital that sufficient care and attention – considerably more than is generally evident – is given to the estimation and interpretation of species numbers so that they are useful and not misleading.

Diversity indices should be used with caution

Diversity indices attempt to collapse measures of both species richness and evenness (the distribution of individuals among species) into a single value, and encompass a bewilderingly large number of alternatives (Magurran, 2004). The aspiration behind such indices is that they represent simplified solutions to the problem of describing the complex and multifaceted concept that is biodiversity (Gaston, 2009). While some diversity indices have been useful in describing biodiversity under particular situations (e.g. Simpson' diversity index is effective at ranking communities at low sampling effort (Lande et al, 2000)), the justification behind such a hope is unfortunately scant.

Instead, it is arguable that diversity indices generally provide very few, if any, advantages over the independent presentation of their component parts (species richness and abundance distributions). Further, they generally represent a highly misleading 'black-box' approach to data analysis, where the ecological significance of changes to index values is almost impossible to interpret because of poorly understood assumptions in their calculation, and the fact that the same result can often be achieved for different reasons. Similarly, indictments and urges of caution in the use of diversity indices have been made repeatedly for at least three decades (Hulbert, 1971; Routledge, 1979), including standard reference books on biodiversity assessment (e.g. Feinsinger, 2001; Magurran, 2004), making it surprising that they have survived for so long. The Shannon index is the most enduring of all, and despite being widely considered as one of the most flawed and uninformative indices (reviewed in Magurran, 2004), it is still commonly encountered in scientific papers, as well as being recommended for inclusion in formal environmental impact assessments of various countries (e.g. Brazil).

Not all single-value 'indices' are deserving of the same criticism as biodiversity indices. There are many examples where other indices play an important role in biodiversity assessment (e.g. in measuring beta diversity – the turnover of species across space, or as calibrated indices of biotic integrity; see below). Rather, the important point is that to be useful indices need to have both a clear ecological meaning and a specific purpose. Diversity indices rarely satisfy either. If the science of biodiversity conservation is to be effective, it is essential that we learn from past mistakes and accept the need for change when concepts become outdated. For the most part, diversity indices are a perfect example of an outdated and unhelpful concept.

A far more effective and simple way of describing patterns of diversity prior to any more detailed analyses on any individual components is through simple graphical summaries, such as SADs. SADs are an intuitive and efficient way of comparing biological communities because they visually combine information on patterns of species identity, relative abundance, richness and evenness (e.g. Figure 15.2), as well as providing the basis for many ongoing investigations in both pure and applied ecology (reviewed in McGill et al, 2007).

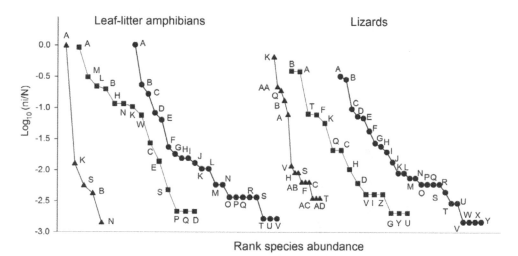

Figure 15.2 Species–abundance plots for leaf-litter amphibians and lizards in primary (circles), secondary (squares) and plantation (triangles) forests of Jari, north-east Brazilian Amazon

Note: Letters represent different species for each taxonomic group. Species are ranked according to total number of individuals sampled of each species (ni) and the total number of individuals of all species (N) for each forest type. This plot integrates a lot of pieces of information. It is immediately clear that the richness is higher and abundance distribution is much more even for both groups in primary forest. It is also clear that there is relatively strong nesting of abundance patterns in different forest types, especially for amphibians, where the most abundant species (A – an unidentified species of *Adenomera*) in primary forest is also the most abundant in secondary and plantation forests
Source: Redrawn with permission from Gardner et al (2007c)

An understanding of spatial scale is essential for describing biodiversity

One of the main principles of ecology is that biodiversity is a strongly scale-dependent phenomenon. As discussed throughout this book, it is therefore essential that great care is taken to accommodate scale considerations in biodiversity monitoring and evaluation work. The employment of appropriate measures of species turnover, or beta diversity, is central to this task.

There is a multitude of ways in which beta diversity can be measured, and a similarly large number of scientific papers and books devoted to the topic. A review of this literature is well beyond the scope of this book but Magurran (2004) and Koleff et al (2003) are recommended as accessible summaries of the main alternatives. The measurement of beta diversity takes two basic forms (Magurran, 2004), namely: (i) comparisons of alpha or site diversity relative to total diversity; (ii) and measurements of difference in the composition and structure of species communities. Often the simplest methods are the most powerful and revealing. The Bray-Curtis index (based on both presence–absence and abundance data) is one of the most reliable measures of the relative distinctiveness of biotic communities (Clarke and Warwick, 2001). For summary measures of the contribution that beta diversity makes to the total (gamma) diversity of region, a simple partitioning of diversity into its alpha (average number of species present in each site) and beta (average number of species absent from each site) components can generate powerful insights and is receiving increasing research attention (for overviews see Veech et al, 2002; Crist and Veech, 2006).

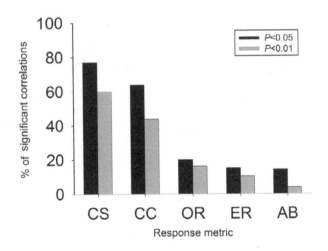

Figure 15.3 The percentage of significant correlations in biodiversity responses between 14 species groups sampled across a gradient of study sites in Jari, north-east Brazilian Amazon

Note: Comparisons were made using community structure (CS), community composition (CC), observed species richness (OR), estimated species richness (ER), and total species abundance (AB)`
Source: Redrawn with permission from Barlow et al (2007a)

Despite a general recognition of the importance of scale in studies of biodiversity there are still many areas of conservation research and monitoring that would bene-fit from a more widespread application of beta diversity analyses. Many multi-taxa studies have demonstrated that indices of community similarity provide a much more sensitive measure of forest disturbance and congruency in cross-taxon responses than assessments of either species richness or abundance (Figure 15.3), and such measures should be included as standard in descriptive analyses of biodi-versity data (e.g. Su et al, 2004; Barlow et al, 2007a; Basset et al, 2008a, 2008b). We still have a poor understanding of patterns of species turnover within any given forest land use and across different spatial scales, and such analyses are urgently needed to help guide landscape-scale conservation planning. For example, there is increasing evidence to suggest that even nearby patches of the same forest type (whether native or modified) can often host distinct sets of species (e.g. due to poorly understood differences in natural environmental gradients, management influences and variation in disturbance histories). In the Jari project, we were surprised to find that beta diversity generally comprises about half of the total sampled diversity for all species groups studied, irrespective of forest types (Figure 15.4).

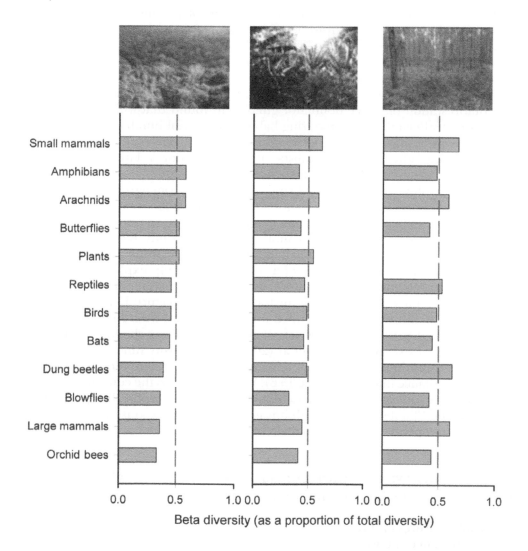

Figure 15.4 Beta diversity as a proportion of total diversity for 12 species groups sampled across different sites of primary, secondary and plantation forest respectively in Jari, north-east Brazilian Amazon

Source: From unpublished data (T. A. Gardner, J. Barlow)

It is often valuable to classify species not just by what they are but also by what they do

Groups of species often interact with their environment in similar ways, for example through similar feeding or reproductive strategies. Such groups are commonly referred to as guilds and are the basis of a long tradition of research in ecology (Simberloff and Dayan, 1991). Guild classifications *sensu lato* provide an intuitive approach to improving our understanding of biodiversity responses to human disturbance as they enable researchers to concentrate on groups that are defined by specific functional roles or life histories. This information can be invaluable in helping to understand why certain species may exhibit contrasting responses, as well as the possible indirect consequences of species losses or gains on key ecological functions and processes (Didham et al, 1996).

For example, Steffan-Dewenter et al (2002) analysed the effect of landscape-scale deforestation on three pollinator guilds and found that while solitary bees were strongly affected by neighbouring habitat loss, social bumble bees were not. The authors infer that this difference in response between key pollinator guilds may have cascading effects on mutualistic plant–pollinator relationships in the landscape. In a similar study in Sulawesi, Klein et al (2002) found a somewhat contrasting result whereby increased land-use intensity from agroforestry had the effect of decreasing the abundance and richness of social bees but increasing the number of solitary bees – demonstrating that functional group responses cannot be readily generalized across systems. In another important study, Becker et al (2007) uncovered how 'habitat split' – human-induced disconnections between habitats used by different life history stages of a species – can help explain why aquatic breeding amphibians declined more than species with direct terrestrial development in the fragmented Atlantic forest of Brazil (Figure 15.5).

Species can also be usefully grouped into what can be described as 'response guilds' (Szaro, 1986) that delineate sets of species exhibiting similar responses to disturbance. For example, Larsen et al (2008) used density functions to classify species based on their population-level responses to fragmentation at different scales of disturbance intensity (Figure 15.6). When applied to the concept of ecological disturbance indicators, this approach can be useful to interpret changes in forest condition (see Chapter 12), although it is important that response guilds are classified from independent data to avoid circularity in analyses.

Assessing conservation priorities associated with individual species can greatly improve the value of data analyses

Almost all measures of diversity (whether to describe individual sites – alpha diversity, or differences between sites – beta) that are commonly used in analyses of biodiversity data, afford equal weighting to the contribution made by different species (Magurran, 2004). This is problematic because not all species have the same level of priority for conservation. By compiling information on the extinction proneness and disturbance sensitivities of individual species, and integrating this

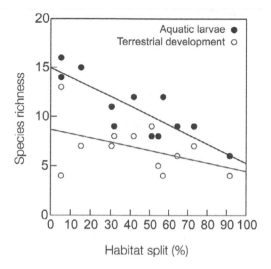

Figure 15.5 The effect of habitat split on species richness of leaf-litter amphibians with either aquatic larvae or direct terrestrial development across 12 Brazilian Atlantic Forest landscapes

Note: Habitat split is calculated as the percentage of the total stream length that does not overlap with areas of natural forest cover
Source: Redrawn with permission from Becker et al (2007)

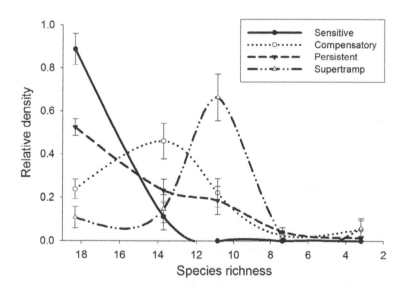

Figure 15.6 Density functions used to classify dung beetle species into four groups based on their response to fragmentation along a gradient of decreasing species richness

Note: Sensitive species rapidly decline with species loss and only occur in one or two richness classes; persistent species also decline but occur in three or more richness classes; compensatory species increase in density in at least one lower richness class; and supertramp species show a peak in density at species-poor sites where all other species were absent or rare. Because species richness is strongly related to fragment area, classifying species on the basis of area instead of species richness as the dependent variable results in similar patterns. Based on data collected across an archipelago of 33 islands created by flooding in Venezuela
Source: Redrawn with permission from Larsen et al (2008)

information into the data analysis process, it is possible to develop a much more nuanced understanding of the biodiversity consequences of human activities and the implications of any observed changes for conservation goals.

For example, to accurately estimate the value of modified landscapes for conserving regional forest biodiversity we need to know the proportion of species inhabiting human-modified systems that were also inhabitants of the original forest landscape (Gardner et al, 2009). Yet, despite the natural priority of these species as a focus of conservation efforts, few studies have been able to identify the extent to which individual species depend on primary habitat. Those studies that have had access to independently collected information on species habitat requirements and sensitivities to disturbance have, somewhat unsurprisingly, reported consistent and marked losses of species with known associations to primary forest following land conversion compared to species with no such known associations, for example, open-area and invasive species (Beukema et al, 2007; Pardini et al, 2009; Figure 15.7). Other work has also shown that it is often those taxa that are of the highest priority for conservation, such as regional forest endemics, which are the most sensitive to forest fragmentation and disturbances at the scale of individual landscapes (e.g. Posa and Sodhi, 2006). However, simply identifying which species comprised part of the original native forest biota for a particular landscape can be challenging, and is dependent on access to suitable control sites or historical species records. It is often impossible to recover this kind of information in poorly studied regions with little forest cover remaining, while the level of forest-specialization exhibited by different species is even harder to ascertain. Primary forest species represent a broad continuum of resource and habitat requirements and life-history strategies, including both highly specialized taxa that are especially vulnerable to forest loss, and disturbance-tolerant species that thrive in more open areas (e.g. edge environments) (Chazdon et al, in press). Understanding the extent to which human-modified systems can support this full range of species requires a more detailed understanding of individual species' habitat requirements and dispersal limitations.

Indices of biotic integrity

Summary metrics that integrate different responses of an ecological system to environmental change can often help disseminate the results of biodiversity analyses to managers and policy-makers (Schiller et al, 2001; Failing and Gregory, 2003). While indices of any kind will inevitably mask some information, if carefully constructed and based on well-grounded ecological principles (rather than mathematical elegance as in the case of many diversity indices) they have the potential to provide more tractable inferences about the effects of management than the complex integration of data from multiple sources (Niemi and McDonald, 2004).

There has been considerable success in the development of indices of biotic integrity based on fish and aquatic invertebrates to assess changes in the integrity of stream systems in the UK, North America and elsewhere (Karr, 1991; Angermeier and Karr, 1994; Barbour et al, 1996; Hughes et al, 1998). In essence, the biotic

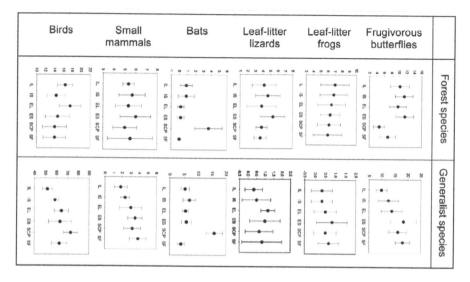

Figure 15.7 Mean and standard deviation of species richness of forest and generalist species of ferns, frugivorous butterflies, leaf-litter frogs and lizards, bats, small mammals and birds along a gradient of forest disturbance in the Una region, South Bahia, Brazil

Key: Interior (IL) and edge (EL) habitat of large remnants; interior (IS) and edge (ES) habitat of small remnants; shade cacao plantations (SCP); second-growth forest (SF)
Source: Redrawn with permission from Pardini et al (2009)

integrity index concept incorporates information on different ecological attributes, including the proportional representation of indicator species that are linked with specific stressors, into a single index to describe the overall condition or integrity of a system. These indices have been so successful in North America that they are now firmly established in the legislation of the US Environment Protection Agency and are employed in almost every state to assess the condition of stream systems (see www.epa.gov/bioiweb1/html/ibi-hist.html).

While corresponding indices of integrity have yet to achieve the same level of acceptance for terrestrial systems, the general attributes of such an index are well recognized and accepted (Noss, 2004). Promising developments in terrestrial indices of biotic integrity (or some named variant thereof) have been made using data on birds (e.g. Canterbury et al, 2000, Glennon and Porter, 2005; Bryce, 2006; Howe et al, 2007), epigeic arthropods (Cardoso et al, 2007), as well as combinations of multiple vertebrate, invertebrate and plant taxa (e.g. Diffendorfer et al, 2007).

The main advantages of the biotic integrity concept are that it provides a more holistic interpretation of overall ecological integrity, and that it may outperform more traditional community level metrics of species richness and community structure when biodiversity responses to anthropogenic influences are complex and variable (Diffendorfer et al, 2007). Nevertheless, the construction of indices of integrity that are technically robust, as well as informative and understandable for non-specialists, is a non-trivial challenge and one remains a priority for future research.

DETECTING CHANGE AND ASSESSING MANAGEMENT PERFORMANCE

The primary objective of effectiveness monitoring is to assess management compliance against minimum performance standards (see Chapter 8, and Figure 8.4). Performance standards are comprised of carefully selected indicators with associated measurable targets or endpoints, and are commonly set by regulatory and certification authorities. They represent an attempt to synthesize what is understood to be best (or at least minimum) practice, and are essential in affording transparency for responsible forestry, as well as providing the starting point in the process of AM and continuous improvement. Although there is ongoing controversy as to the type of indicators that should be used to set performance standards, I argue elsewhere in this book (in particular in Chapters 8 and 11) that indicators of forest structure (and to a lesser extent disturbance processes) provide the most suitable measures, while species-based indicators and targets are best interpreted as *evaluators* of standards rather than standards themselves.

The analytical challenge at the heart of effectiveness monitoring is to judge whether performance indicators are maintained above minimum target levels or within acceptable limits of variability. The way in which science has traditionally approached this challenge is through a hypothetico-deductive framework and NHT. Data on indicators are collected and used to test whether indicator levels are being maintained above (or below in the case of a negative indicator) a predetermined minimum value, and/or whether indicator values exhibit any change compared to a natural baseline (Di Stefano, 2001). In Chapter 6, I argued that this approach is not only fraught with technical difficulty but is also naive to the practical challenge of developing responsible forestry (see Chapter 2; Ghazoul and Hellier, 2000; Prato, 2005; Nichols and Williams, 2006).

Proposed solutions to the problem of measuring compliance against management standards include the notion of setting 'acceptable limits' to changes in indicator values rather than imposing an unrealistic expectation of maintaining a management system within natural limits of variability (see Chapter 11), and using power analyses to ensure that statistical tests are sufficient to detect an important effect where one exists (see Chapter 14; Di Stefano, 2001; Lindenmayer and Burgman, 2005).

While formal power analyses can help to develop more effective sampling designs, I agree with Walshe et al (2007), who suggested that they fall short of a satisfactory approach for assessing compliance against management performance standards. Using *a priori* power analyses, the minimum sampling effort necessary to reduce the risk of a Type II error below some acceptable level is calculated. With this assurance in place, an assessment is then made as to whether compliance against minimum standards has or has not been achieved. However, while clarity in auditing is desirable, it is neither reasonable nor constructive to view compliance in natural resource management as a black and white issue (Walshe et al, 2007). For reasons discussed throughout this book, management standards are inherently subjective and evolving constructs rather than absolute assurances of 'sustainability'. Moreover, even if absolute standards could be set, definitive tests of compliance are not possible due to irreducible sources of bias in the process of data collection, and the arbitrariness of significance thresholds in NHT (see discussion in next section). A further problem lies in communicating the results of monitoring to non-expert stakeholders and interested parties. Here transparency and intuition are key to success and statistical rigour can be a difficult concept to translate.

As a practical alternative to the use of formal power analyses and hypothesis testing to assess management compliance against performance standards, Walshe et al (2007) propose the use of confidence intervals (Box 15.1). Comparable approaches are possible under a Bayesian framework that employs Bayesian-credible intervals and allows the inclusion of prior information on the probability of a certain outcome (see below and McCarthy, 2007).

Box 15.1 Use of confidence intervals in demonstrating performance against management standards

Confidence intervals support a less rigid interpretation of statistical power and variability in indicator values, as well as providing an intuitive visual basis for discussing issues of compliance (Gardner and Altmann, 2000; Burgman, 2005). The meaning of confidence intervals is often the source of much confusion in ecology. In precise language, a 90 per cent confidence interval is the interval which will contain the true value on 90 per cent of occasions if a study were repeated many times – theoretically an infinite number – using samples from the same target population. In practice, however, the interval is interpreted to be the plausible range for the true parameter value (Walshe et al, 2007).

The graphical expression of confidence intervals emphasizes the effect size and uncertainty around the effect size that is associated with sampling and measurement error (Di Stefano, 2001). Confidence intervals only communicate uncertainty associated with sampling error and do not control for more systematic errors in the design and execution of monitoring programmes (see later section in this chapter). The size of confidence intervals depends on four factors: sample size, natural variability in the indicator being sampled, the degree of confidence required (e.g. 90, 95 per cent) and the underlying statistical distribution (Gardner and Altman, 2000). As in the case of traditional NHT, there is no *a priori* justification for using the conventional size of 95 per cent for a confidence interval. Instead the decision should be based on the level of precaution that is desired by stakeholders of the management system (Newton and Oldfield, 2005).

Walshe et al (2007) demonstrate the intuitive nature of confidence intervals through a hypothetical example of a forestry operation that is obliged to limit increases in stream turbidity resulting from soil erosion because of the negative impact on local frog populations. A threshold increase in turbidity above background levels is set as the management standard based on prior research demonstrating harmful levels to frogs. Figure 15.8 illustrates the confidence intervals that are calculated from three different monitoring scenarios, and how their interpretation sheds light on management performance and provides recommendations for future monitoring. In scenario (a) monitoring is inconclusive; forest management could result in either a decrease in turbidity or an increase large enough to cause damage to the frogs. Additional monitoring is required to evaluate compliance, during which time it may be advisable to enforce precautionary mitigation measures. In scenario (b), management increases the turbidity but the magnitude of the increase is too small to affect the frogs substantially. The assessment is conclusive as the entire confidence interval falls beneath the indicator threshold and compliance is awarded. In scenario (c), stream turbidity exceeds threshold levels and managers are required to change their practices or implement a mitigation strategy to achieve standard compliance. However, arguments could be made that monitoring is inconclusive (lower confidence interval falls beneath the threshold) and the auditor may decide that further monitoring is required.

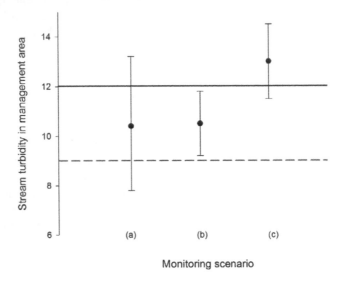

Figure 15.8 Confidence intervals for changes in stream turbidity following forest management activity under three monitoring scenarios

Note: The dashed horizontal line refers to background turbidity levels (i.e. no effect), while the solid black line is the threshold indicator level imposed by the management standard
Source: Theoretical example adapted from Walshe et al (2007)

In cases where it is not possible to set a clear indicator threshold (e.g. due to uncertainty in threshold levels and or variability among stakeholders about what constitutes acceptable change), additional flexibility can be accommodated by setting a lower and upper bound on performance indicator levels (Walshe et al, 2007).

Temporal trend analyses and surveillance monitoring

An additional area of data analysis that is concerned with detecting change is the analysis of temporal trends in indicator values or target species at a single site. Trend analyses are explicitly concerned with finding out whether some attribute of interest is stable, declining or increasing over time, and generally fall under the banner of what is called surveillance monitoring (see Chapters 4 and 5). Although surveillance monitoring can provide an important function (e.g. to provide a warning signal of unexpected declines in an endangered species and raise awareness of the state of the environment at regional and national levels, see Chapter 4) it is purely observational and not linked to the assessment or evaluation of the management system in any specific way (unless trends are compared between different management regimes). In addition, the detection of significant trends over time can be very challenging (Chapter 6). Readers interested in this literature are recommended to look at Dixon et al (1998), Urquhart et al (1998) and Schmidt and Meyer (2008). The excellent book on analysing ecological data by Zuur et al (2007) has a comprehensive section on time series analyses in general and illustrates how standard statistical methods can be adapted to deal with time series data.

EVALUATING CHANGE AND VALIDATING MANAGEMENT PERFORMANCE

The purpose of validation monitoring is to relate observed changes in biodiversity to management in order to evaluate the adequacy of existing performance standards, and identify where cost-effective improvements can be made. In essence, it represents the research effort needed to identify and understand the empirical and theoretical relationships that link human impacts, through changes in measurable and manageable elements of forest structure and function, to outcome measures of biodiversity and ecological integrity (Possingham and Nicholson, 2007; and see Chapter 8). Statistics and statistical modelling represent an essential element of this process – the glue that connects the different components. In providing a means of formalizing our conceptual understanding of a human-modified ecosystem – which factors we consider to be important in driving observed changes in biodiversity, their relative importance and the ways in which they may interact (Chapter 13) – statistical models allow us to identify the extent to which our predictions about how humans influence nature are supported by sample data. Armed with such evidence, we can then make defensible recommendations concerning if and how management may need to change in order to mitigate biodiversity loss and encourage progress towards long-term conservation goals, while also identifying priorities for future research.

The application of inferential statistics to ecology is an enormous field of research in itself. Fortunately, however, there are some truly excellent textbooks available to guide the judicious selection of techniques and ways to extract the most information from hard-won sample data (while also avoiding misuse). An annotated list of particularly recommended titles is given in Table 15.2. A number of these books recommend using the free R statistical language to perform analyses (Box 15.2).

A detailed review of statistics and their application to the process of validation monitoring and research in managed forests is far beyond the remit of this book. However, the following sections provide a brief overview of some key principles and concepts, the distinction between different approaches and their relative benefits

Table 15.2 Some of the most popular textbooks on the application of statistics and statistical modelling in ecology

Book	Description
Statistics: An Introduction Using R. (Crawley, 2005)	A very readable introduction to both statistics and the R statistical language
The R Book. (Crawley, 2007)	A thoroughly comprehensive and accessible reference manual for both a huge range of statistical techniques and the R statistical language
Modern Statistics for the Life Sciences, Grafen & Hails. (2002)	A general statistics textbook with a novel and insightful approach to explaining key concepts and model formulae based on the General Linear Model
Analysing Ecological Data. (Zuur et al, 2007)	Comprehensive coverage of a wide range of statistical techniques written specifically from the perspective of an ecologist and employing many thought-provoking and clearly presented examples using real data.
Ecological Models and Data in R. (Bolker, 2008)	An accessible introduction to likelihood and Bayesian statistical methods in ecology. Focused on understanding how to combine models with data to answer ecological questions. Careful worked examples provided on how to implement different techniques using the R statistical language
Model Selection and Multi-Model Inference, (Burnham and Anderson, 2002)	A comprehensive reference on the use of likelihood-based information-theoretic approaches to achieving statistical inference from multiple models. Accessible while also providing access to underlying theory for interested reader
Bayesian Methods for Ecology, (McMcarthy, 2007)	Accessible introduction to Bayesian statistics, including a useful and practical comparison of the major differences between Bayesian, likelihood and NHT statistical frameworks. Also provides a guide to WinBUGS, the popular software program for Bayesian analyses.
The Ecological Detective: Confronting Models with Data, (Hilborn and Mangel, 1997)	A now classic introduction to the idea of combining models with real-world data and taking analyses beyond Popperian hypothetico-deductive methods of falsification. An accessible account of ecological models, maximum likelihood estimation and Bayesian statistics
Numerical Ecology. (Legendre and Legendre, 1998)	A classic reference book for many multivariate analyses in ecology
Structural Equation Modelling and Natural Systems (Grace, 2006)	A good introduction to structural equation modelling of ecological data

for the analysis of biodiversity monitoring data. I first give an overview of the three major approaches for statistical inference (i) classical frequentist (i.e. NHT), (ii) likelihood-based information-theoretic and (iii) Bayesian methods. I then summarize some of the commonly cited criticisms and limitations of NHT (the most popular method), and highlight why likelihood and Bayesian methods may often be more appropriate for the analysis of biodiversity monitoring data (while also cautioning that there are no silver-bullet solutions). I then briefly introduce the use of statistics to develop spatially explicit models of biodiversity in human-modified landscapes. I end with a note of caution: although statistics provide an essential and powerful tool for biodiversity conservation research, we should remember that they are just one piece of the puzzle and their usefulness can be readily compromised by limitations in the quality of sample data or our understanding of the ecology of the study system.

Box 15.2 The open source statistical language and environment of R

R is technically termed a 'language and environment for statistical computing and graphics'. This means that it is simultaneously a software package and a programming language, while also being capable of producing highly professional graphics. To an increasing number of people around the world, R has revolutionized the teaching and application of cutting-edge statistics in ecology. While it requires an initial period of hard work to learn (especially for those not used to a command-line, rather than a Windows-based interface) it is very much worth the effort and is strongly recommended to any readers interested in analysing their own data. Bolker (2007) summarized why it is so popular:

- It is freely accessible to download from the internet: www.r-project.org/.
- It is open source – the code behind all the statistical methods provided can be easily viewed and changed.
- It is written, and constantly added to (in the form of method-specific packages), by some of the world's leading statisticians. The statistical routines are therefore generally more robust and up-to-date than many commercial packages. An increasing number of packages have been developed specifically for the analysis of ecological and biodiversity data (e.g. Vegan: www.cc.oulu.fi/~jarioksa/softhelp/vegan.html).
- It can run easily on any operating environment.
- There are a number of excellent textbooks, including those focused on ecology, that use R as their main method of teaching statistics (see Table 15.2).

A (very) brief primer on the major approaches in statistical inference

When analysing ecological data, it is important to be aware that there are three quite different frameworks or approaches for making statistical inferences: (i) traditional or Neyman–Pearson NHT, (ii) likelihood-based information-theoretic methods, and (iii) Bayesian statistics. In addition to their technical differences, these different approaches have different philosophical underpinnings that often generate considerable controversy as to which is more appropriate.

NHT

The most common application of statistics in the analysis of biodiversity data is null-hypothesis testing (NHT; also called Neyman–Pearson NHT). NHT is essentially concerned with testing whether a given result is unusual. The basic steps for NHT are (McCarthy, 2007):

1 A null hypothesis is defined, along with a single alternative hypothesis.
2 Data are collected.
3 The probability of collecting the data or more extreme data given that the null hypothesis is true is calculated. This is the infamous *P* value.
4 If this probability is sufficiently small, a conclusion is reached that the data are unusual given the null hypothesis. The traditional (and entirely arbitrary) cut-off is if *P* is less than 0.05.
5 If the probability is not below the critical threshold, the data fail to reject the null hypothesis. The null hypothesis is then accepted, although cannot be proven as NHT only falsifies hypotheses.

NHT is commonly referred to as classical frequentist statistics – frequentist because it is based on the probability of a particular outcome, defined as the average frequency of that outcome if the study was repeated a very large number of times. NHT is interpreted in terms of Type I (the probability of falsely rejecting the null hypothesis when no effect exists – determined by the *P*-value threshold) and Type II (the probability of falsely accepting the null hypothesis when an effect does exist – the inverse of the power of the test) errors. Determining the risk of a Type II error requires *a priori* power analyses (see Chapter 14). Guidance for NHT can be found in any standard statistics textbook.

Likelihood-based information theoretic methods

Information-theoretic (IT) methods also fall under same broad paradigm of frequentist statistics and use what is called maximum-likelihood estimation to find the set of parameters that make the observed data most likely to have occurred. The basic steps for IT methods are (McCarthy, 2007):

1 A set of candidate models are selected that represent different hypotheses for explaining variability in the variables of interest (e.g. some measured element of biodiversity).
2 Data are collected.
3 Data are used to assess relative support for the different models, by estimating the amount of information about the sample data that is lost when using each model. The best model within the initial set is then selected as the one that is estimated to lose the least amount of information.
4 If required, a predictive model can be developed based on model parameters averaged across a suite of the most likely models.

The model selection process – assessment of relative likelihoods – is commonly based on Aikaike's information criterion (AIC, or AIC_c to correct for small sample sizes that are common in ecology), which is a function of the likelihood of the model and the number of independent variables used. AIC values are commonly transformed into AIC weights (scaled to sum to 1), which give a measure of the *relative strength of evidence* in support of each candidate model. Pairs of candidate models can be compared using evidence ratios of AIC weights. Key to IT methods are the notions that all models are false because they represent incomplete approximations of the truth (Elliott and Brook, 2007), and that attempts to identify a single 'best' model (e.g. as in multiple regression based on NHT) are futile and unjustified as real data invariably give support to more than one model. Burnham and Anderson (2002) provide a clear and thorough breakdown of the background, justification and application of IT methods in ecology (see also Johnson and Omland, 2004). Most IT methods are best implemented in R (Box 15.2).

Bayesian methods

Null hypothesis and information-theoretic approaches (both types of frequentist statistics) both assume that there is a 'true' state of the world, and attempt to answer the same type of underlying question – *what is the probability of observing the data given that the assumed model/hypothesis is true?* Bayesian statistics take a different view and ask *what is the probability the model/hypothesis is true given the sample data?* They are the only form of statistics that can calculate the probability of a given hypothesis being true – and this is their most defining feature (Hobbs and Hilborn, 2006; McCarthy, 2007). The other unique feature of Bayesian methods is that relevant prior information can be incorporated into analyses to improve precision of estimates. The basic steps for Bayesian analyses are (McCarthy, 2007):

1   A set of candidate models are selected that represent different hypotheses for explaining variability in the factors of interest (same procedure as for IT methods).
2   Prior probabilities are assigned to these different models.
3   Data are collected.
4   Bayes' rule is used to combine the prior probabilities with the information contained in the sample data to generate predictions.

The Bayesian framework solves many of the conceptual problems of frequentist statistics: answers depend on what we actually saw and not on a range of hypothetical outcomes, and it is possible to make legitimate statements about the probability of different hypotheses or parameter values (Bolker, 2008). Because it is possible to include prior information, Bayesian methods are particularly suited to address problems where our expectation of a particular outcome changes with the accumulation of evidence – for example, the prediction of whether a species is absent from a site when it has not been sampled (McCarthy, 2007). That said, Bayesian methods have been the source of much criticism because biased prior probabilities can be used to drive particular outcomes. However, it is of course possible to cheat with any statistical framework (Bolker, 2008).

Why information theoretic and Bayesian methods are invariably more appropriate for analysing biodiversity monitoring data

NHT has received an increasing amount of criticism in ecology (for reviews, see Hilborn and Mangel, 1997; Burnham and Anderson, 2002; Johnson and Omland, 2004; Prato, 2005; Hobbs and Hilborn, 2006). The two principal shortcomings are:

1   NHT is commonly used to test a single hypothesis of no change or no effect that is known to be false from the outset. Such null hypotheses are often known as silly nulls or straw man hypotheses (Burnham and Anderson, 2002; McCarthy, 2007).
2   P values do not provide direct statements about the reliability of a hypothesis because they conflate effect size and sample size (larger sample sizes always result in smaller P values). Moreover, P values are usually interpreted against a completely arbitrary threshold significance level of 5 per cent.

Defendants of NHT point out that the fault lies more in the abuse of the method rather than the method itself, arguing that ecologically meaningful null hypotheses should be used and inferences should not depend solely on P values but also be supported by power analyses and confidence estimates around effect sizes. Other statistical frameworks can be similarly abused (IT methods can still employ silly hypotheses, AIC is open to inappropriate interpretation and Bayesian analyses can be misleading if inappropriate priors are used). However, NHT methods have an underlying limitation in that they are explicitly designed for manipulative experimental studies. As discussed in Chapter 10, replicated and randomized experiments that typically deliver 'yes or no' type answers regarding a particular hypothesized effect are difficult to conduct in the real world. Instead, ecological inferences are usually derived from a careful balance of accumulated evidence regarding multiple working hypotheses, and rarely succeed in identifying a single 'best' model for explaining nature (Box 8.2; Hobbs and Hilborn, 2006; and see Chapters 13 and 14 for discussion on the value of multiple working hypotheses in ecological research). This distinction means that the interpretation of ecological monitoring data is often best served through approaches that explicitly accommodate uncertainty and the evaluation of a range of alternatives. IT and Bayesian methods are both naturally suited to this task because while they can also easily be applied to experimental data, they have the advantage of weighing up differences in the strength of evidence among alternative explanations rather than attempting to find a single answer to a narrowly constrained problem (Walters and Holling, 1990; Hilborn and Mangel, 1997; Walters and Green, 1997; Hobbs and Hilborn, 2006). These advantages allow for a more balanced and transparent assessment of the costs and benefits of different conservation strategies and underpin repeated calls for their application to natural resource management problems (Ghazoul and McAllister, 2003; Prato, 2005; Hobbs and Hilborn, 2006; Figure 15.9).

Despite these advantages, we should be careful in viewing IT and Bayesian approaches – what Hobbs and Hilborn, 2006 term 'new statistics' for ecology – as a panacea. Part of the real strength of these 'new statistics' is that they strongly

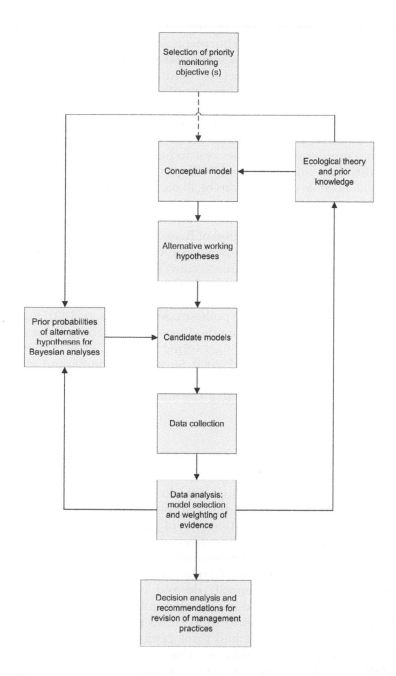

Figure 15.9 Data collection and analysis as part of an iterative monitoring cycle driven by the formulation and evaluation of alternative working hypotheses and associated models

Note: The strength of support for alternative hypotheses (and any associated management recommendations) is tested following repeated confrontations with empirical data during the monitoring process. Prior knowledge can enter the process both informally in developing the conceptual model as a basis for identifying candidate hypotheses, and formally by updating prior probabilities in Bayesian analyses. See also Figure 8.5 and Chapter 8 for a wider summary of the process of validation monitoring and its role in the management system encourage careful prior thinking in order to specify ecologically meaningful hypotheses and associated models (Burnham and Anderson, 2002). However, this same strength can also be a weakness as the *a priori* identification of plausible models can often be very difficult to achieve (Eberhardt, 2003; Taper and Lele, 2004) – this is especially so in tropical forests, where we often lack the most basic information about a species habitat preferences, natural history and life cycle. Evaluating the strength of evidence in support of competing models is only a useful exercise if some of the models represent meaningful candidates.

Although considered by some researchers to represent an 'unthinking approach' (Burnham and Anderson, 2002), full subsets regression (i.e. selecting models from all possible combinations of all parameters) coupled with model averaging offers some protection against ignorance and potential bias when selecting candidate models for analysis (provided sufficient data are available). Thomson et al (2007) present a detailed application of Bayesian model averaging to predict bird species distributions in restored landscapes in Australia, and show how it can be an effective way of dealing with model uncertainty when confronted with a large set of candidate variables. Stephens et al (2005, 2007) also suggest that careful exploratory data analysis and graphical analyses can play a vital role in establishing the patterns necessary to generate meaningful hypotheses (see also Peters, 1991; Stauffer, 2002). For example, variance partitioning methods (such as hierarchical partitioning) can make a useful contribution towards uncovering the complexity of species–environment relationships (e.g. Oliver et al, 2000; Smyth et al, 2002), and identifying the relative importance of different sets of candidate explanatory variables – for example, spatial and environmental (e.g. Cushman and McGarigal, 2004; Heikkinen et al, 2005).

Predicting biodiversity across human-modified forest landscapes

In addition to their role in evaluating individual elements of forest management, biodiversity data can also help to address whole-landscape or regional-scale planning problems (including the identification of priority areas for conservation and restoration activity) through spatially explicit species and habitat models.

In their broadest sense, species distribution and habitat models are empirical models that relate biodiversity sample data to environmental predictor variables based on statistically or theoretically derived response surfaces (Guisan and Zimmermann, 2000). A large and increasing number of prediction and modelling techniques have been developed to address this problem depending on specific objectives, data availability, scale and prior ecological knowledge of the species involved. At the simplest level, this includes models based on habitat suitability indices (HSI) that scale key environmental variables (e.g. forest type, edge effects) to expected differences in habitat value for target species or species groups. Habitat suitability modelling (often based on HIS) has a long history in forest conservation and management, especially in North America (e.g. O'Connell et al, 2000; Larson et al, 2004; Edenius and Mikusinski, 2006; McComb et al, 2007). The last two decades have seen a rapid rise in the use of statistical modelling methods as well as other options, such as neural networks. Comprehensive reviews of the main methods and their underlying principles and assumptions are given by Guisan and Zimmerman (2000), Guisan and Thuiller (2005), Burgman et al (2005) and Elith et al (2006).

Predictive models of biodiversity are only useful for management if they can be reliably parameterized using easy-to-measure forest structure variables (i.e. structural indicators that are routinely collected during forest inventories and auditing), available land-cover maps (e.g. soil and topographic maps) and/or remote sensing data on landscape structure and composition. Larger-scale species mapping exercises often also use climate data (e.g. Thomson et al, 2007). Where possible, model specifications can be improved by the inclusion of taxon-specific life history information. The majority of model developments and applications to date has been focused on single-species models. However, recent advances have seen an increase in applications of spatial modelling of biodiversity at the community level, including patterns of species richness, community types and patterns of dissimilarity in species composition (Ferrier and Guisan, 2006; Table 15.3). Such techniques are vital to the kind of whole-system biodiversity monitoring advocated in this book, and offer a greatly enhanced capacity to synthesize complex data into a form more readily interpretable by scientists and decision-makers. One of the most promising developments is generalized dissimilarity modelling (GDM) (Ferrier, 2002; Ferrier et al, 2007), a form of non-linear matrix regression that models compositional dissimilarity observed between pairs of surveyed locations as a continuous function of multiple environmental gradients. Spatial mapping of community data has thus far seen little application to conservation planning and management problems but the approach holds considerable promise (Ferrier and Gusian, 2006; Ewers et al, 2009).

To make biodiversity mapping more attractive for forest management and planning, an important next step is to combine species-based models with scenario models of landscape change (Edenius and Mikusinski, 2006). McComb et al (2007) provide an interesting example of this approach, linking biodiversity models to a 'landscape management policy simulator' to evaluate potential future patterns of habitat availability for three focal species in Oregon (Figure 15.10). This kind of integrated modelling continues to be an active area of research.

Finally, it is vital that biodiversity models are validated using independent data (i.e. not those used in the parameterization process) to ensure that inferred relationships are robust, and predictions reliable. To date, relatively few modelling studies have conducted independent validations of model predictions, although those that have demonstrate that models based on data from a limited area often perform poorly when transferred to other regions (e.g. Rhodes et al, 2008).

Going beyond the numbers: Dealing with uncertainty in biodiversity data

Efforts to understand the relationship between human activities and biodiversity are fraught with uncertainty. Understanding where this uncertainty originates from, and the ways in which it can be reduced is necessary to understanding the reliability that can be attached to any management recommendations, as well as ways in which the monitoring process can be further improved.

Table 15.3 The six main types of spatial output that can be generated using community-level modelling

Spatial output	Description	Structure of derived grid layer(s)
Individual species	Predicted distributions of multiple species, as for species-level modelling	A separate layer for each species, indicating the predicted probability of occurrence or abundance of that species in each cell
Community types	Each 'community type' defined as a group of locations (grid cells) that closely resemble one another in terms of predicted species composition. Grouping normally achieved through some form of numerical classification	Either (i) a single layer with each cell assigned exclusively to one community type (depicted as a map with different colours indicating different types or (ii) a separate layer for each community type, indicating the probability of that type occurring in each cell (depicted as multiple grey-scale or colour-ramp maps)
Species groups	Each 'species group' defined as a subset of species with similar predicted distributions. Grouping again achieved through numerical classification, but in this case the objects classified are species rather than locations	A separate layer for each species group, indicating the predicted prevalence or abundance of that group in each cell (depicted as multiple grey-scale or colour-ramp maps)
Axes of compositional variation	A set of continuous axes (or gradients) representing dimensions of a reduced space that summarizes the compositional pattern exhibited by multiple species. These axes are most commonly derived through some form of ordination	A separate layer for each axis, indicating the predicted score for that axis in each cell (depicted either as multiple grey-scale or colour-ramp maps, or as a single map by assigning each of the first three axes to a different colour dimension, e.g. red, blue, green)
Levels of compositional dissimilarity between pairs of cells	The predicted level of dissimilarity in community composition between all possible pairs of grid cells in a region	In theory, a complete matrix of pair-wise dissimilarities, but, in practice, these values are usually predicted dynamically as required by the application of interest (difficult to depict spatially without prior conversion to community types or axes of compositional variation)
Macro-ecological properties	Most commonly modelled property is species richness, either of a whole group (e.g. all vascular plants) or of a functional subgroup (e.g. annuals and trees). Many other macro-ecological properties (e.g. mean range size and endemism) can potentially be modelled	A separate layer for each macro-ecological property (depicted as a grey-scale or colour-ramp map)

Source: From Ferrier and Guisan (2006)

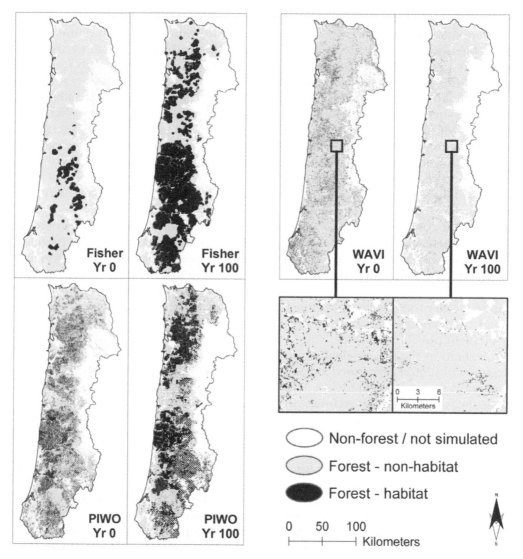

Figure 15.10 Patterns of habitat change over a 100-year simulation in the Oregon Coast Range under current land management policies on federal, state and private forest lands for the Pacific Fisher (*Martes pennanti*, Fisher), warbling vireo (*Vireo gilvus*, WAVI) and pileated woodpecker (*Dryocopus pileatus*, PIWO)

Source: Redrawn with permission from McComb et al (2007)

As emphasized at the start of this chapter, it is important to recognize that the interpretation of biodiversity data is an entirely context-dependent process. Forests are known to be complex ecosystems, and the ways in which human activities influence the composition, structure and function of forest biodiversity even more so. The process of studying nature is essentially comprised of a series of steps that attempt to simplify and abstract this complexity in meaningful ways. In other words, the depth of our understanding is determined not only by the complexity of the system itself, but also the inevitably restricted and biased lens through which we view the world (Peters, 1991). This point may seem blatantly obvious but we rarely stop what we are doing to give it sufficient attention.

The definition of a monitoring programme starts during the initial scoping stage in setting specific goals and objectives, before passing through a series of choices relating to indicator selection, sampling design and data analysis. At each stage the choices we make are a product not only of our prior knowledge about the ecology of the system but also limitations imposed by resource constraints and access to technical expertise, as well as a variety of cultural traditions and norms that dictate how science should be conducted. All of these factors contribute uncertainty to the monitoring process and undermine our ability to draw reliable and ecologically meaningful conclusions from sample data. How best to deal with this problem? There are no easy solutions; uncertainty will always pervade attempts to ascertain the truth about the environment. Nevertheless, a lot of progress can be made by giving more explicit recognition to the multi-faceted nature of uncertainty (Regan et al, 2002; Burgman et al, 2005). By revealing that uncertainty enters the monitoring and research process in various guises, many of which are cryptic and poorly understood, it becomes easier to identify the areas of greatest concern as well as implement possible solutions.

To help in the task of dealing with uncertainty in a more systematic manner, Regan et al (2002) developed a typology of the different sources of uncertainty in ecological research (Table 15.4). While these categories should not be thought of as clear-cut and mutually exclusive, they do provide a very useful starting point for thinking about the main issues, including many that are frequently ignored in science. In particular, Regan et al (2002) highlight the distinction between epistemic uncertainty (uncertainty in determinate facts) and linguistic uncertainty (uncertainty in the language we use to specify the way in which we work), both of which can be very important.

Some sources of uncertainty are more easily dealt with than others. For example, an understanding of measurement error can be improved by estimating confidence intervals around parameter values. Context dependence and ambiguity can be reduced by using more careful language, and little-used techniques like fuzzy logic can help address borderline cases that come from vagueness (Regan et al, 2002). Often, however, uncertainty is difficult to deal with because it is not recognized as being a problem to start with. For example, a common criticism of much of landscape ecology to date is that researchers frequently approach problems with a predetermined (anthropogenic) view of what constitutes viable species habitat, and what processes are responsible for driving observed patterns (Manning et al, 2004; Fischer and Lindenmayer, 2006; and see Chapter 13). This problem has led to a strongly patch-centric view of human-modified landscapes (a legacy of island biogeography theory) and a focus of data collecting efforts in forest remnants at the expense of sampling in the wider landscape matrix (which in many cases may contain important species habitat (Gardner et al, 2009)). Another poorly appreciated source of epistemic uncertainty is the dominant influence that the choice of modelling framework

Table 15.4 A typology of epistemic and linguistic uncertainty in ecological and conservation research

	Type of uncertainty	Description
Epistemic uncertainty	Measurement error	Results from imperfections in measuring equipment and observational techniques including observer and instrument error. This is one of the easiest forms of uncertainty to deal with
	Systematic error	Results from bias in the design and/or implementation of a sampling and/or data analysis strategy. Systematic error can be introduced both deliberately (exclusion of certain sampling locations, species groups, outlier data points, etc), and sub-consciously through the use of recording equipment or sampling methods that have unknown biases and assumptions. This kind of uncertainty is commonly found in studies of biodiversity in human-modified landscapes when researchers do not have access to undisturbed control sites or if sampling is only conducted in certain portions of the landscape (e.g. forest remnants). Many sources of systematic error are legacies of flawed ecological theories that dictate much about the way in which we work
	Natural variation	Results from natural heterogeneity (spatial and temporal) in variables of interest such that sample measurements will always provide a varyingly biased estimate of the truth. This kind of uncertainty is what confounds efforts to identify what could be called a 'significant; population decline in the analysis of trend data
	Inherent randomness	Results from the fact that many ecological and human processes are so complex that (for all practical purposes) they are unpredictable and are therefore very difficult, if not impossible, to incorporate into models
	Model uncertainty	Results from the simplified and biased models (both conceptual and mathematical) that we use to represent our study system. Model uncertainty arises in two main ways: through the study of only a subset of a much larger pool of variables and processes; and through the use of simplified constructs that are used to characterise observed patterns in data (e.g. choice of statistical methods and functional form of equations)
	Subjective judgement	Results from subjective interpretation of sample data. No matter how much care is taken in the collection and analysis of biodiversity data, the interpretation of any findings is very rarely unequivocal and instead relies upon the subjective judgement of responsible individuals – which in turn depends on differences in personal experience and knowledge of the study system. Bayesian statistics offer a formal way of incorporating subjective estimates of model predictions into the modelling process
Linguistic uncertainty	Vagueness	Results from the fact that unclear language and terminology can allow for borderline cases. A classic example of this is the use of the term 'endangered' to describe species that suffer a particular level of threat. The large number of definitions used to define what constitutes an endangered species means that it can be difficult to identify management priorities
	Context dependence	Results from a failure to specify the context within which the finding of a study should be interpreted and understood. This is a common problem in landscape ecology where it is becoming increasingly recognised that landscape context and regional variability in patterns of deforestation can have an over-riding effect on local biodiversity dynamics
	Ambiguity	Results from the fact that the language of science is often so unclear that meaning can be lost and the same terms are used to describe very different phenomena, resulting in needless confusion and contradictory outcomes. This is a serious problem in ecology where generic terms such as 'habitat fragmentation' and 'disturbance' are commonly employed to describe a variety of specific threatening processes
	Indeterminacy of theoretical terms	Results from the fact that the future usage of theoretical terms is not completely fixed by past usage – i.e. theory is continuously evolving. A specific example of this is the species concept where the name of a particular taxon may take on new meaning following revision of the species group. More generally, terms such as 'biodiversity' reflect this kind of linguistic uncertainty because their meaning is constantly evolving in the face of new ideas and research
	Underspecificity	Results from unwanted generality and a failure to provide sufficient details to specify the meaning of a statement about research findings

Source: Based on Regan et al (2002)

can have on our ability to draw reliable inferences from sample data (see above and Burgman et al, 2005).

Perhaps the most effective way of reducing uncertainty is to ensure that careful reflection of the implications of individual design choices remains a constant and pervasive component of the entire monitoring and research process (Feinsinger, 2001). Above all, the design and implementation of a monitoring programme needs to be strongly grounded in what we already know and understand about the ecology of species and ecosystems concerned. No amount of statistical wizardry can compensate for a poor study design, inadequate or inappropriate measurements, or poor judgement regarding the relationship between any findings and existing theory (Manly et al, 2002; Stauffer, 2002). The development of a conceptual model can be helpful in guiding the design process, as well as helping to interpret findings in the context of current knowledge (see Chapter 13). We should also be wary of always adopting the same approach to data collection and analysis, especially where we have a poor understanding of how alternative approaches may influence results. Where possible, multiple, carefully selected techniques should be used and outcomes compared, with defensible reasons given for the final preferred technique. The importance of this kind of bet-hedging strategy lies behind arguments for selecting multiple indicator taxa for monitoring, sampling across multiple spatial and temporal scales and not being dogmatic about the use of a single preferred approach to analysing data (see Chapters 11–14). Finally, it is worth remembering that decision-making under uncertainty is where science can make the most substantial contribution to management. Therefore uncertainty is no excuse for inactivity.

ANALYSING BIODIVERSITY DATA IN CONTEXT: THE IMPORTANCE OF MULTIPLE MANAGEMENT OBJECTIVES AND TRADE-OFFS

Deciding what represents an acceptable balance of environmental, economic and social values for responsible forest use is the task of the wider stakeholder community (McDonald and Lane, 2004). However, for biodiversity researchers to make an effective contribution to this debate, it is essential that the monitoring process gives explicit recognition to this wider management context, and actively seeks to identify ways to integrate competing objectives.

Biodiversity conservation in human-modified forest ecosystems will not be successful unless it first recognizes that although some synergies may be possible (Rosenzweig, 2003), fundamental societal trade-offs between competing land uses are often unavoidable (DeFries et al, 2004; Maness, 2007). These trade-offs exist because more intensive forms of land use invariably generate greater profits (ignoring here any market value placed on ecosystem services) while at the same time incurring greater losses to biodiversity. Once trade-offs are recognized, careful planning can often identify significant opportunities for achieving multiple objectives, especially when evaluating different forest management and restoration projects (e.g. Lamb et al, 2005; Figure 15.5).

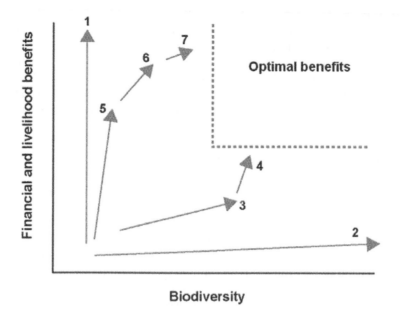

Figure 15.11 Balancing trade-offs between human livelihoods and biodiversity conservation in reforestation projects

Note: Arrows represent alternative reforestation methods. Traditional monoculture plantations of exotic species (arrow 1) mostly generate just financial benefits, whereas restoration using methods that maximize diversity and enhance biodiversity (arrow 2) yields few direct financial benefits to landowners, at least in the short term. Protecting forest regrowth (arrow 3) generates improvements in both biodiversity and livelihoods, although the magnitude of the benefits depends on the population density of commercially or socially important species; these can be increased by enrichment of secondary forest with commercially attractive species (arrow 4). Restoration in landscapes where poverty is common necessitates attempting both objectives simultaneously. But, in many situations, it may be necessary to give initial priority to forms of reforestation that improve financial benefits, such as woodlots and agroforestry systems (arrow 5). In subsequent rotations, this balance might change over time (moving to arrow 6 and later to arrow 7 by using a greater variety of species). There may be greater scope for achieving multiple objectives by using several of these options at different locations across the landscape
Source: Reproduced with permission from Lamb et al (2005)

Dealing with trade-offs and competing objectives is a normal part of any decision-making process. However, the majority of decision-making processes take place on an ad-hoc basis, involving a limited number of alternatives and a subjective appraisal of their relative benefits by a subset of stakeholders (Lindblom, 1959). Forest policy and management offers no exception to this general situation, yet if we are to guide progress towards more sustainable management systems for the world's remaining areas of forest we need to adopt a more coherent analytical framework for balancing conservation, social and economic values. A large number of formal scientific methods that can help develop such a framework are already well established in other disciplines outside ecology and conservation science, including economics, applied mathematics and engineering. These methods are founded in decision theory and help guide more cost-effective decision-making by clearly stating the objectives, constraints and underlying assumptions and uncertainties that pertain to a given problem (Possingham et al, 2001; Possingham and Nicholson, 2007). Decision theory approaches have seen increasing success in tackling conservation planning problems around the world (Wilson et al, 2009), and they offer considerable promise in helping to resolve competing objectives of economic production and biodiversity conservation in managed forest landscapes (e.g. Rempel and Kaufmann, 2003; Polasky et al, 2008).

Of course, science does not make policy and management decisions, it only informs them. To be most effective, the process of analysing and modelling conservation–development trade-offs needs to be embedded within a broader (and largely qualitative) analysis of the social-ecological context of the management system (Fischer et al, 2009), and the values of local stakeholders who will ultimately decide whether to accept or reject any conservation recommendations from monitoring (Knight et al., 2006). Techniques such as multi-criteria decision-making (Kiker et al, 2005; Mendoza and Martins, 2006) and conflict resolution management (Niemalä et al, 2005) can help in providing a constructive environment for participatory decision-making and transparent and easily understood methods to evaluate priorities and develop lasting solutions (see also discussion in Chapter 16).

Putting Forest Biodiversity Monitoring to Work

Synopsis

- Many biodiversity monitoring programmes either fail or fall short of their original intentions because insufficient attention is given to the practical factors that determine viability in the real world.
- The outputs of biodiversity monitoring will only be as good as the people who are striving for them. Deciding on the appropriate blend of people to be responsible for designing and running a biodiversity monitoring programme, depends on both the desired level of detail as well as who the data are intended to benefit.
- In many situations, an integrated approach to monitoring that combines expert guidance and management from professional scientists with a close involvement of local people is likely to provide the most attractive solution. The contribution of professionals ensures scientific rigour in programme design and data analysis, while the involvement of local people facilitates the process of implementing any management recommendations, while also providing a cost-effective and sustainable means of data collection and a potentially rich source of local knowledge to aid interpretation of results.
- Many monitoring programmes are highly wasteful of resources, and significant improvements in return on investment are possible through the judicious selection of cost-effective objectives, indicators, and sampling methods, as well as the involvement of local people in data collecting and processing.
- The viability of monitoring can be further enhanced by increasing the relevance and utility of monitoring products to as wide an audience as possible, including forest management authorities responsible for standard development, government agencies responsible for national biodiversity assessments, the scientific community and environmental educators.
- Instead of being developed in isolation, monitoring programmes should be conceived and nurtured within an interactive learning environment that brings together researchers, managers and regulators from different places and institutions to share data, expertise and ideas.
- Without a clear recognition of the broader societal context within which the monitoring process is situated and the underlying conservation values that define the ultimate purpose of monitoring, even the most technically robust monitoring programmes will be committed to failure.

A SUMMARY SO FAR

The three main sections of this book have: discussed the importance and origins of monitoring forest biodiversity (Chapters 1–4); identified some of the key challenges that have confounded efforts to implement monitoring to date (Chapters 5–7); and outlined an operational framework for how future programmes can be developed to overcome some of these difficulties, and make a genuine contribution towards progress in biodiversity conservation (Chapters 8–15). It is, hopefully, clear from reading these chapters that biodiversity monitoring has an enormous contribution to make towards practical conservation. However, it should also be evident that the success of any such contribution is far from assured. Despite many popular impressions, monitoring is about far more than collecting data. Without careful consideration of the underlying goals and objectives, how these objectives guide the development of an effective and meaningful strategy for data collection and analysis, and the ways in which the entire monitoring process relates to management, monitoring is likely to result in little more than a waste of precious resources and time.

While differences in environmental and social context, including local conservation priorities, mean that it is not possible to create a universal how-to manual for biodiversity monitoring, a lot can be learnt from reflecting on past experiences. In stepping back from the technical details that often confound the planning process, and thinking about broad questions of purpose, design and implementation, a number of general guidelines emerge (see Chapters 8–15). Many of these principles relate to three overarching themes that characterize all biodiversity monitoring programmes:

1 **Interconnectedness of elements**. No single element of a monitoring programme can or should be considered in isolation from others to which it is naturally related. This applies to the entire monitoring process itself as an integral part of the wider management system, as well as the relationship between different approaches and indicators for monitoring, and the interdependencies between the processes of collecting, analysing and interpreting sample data. A key lesson in thinking about the interconnectedness of monitoring is the need to be clear about the *purpose* of individual elements, and ways in which these elements contribute towards the wider programme. This especially concerns the choice of indicators, and the need to be clear about the extent to which species-based, structural and management indicators serve complementary functions.

2 **Uncertainty**. Understanding how humans impact biodiversity and the ecological integrity of forest ecosystems is an extremely complex problem, and one that is characterized by considerable uncertainty. The pervasive nature of this uncertainty requires it to be given explicit recognition as an inherent component of any management system, rather than being denied, ignored or trivialized. This observation lies at the heart of AM. Recognition of uncertainty also exposes the fact that there do not (nor could not) exist any absolute or precise targets for sustainability. Rather, a responsible management approach needs to be based on

minimum standards and a formal commitment to continuous improvement which holds sustainability not as an absolute goal, but as a guiding vision. Viewed this way, monitoring becomes an *integral part of the management system* and not an external surveillance tool, as is so commonly the case. Finally, recognition of uncertainty underlines the fact that we need to be strategic about the choice of monitoring objectives. Only by making alternative choices explicit, is it possible to understand which areas of uncertainty matter the most, and how they can best be reduced through monitoring.

3 *Subjectivity.* The monitoring process involves a series of choices, starting from decisions about overarching conservation goals and moving through specific objectives, indicator selection, sampling design and data collection method-ologies, description of biodiversity, data analysis and interpretation of findings. All of these choices are at least partly subjective and are influenced by the personal experiences, expertise and underlying agendas of those involved in developing the programme. In combination, such factors can greatly constrain the breadth of outcomes that are possible from monitoring, mask important information and lead to significant bias in the results. Subjectivity is best dealt with by working to increase transparency in the monitoring process and, wher-ever possible, making assumptions explicit (see Chapter 13).

Having established the basic elements of the conceptual framework for biodiversity monitoring, the focus of this final chapter turns to address some of the challenges of making biodiversity monitoring work effectively in the real world. In the first section, I deal with the fundamental importance of the *people* that are involved in designing and implementing the monitoring process on the ground, and consider the advantages and limitations associated with different options. Next, I consider how a biodiversity monitoring programme can be developed to ensure long-term viability, both by reducing costs and enhancing collateral benefits. In the final section, I reflect on the nature of the challenge ahead and the fundamental changes in behaviour that are needed if biodiversity monitoring is to meet this challenge.

THE IMPORTANCE OF PEOPLE

In all the debate about the purpose and practicalities of monitoring biodiversity, it is disturbingly easy to forget the importance of people in every part of the equation. I cannot do better than quote Duncan Poore, in his book *Changing Landscapes*, which charts the influence of the ITTO on tropical forest management (Poore, 2003):

> *'Results will only be as good as the people who are striving for them. Good management of forests depends not only on sound science but also on accurate observation and sensitive judgement. Sustainable forest management will only succeed when there are many people with such talents in positions where they are able to shape the future forest landscape.'*

It is people who design and implement monitoring activities, people who draw conclusions from sample data, people who choose which findings will be incorporated into new management approaches on-the-ground and which will be discarded, and people who decide whether monitoring programmes are ultimately worthwhile and should be sustained in the long term. Given these obvious facts it is important to recognize the roles played by different actors, and how a better understanding of their relative contributions can help in developing the most cost-effective and justifiable approaches to monitoring.

The role of different people in biodiversity monitoring: A typology of approaches

A wide variety of people are involved in biodiversity monitoring programmes in different forests around the world, and they generally separate into two groups. In one group are those who live outside the place where monitoring is conducted and who appear on occasional visits to work or provide advice and guidance to others. This includes almost all professional scientists as well as advisors and auditors from regulatory or certification authorities. In the other group there are people who live and work within the study area, including forest management employees and/or locals who are paid for their involvement or engage on a volunteer basis. This distinction is important as it often underpins fundamental differences in the motivations and relationships that different people have with the monitoring process.

Discussion of the relative benefits and disadvantages of scientist-run and locally based biodiversity monitoring often focuses on these two extremes, whereas in reality they simply form the ends of a spectrum of possible monitoring protocols (Danielsen et al, 2009). To facilitate thinking about the relative contribution of different people, and what combination is most appropriate for different situations, Danielsen et al (2009) proposed a typology of monitoring protocols, ranging from no local involvement with monitoring undertaken by professional researchers to an entirely autonomous monitoring system run by local people (Table 16.1).

Choosing the appropriate blend of people to be responsible for designing and running a monitoring programme depends on both the desired level of detail, as well as who the data are intended to benefit. Central to this decision-making process is recognition of a number of inherent trade-offs amongt different groups of people regarding cost, precision and accuracy, and the interest and ability of individuals to integrate the results from monitoring into revised management.

This book is aimed primarily at the community of conservation scientists and professionals who have an interest in, or are directly involved with, the development of forest biodiversity monitoring programmes. Given this, the implementation of many of the more detailed proposals presented in Chapters 8–15 will invariably require a minimum degree of guidance from experts. However, this does not necessarily mean that all monitoring can or should be designed and implemented solely by teams of academic experts. Professional scientists, whether they are consultants, university researchers or even doctorate students, invariably come with a very high price tag and limited time commitments. Perhaps of greater concern is the

Table 16.1 A typology of biodiversity monitoring approaches based on the relative participation of different actors

Category of monitoring	Primary data gatherers	Primary evaluators and users of data	Description
1. Externally driven, professionally executed	Professional researchers	Professional researchers	Design and implementation conducted entirely by professional scientists who are funded by external agencies and generally reside elsewhere. Little to no local stakeholder involvement in the planning of the programme
2. Externally driven with local data collectors	Professional researchers, local people	Professional researchers	Local people involved only in data collection stage, with design, analysis and interpretation of monitoring results for decision-making being undertaken by professional researchers from elsewhere. Local participants may include volunteers (especially in more developed countries), as well as parataxonomists
3. Collaborative monitoring with external data interpretation	Local people with professional researcher advice	Local people and professional researchers	Local people involved in data collection and monitoring-based management decision-making, but the design of the scheme and the data analysis and interpretation are undertaken by external scientists. Local people may either be paid for their time or contribute their time freely
4. Collaborative monitoring with local data interpretation	Local people with professional researcher advice	Local people	Locally based monitoring involving local stakeholders in data collection, interpretation or analysis, and management decision-making, although external scientists may provide advice and training. The original data collected by local people remain in the area being monitored, which helps create local ownership of the scheme and its results, but copies of the data may be sent to professional researchers for in-depth or larger-scale analysis
5. Fully autonomous local monitoring	Local people	Local people	The whole monitoring process, from design, to data collection, analysis and finally to the use of data for management decisions is carried out autonomously by local stakeholders. There is no direct involvement of external scientists or agencies, except possibly through an advocacy role

Source: Based on Danielsen et al (2009)

fact that performance assessments in academic and research institutions are commonly based not around the delivery of findings that can help address the most pressing practical problems, but on generating scientific publications and winning grant money (see Chapter 7).

There are very few, if any, arguments against involving local people in biodiversity monitoring (Sheil and Lawrence, 2004). Beyond any other reason, they are often the ones who are ultimately responsible for the on-the-ground monitoring and management of forests and forest biodiversity, whether they are representatives of local communities or forestry officials. Local people are often those with the greatest opportunity and vested interest to ensure the long-term success of monitoring, and often contribute a unique set of field skills (e.g. surveying of cryptic and nocturnal species) that is difficult (or very expensive) to find in external experts. For more technical elements of monitoring, it is almost always possible to train local people to a good standard to collect data and perform experiments for a mere fraction of the cost needed to pay outside professionals to do the same work (Basset et al, 2004b; Janzen, 2004; Sheil and Lawrence, 2004; Box 16.1). In addition, there are major economies of scale in time and cost that can be exploited (yet rarely are) by harmonizing biodiversity sampling work with standard forest inventories that are regularly conducted by forestry companies to quantify timber stocks (Newton and Kapos, 2002; Winter et al, 2008). In many parts of the more developed world, further cost savings and opportunities for mainstreaming monitoring activities are possible through the use of voluntary data collectors (e.g. Schmeller et al, 2009).

The last decade has seen an increasing level of interest in biodiversity monitoring that is either mostly or entirely under the control of local people (Sheil and Lawrence, 2004; Danielsen et al, 2005a, 2005b, 2007; Evans and Gauriguata, 2008). These more autonomous approaches are reflected in Categories 4 and 5 of Danielsen et al's (2009) typology (Table 16.1) and are commonly referred to as participatory or collaborative monitoring (Box 16.2). The principle advantage of participatory monitoring is that it transfers responsibility and ownership of monitoring into the hands of the very same people who are responsible for making management decisions. Increasing local ownership of monitoring helps ensure the sustainability of monitoring activities and can also greatly increase the speed and political acceptability of the decision-making process (Box 16.2).

Work to date suggests that participatory approaches are particularly useful for monitoring of key natural resources (e.g. hunted species and other forest products) in forests that are de facto managed by the communities who live in or near the area and who, on a daily basis, are using the resources in the area (e.g. Danielsen et al, 2007; Constantino et al, 2009; and see Box 16.2). In many situations, and especially where conservation goals are focused on both the intrinsic and the utilitarian value of biodiversity, and data is being collected on entire assemblages of species, an integrated approach that combines expert guidance and management from professional scientists with a close-involvement of local people (i.e. categories 3 and 4 in Table 16.1) is likely to provide the most attractive solution for monitoring (Evans and Gauriguata, 2008; Garcia and Lescuyer, 2008; Holck, 2008). This integrated approach can deliver significant benefits. The contribution from professionals ensures a necessary degree of scientific rigour in study design and data analysis while the involvement of local people helps facilitate the process of implementing any management recommendations, as well as providing a cost-effective and sustainable means of collecting data and a rich source of local ecological and historical knowledge to aid the interpretation of monitoring results (Sheil and Lawrence, 2004; Kainer et al, 2009; Box 16.1).

The important point here is that even when biodiversity monitoring requires significant input from professional scientists, the contribution of local people can go way beyond the provision of cheap labour. Local stakeholders, whether they are community representatives, reserve guards, local environmental groups, or employees of a forest management company, can play a central role in ensuring that a monitoring programme serves its purpose and survives into the long term. The contribution made by scientists to conservation is almost always more effective when local people are actively involved in as many stages as possible of the monitoring process, from thinking about goals and indicators to day-to-day field collection and discussion of findings, than when they simply have research findings dictated to them.

Figure 16.1 Sharing the experience of biodiversity monitoring: Joint participation of research technicians and forestry company employees in a biodiversity monitoring programme in a plantation forestry landscape, Jari, Brazilian Amazonia

Source: Photograph courtesy of Jos Barlow

Box 16.1 Employing local people in expert-led biodiversity monitoring

Scientists the world over have long employed local people to help conduct field research and run monitoring programmes, especially in the developing world where labour is comparatively cheap. However, through careful investment and training, local people can provide far more than just manual labour and assistance with menial tasks. Rather they can act as 'parascientists', in the same tradition as paralegals and paramedics, and provide a powerful supporting role for a much smaller cadre of professionally trained scientists. This concept was first trialled in the late 1980s in Costa Rica's Area de Conservación Guanacaste by Daniel Janzen, who coined the term 'parataxonomist' as a:

> *resident, field-based, biodiversity inventory specialist who is largely on-the-job trained out of the rural work force and makes a career of providing specimens and their natural history information to the [wider scientific community], and therefore to a multitude of users across society.'* (Janzen, 2004)

A similar label of 'parabiologists' has been applied to local people trained to conduct essential research tasks, such as observational and experimental work (Sheil and Lawrence, 2004; Figure 16.2).

Parataxonomists and parabiologists can make an enormous contribution to biodiversity monitoring as there is not, nor ever will be, a sufficient number of professional scientists, to sample and study the world's forest biodiversity, especially regarding hyper-diverse groups such as tropical invertebrates (Basset et al, 2004b). Their worth is well demonstrated by the central role that parataxonomists have played in some of the world's most ambitious biodiversity research projects (Table 16.2). Aside from reducing costs, parataxonomists are also extremely valuable because they are often uniquely skilled to conduct surveys of cryptic species (e.g. mammals) and live in the area where the research and monitoring is being conducted, allowing them to manage long-term programmes that require frequent collections (e.g. sampling of highly seasonal taxa, leaf-litter decomposition experiments and phenology

data). Beyond their professional capacity, locally trained people can become important advocates for long-term investments in conservation through the development of a keen sense of ownership over their natural resources. These individuals may ultimately move on to jobs in conservation organizations and government agencies elsewhere, helping to improve representation of local voices in regional or national decision-making (Janzen, 2004; Basset et al, 2004b).

Figure 16.2 Parabiologist Edivar Dias Correia, from Tefé, Brazil

Note: Edivar is single-handedly responsible for conducting literally thousands of hours of line-transect surveys of vertebrates across the Amazon, as well as managing and implementing the collection of sample data on dung beetles, orchid bees, carrion flies and vegetation as part of a long-term monitoring project in Jari
Source: Photograph by the author

Table 16.2 Examples of scientific and educational biodiversity projects around the world that have relied heavily on parataxonomists

Project/course	Description and country	Website
INBio	Instituto Nacional de Biodiversidad, Costa Rica	www.inbio.ac.cr
National Museums of Kenya	Plants, insects and birds, Kenya	www.museums.or.ke
All Taxa Biodiversity Inventory	All taxa biodiversity inventory, USA National Parks	www.dlia.org
Manual de Plantas de Costa Rica	Botanical inventory in Costa Rica	www.mobot.org/MOBOT/research/edge
Macrofungi of Costa Rica	Inventory of fungi in Costa Rica	www.nybg.org/bsci/res/hall
Biodiversity inventory of the caterpillars and adult *Lepidoptera* of the Area de Conservación Guanacaste	Biological inventory of Lepidoptera in Costa Rica	http://janzen.sas.upenn.edu/
Arthropods of La Selva	Biological inventories at La Selva, Costa Rica	http://viceroy.eeb.uconn.edu/ALAS/ALAS.html
The Parataxonomist Training Centre	Training of parataxonomists in Papua New Guinea	www.entu.cas.cz/png/parataxoweb.htm
Faunistic survey of *Dynastinae* beetles	Inventory of scarab beetles in Central America	www.museum.unl.edu/research/entomology/scarabcentral.htm
SIMAB Gamba Project	Biological monitoring in Gamba, Gabon	www.si.edu/simab/Gabon.htm

Source: From Basset et al (2004b)

Despite the undisputable benefits offered by parataxonomists, they have seen relatively little involvement in biodiversity monitoring programmes to date. This is partly because there is not a pool of cheap, well-trained local people out there waiting to be contracted for scientific research. Parataxonomists require careful training and nurturing to develop not only expertise but also pride and confidence in their work – something that is an essential ingredient for long-term monitoring (Janzen, 2004). Scientists are also often wary of trusting the quality of data provided by parataxonomists, especially with regard to morphospecies identifications for areas where identification guides are lacking. Indeed, comparative work has shown that error rates can often be both high and unpredictable (Basset et al, 2004b; Krell, 2004). That said parataxonomists and parabiologists cannot operate in isolation from the wider scientific community that is responsible for providing a long-term source of guidance and training (Janzen, 2004). By developing an effective working relationship between local people and experts, it is possible to greatly improve data quality and hence the long-term integrity of monitoring programmes.

Box 16.2 Participatory monitoring of biodiversity

The term 'participatory monitoring' has been loosely interpreted, but generally refers to monitoring that involves significant, if not exclusive, participation by local people (Danielsen et al, 2009; Table 16.1). Largely synonymous terms include locally based monitoring, collaborative monitoring, participatory assessment, self-monitoring and community-based monitoring (Danielsen et al, 2005a; Evans and Gauriguata, 2008). A detailed review of these approaches is provided by Danielsen et al (2005a) encompassing both fully participatory monitoring of biodiversity and natural resources by local users, as well as censuses by local rangers and inventories by amateur naturalists (see also: www.monitoringmatters.org/). Most participatory monitoring programmes are directly linked to resource management and locally important ecosystem goods and services (including harvestable species), and are distinguished by the fact that monitoring is conducted at the local scale by individuals with little formal education, with the same people also involved in data processing and decision-making (Danielsen et al, 2005a).

Participatory monitoring is particularly relevant in developing countries, where it can help empower local people to be directly involved in the management of their own natural resources. It can also encourage a culture of learning, which is key to the success of AM (Cundill and Fabricius, 2009). In turn, this empowerment and enhanced proximity between monitoring and management activities can often lead to more rapid and effective decision-making, as demonstrated by Danielsen et al (2007) in a study of protected area management in the Philippines (Figure 16.3).

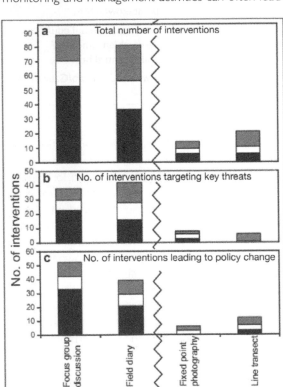

Figure 16.3 Effectiveness of participatory versus conventional scientific biodiversity monitoring methods in generating conservation management interventions intended to improve the way local people (black shading), outsiders (open bars), and both (grey shading) manage Philippine protected area resources

Note: (a) The total number of interventions generated by each method for the same recurrent investment; (b) The number of interventions that targeted the three most serious threats to the biodiversity of each site; (c) The number of interventions that led to policy change within local government and community institutions
Source: Redrawn with permission from Danielsen et al (2007)

Nevertheless, the conditions under which truly participatory monitoring (i.e. Categories 4 and 5 in Table 16.1) is likely to be viable in the long run are quite specific. Danielsen et al (2005a) identified six principles that contribute to the ability of participatory monitoring to be sustained in the long run:

1   Locally based monitoring has to identify and respond to the benefits that the community derives from the habitat or population being monitored.
2   The benefits to local people participating in monitoring should exceed the costs.
3   Monitoring schemes must ensure that conflicts and politics between government, managers and communities do not constrain the involvement of local stakeholders in the monitoring process.
4   Monitoring should build on existing traditional institutions and other management structures as much as possible.
5   It is crucial to institutionalize the work at multiple levels, from countrywide policies down to the job descriptions of local government officers.
6   Data should be stored and analysed locally, even if this means some loss of quality. It should also remain accessible to local people.

MAKING BIODIVERSITY MONITORING PROGRAMMES VIABLE AND EFFECTIVE IN THE LONG TERM

Very few biodiversity monitoring programmes survive in the long term or come close to delivering on their initial promises (Chapters 5–7). As discussed throughout this book, a key reason for the failure of many monitoring projects is that they fail to deliver results that are meaningful (or delivered over meaningful time frames) for management. This disconnection between the activity and purpose of monitoring can lead to an erosion of credibility in the entire process and the loss of enthusiasm among both participants and funders (see Chapter 5). However, even well-meaning monitoring programmes can fail. In many cases work is cut short prior to the generation of the main results due to a lack of funds, often precipitated by insufficient contingency to cover the costs of repairing vital equipment and project vehicles. It is also common for programmes to grind to a halt due to the departure of personnel, whether they be field or managerial staff or expert advisors. In addition to these operational problems, monitoring programmes often fail to address the most important questions because of a mismatch between the scale at which monitoring programmes are conducted, and the scale at which many ecological processes and human activities operate (Hobbs, 2003). Because the vast majority of monitoring programmes are conducted in isolation from one another, there is very limited scope for addressing questions about large-scale processes and context-dependency that require integrating data from surrounding areas and across multiple landscapes (Hammond and Zagt, 2006; DeFries et al, 2009; and see Chapter 13).

Past experience indicates that none of these problems are easily dealt with. Nevertheless, there are many areas where improvements can be achieved and the longevity of monitoring programmes enhanced. These opportunities fall into three broad categories: (i) reducing costs; (ii) enhancing benefits; and (iii) improving interaction and cooperation among individual monitoring initiatives.

Reducing the costs of monitoring

Biodiversity monitoring is commonly perceived to be an extremely expensive operation and beyond the limited budget of many conservation projects and management organizations (Margules and Austin, 1991). Although funding is almost always a limiting factor, an often deeper problem is the fact that many programmes are very wasteful (often unintentionally) in the way that available funds are spent. Indeed, cost considerations are often ignored in the planning process of conservation activities in general – a situation that threatens the sustainability of many otherwise well-meaning projects and can result in significant lost opportunities for biodiversity (Possingham, 2001; Naidoo et al, 2006).

Even a superficial inspection of the design and implementation strategy of most monitoring programmes invariably reveals ways in which costs can be significantly reduced. Many such opportunities are discussed in earlier chapters of this book, and include:

- a strategic choice of monitoring objectives. Instead of basing monitoring priorities on an ad hoc decision-making process, resources need to be channelled towards addressing the problems that are most likely to deliver real benefits for conservation and encourage lasting investment and participation in monitoring (see Chapter 10).
- the selection of cost-effective indicator taxa. Remarkably few biodiversity studies incorporate considerations of cost in the indicator selection process, although comparative work clearly demonstrates the significant benefits to be gained by doing so (Gardner et al, 2008a; and see Chapter 12).
- the use of a cost-effective sampling design. It is common in biodiversity work to see an over-emphasis of sampling effort on the exhaustive surveying of a small number of locations, invariably at the expense of securing sufficient replication of independent samples. However, most conservation objectives are better served by collecting less comprehensive data from a larger number of sampling locations, thereby encompassing scales that are more relevant to dominant ecological and management processes (see Chapter 14).
- the choice of a cost-effective sampling methodology. There are usually a variety of ways in which a given species group can be sampled, and careful examination of the alternatives can lead to the identification of methods that allow significant cost savings, while also maintaining data integrity (Franco et al, 2007; Garden et al, 2007; and see Chapter 14).
- the participation of local people in data collecting and processing. As seen earlier in this chapter, parataxonomists and parabiologists can provide

enormous cost-savings in executing many parts of the monitoring process compared to when relying solely upon postgraduate students or professional scientists.

Although money and time will always be limiting factors, resource limitations cannot justifiably be used as an excuse for failing to monitor biodiversity until a genuine effort has been made to evaluate the cost of alternative approaches.

Enhancing the benefits of monitoring

Other than reducing costs, another way of enhancing the viability of biodiversity monitoring programmes is to maximize their relevance and utility to as wide an audience as possible. This means identifying additional 'customers', beyond those responsible for the management of the study system itself, who may be interested in offering some kind of investment in return for access to monitoring data and outputs. Such investments may be financial but could just as likely be through improved access to outside experts, technical facilities (e.g. museums, soil laboratories) and institutional support, all of which can help strengthen the long-term viability of a monitoring programme. The following sections outline a number of ways in which the benefits of monitoring can be sold more widely.

The development of regional management standards

Ideally, each forest landscape would be managed under a tailored management plan, and the ongoing refinement of management strategies would be supported by a fully comprehensive and long-term validation monitoring programme. Of course, the reality is very different, and the majority of forest management organizations lack the financial and technical capacity or political will to dedicate sufficient funds or time to monitoring. This discrepancy means that scientists responsible for advising forest managers and certification authorities on the development of management standards frequently have to rely upon outside experts and findings from studies conducted elsewhere (Bormann et al, 2007). This level of dependency is unsatisfactory because the responses of biodiversity to forest disturbance and management strategies (e.g. logging cycles, corridor design, etc.) are highly context and scale dependent (Bunnell and Huggard, 1999; Hobbs, 2003; Gardner et al, 2009). Indeed, many important management questions can only be addressed by combining data from across multiple landscapes (see Chapter 13). The way to improve this situation and develop standards more appropriate for specific regions is through improved site-level monitoring, and closer harmonization of protocols for data collection and analysis. This would allow the results of individual programmes to be more effectively aggregated to a level that is meaningful to policy makers and the certification process (Holvoet and Muys, 2004; McAfee et al, 2006).

The development of regional and national biodiversity assessments

There is an urgent need for biodiversity data at larger spatial scales to evaluate progress towards national and international conservation targets. The most prominent such example is the Convention on Biological Diversity's (CBD) target to significantly reduce the loss of biodiversity by 2010 (UNEP, 2002; Balmford et al, 2005; McAfee et al, 2006) that was recently incorporated into the Millennium Development Goals in recognition of the impact of biodiversity loss on human well-being (Walpole et al, 2009). Just two indicators relating to sustainable forest use were identified at the seventh meeting of the Conference of the Parties of the CBD in February 2004: the area of forest under sustainable management and the proportion of products derived from sustainable sources. Furthermore, the agreed indicators for assessing the status and trends of biodiversity within the CBD 2010 initiative are currently limited to the Global Wild Bird Index, the Red List Index and the Living Planet Index (2010 Biodiversity Indicators Partnership; www.twentyten.net). Consequently, the contribution of species data to the 2010 indicators is severely biased towards a relatively small number of taxa (mostly charismatic vertebrates, such as mammals, birds and amphibians), and particular areas of the world (mostly Europe and North America) (Dudley et al, 2005; Walpole et al, 2009).

A number of global-scale proposals for the systematic monitoring of selected indicator taxa have been proposed (e.g. Niemelä, 2000; Pereira and Cooper, 2006; Teder et al, 2006). However, such programmes are unlikely to be realized because they depend on independent core funding to cover the substantial implementation and running costs (e.g. Pereira and Cooper (2006) estimate an annual budget of $10 million), as well as posing formidable logistic, administrative and governance challenges (Henry et al, 2008). Hence, the only realistic option is to exploit a bottom–up approach that collates information from *existing* monitoring programmes whose primary raison d'être is not to provide a national or global-scale status assessment of biodiversity (Gregory et al, 2005; Henry et al, 2008). Even when comprised of data collected using different techniques, regional syntheses of monitoring programmes can be extremely effective in providing large-scale summaries of the status and trends of entire ecosystems for policy makers (e.g. Gardner et al (2003), who compiled data from nearly 200 separate monitoring sites to assess the status of Caribbean coral reefs). If appropriately coordinated and standardized, biodiversity monitoring programmes in forest management systems provide a major opportunity to enhance our understanding of the condition of forest ecosystems worldwide.

Tackling large-scale science questions

Modified forest landscapes in different regions of the world are distinguished by their evolutionary, biophysical, geographic, historical and current socio-economic context. These differences provide both challenges and opportunities for biodiversity conservation (Whittaker et al, 2005; Gardner et al, 2009). For example, the conservation challenges in a given region can be strongly influenced by the underlying resilience of forest biota and regional variability in the historical legacy of anthropogenic disturbance. This complexity is further exacerbated by the spectre of global climate and environmental change and the ever tighter coupling of ecological and socio-economic systems (Liu et al, 2007; Rockstrom et al, 2009). Untangling the complexities of these large-scale couplings, such as feedback and threshold effects, could play a critical role in developing more SFM strategies. The only way of effectively addressing such large-scale questions in the short term is through the integration of independent datasets from multiple study sites.

Capacity building and education

Well-designed biodiversity monitoring programmes can make a significant contribution towards the much needed capacity building of biodiversity scientists and research assistants, and generally help build support for conservation. Developing a skills base is essential for promoting biodiversity conservation research, especially in the developing world where the success of any new project is often dictated by the availability of a small number of desperately overstretched experts. The situation is particularly dire in the case of taxonomy, which has seen a continued decline in investment in many parts of the world, even in developed countries with a long tradition of natural history, such as the UK (Hopkins and Freckleton, 2002). For many of the world's tropical rainforests, there are only a mere handful of expert taxonomists available for a given species group to provide the vital support necessary for processing the specimens that come from biodiversity monitoring programmes (Table 16.3).

It is often possible to establish mutually beneficial partnerships between forest management organizations and local universities and research institutions where students and technicians are supplied free of charge in return for in-kind logistic support and help in securing research permits. In this way, long-term monitoring programmes can provide a reliable source of fieldwork and laboratory experience, as well as specimen material to support the training of a new generation of conservation scientists and professionals.

The development of emerging markets for ecosystem services

In addition to existing regulatory requirements and incentives from certification, payments for environmental services offer a promising new way to promote the conservation of forest biodiversity (Wunder, 2007; Wunder et al, 2008). In particular, there are strong hopes that the rapidly growing forest carbon markets may be adapted to offer 'biodiversity rich' carbon credits (Miles and Kapos, 2008). Biodiversity monitoring is critical to the success of such payment schemes (Meijerink, 2008), which could offer a significant new source of finance to support monitoring programmes in forests around the world.

Table 16.3 Availability of taxonomic experts, both within Brazil and globally, who are deemed capable of separating specimen collections from a site in the north-east Brazilian Amazon without extensive consultation with reference materials or other scientists

Taxonomic group	Approximate number of experts globally	Approximate number of experts in Brazil
Leaf-litter amphibians	20	10
Lizards	20-30	8
Large mammals	>250	>150
Small mammals	20	10
Bats	30	10
Birds	>50	15-20
Scavenger flies (Diptera: *Calliphoridae, Mesembrinellidae* and *Sarcophagidae*)	NA	5
Fruit-feeding butterflies (Lepidoptera: *Nymp-Halidae*)	>50	20
Dung beetles (Coleoptera: *Scarabaeinae*)	3	1
Epigeic arachnids (orders *Amblypygi, Araneae, Opiliones, Scorpiones* and *Uropygi*)	30	20
Fruit flies (Diptera: *Drosophilinae*)	4	3
Orchid bees (Hymenoptera: *Apinae and Euglossini*)	30	10
Moths (Lepidoptera: *Arctiidae, Saturniidae and Sphingidae*)	15	3

Source: Data from Gardner et al (2008a)

Working towards a more interactive and cooperative monitoring environment

An individual biodiversity monitoring programme cannot be developed in isolation of other programmes if is to be successful in the long term. Instead, programmes need to be conceived and nurtured within an interactive learning environment that brings together researchers, managers and regulators from different places and institutions to share data, expertise and ideas (Salafsky and Margoluis, 2003). For example, many of the opportunities for cost saving in monitoring rely upon a high level of familiarity with previous research and the latest developments in monitoring methods and techniques (many of which may be unpublished), which is impossible to achieve without close communication and cooperation with workers involved in other programmes.

Although there is a multitude of ways in which learning can be achieved, semi-formal learning or research networks have been presented as a proactive way in which we can increase the effectiveness of conservation research and action (Salafsky and Margoluis, 1999a). Learning networks (also termed portfolios) were developed to facilitate stakeholder interactions in AM systems for integrated conservation and development projects (ICDPs) but are broadly applicable to all forms of conservation activity (Salafsky and Margoulis, 1999a, 2004; and see Foundations of Success www.fosonline.org). They represent varyingly formalized structures for facilitating collaborative learning and action, and ensure that the latest ideas, practical and technical experience and research findings can be readily exchanged and used to benefit conservation. Learning networks are conceptually analogous to the 'communities of practice' model that is used by many businesses to bring together groups and individuals with a common interest (Wenger et al, 2002).

As described by Salafsky and Margoulis (1999a, 2004), the basic unit of a learning network is the individual management system, which is linked to other systems through a common sets of goals and objectives (see Chapters 9 and 10). The first task in developing a learning network is to build a multi-stakeholder network coordination team. In the case of large and complex forest monitoring programmes operating across multiple commercial enterprises, this is best mediated through an independent research body such as a university or research institution that works in collaboration with certification and/or government authorities (Yamasaki et al, 2002; Higman et al, 2005). A key task of the coordination team is the development of a learning framework – a written document combined with other learning–teaching materials that can act as a guide for the full process of planning, designing, implementing and analysing biodiversity monitoring programmes (see Chapters 8–15 of this book). An important part of any such guidance document is the establishment of a common set of technical terms that is needed to avoid miscommunication (Stem et al, 2005; Salafsky et al, 2008; and see Chapter 5). In addition to providing a mechanism for harmonizing the gathering, evaluating and disseminating monitoring data, a learning framework can also serve as an appropriate and effective vehicle for reviewing and synthesizing priority objectives for validation monitoring (see Chapter 10).

Perhaps the greatest challenge facing the development of an effective research network, is the political and cultural resistance to data sharing that is commonly found within both forestry companies and the scientific research community (see Chapter 7). While formal requirements for interaction and knowledge sharing can be incorporated into certification requirements, the only way to really overcome this barrier is through the development of incentive schemes and informal trust-building initiatives that demonstrate shared benefits for all parties.

The need for a more formal institutional framework for integrating the results of monitoring programmes and helping to guide responsible management protocols at national and regional scales is well recognized in the forestry sector (Davis et al, 2001b; Allen et al, 2003; Curran et al, 2005). Support for such an endeavour could be provided from the existing Global Forest Information Service (GFIS) that was established in 2001 in cooperation with the FAO, CIFOR and the European Forest Institute (EFI) to provide an online meta-database of forest information (Buck and Rametsteiner, 2003). Ultimately, any such coordinating institution or set of institutions needs to adopt a radically more democratic approach to the way that science is conducted, and scientific knowledge shared, than the one that currently dominates the academic community (Sayer and Campbell, 2004). An excellent example of such an approach that is relevant to forest conservation is the Global

Restoration Network which operates to 'link research, projects and practitioners in order to foster an innovative exchange of experience, vision and expertise', and makes available a wealth of information on all aspects of restoration through its website (www.globalrestorationnetwork.org/). Nevertheless, resource management regimes are, like all other social institutions, prisoners of history and precedent, and the task of forging new collaborative institutions can be expected to be neither trivial nor short (Connor and Dovers, 2004; Dovers, 2005).

THE WAY AHEAD

In drawing upon a wealth of proposals made by scientists and conservation practitioners working throughout the world, this book presents what can best be described as a new paradigm for monitoring forest biodiversity. I have attempted to lay out the arguments and opportunities for moving away from a static and reactive framework, whose primary purpose is to satisfy auditing requirements, towards a system that is designed to enable a process of continued learning and improvement in management. Achieving such a change in philosophy is far from easy. Indeed the notion of 'sustainable forest management' (SFM), which provides the guiding vision for monitoring and AM, has been described as a quintessentially 'wicked problem', characterized by high levels of uncertainty and distrust and a low stakeholder consensus (Allen and Gould, 1986; Ludwig, 2001). Nevertheless, after decades of research and increasing political and public concern about the state of the world's forest and biodiversity, we have an unprecedented opportunity to instigate real change. We cannot allow past failures to set the precedent for the future, nor can we permit the urgency of the conservation crisis to dishearten or blind us as to the importance of rigorous and meaningful monitoring programmes. If we don't effectively monitor the condition of the world's forest, we will also fail to understand the relative success and failure of our actions, and how limited conservation resources should be invested in the future (Nichols and Williams, 2006).

The path to success is perhaps most clearly defined, not simply by overcoming limitations in technical capacity, scientific knowledge or funds for research and monitoring, but rather by the behaviour of the people involved. In seeking a more constructive and rewarding approach to biodiversity monitoring, it is useful to hold in mind four overarching principles that run throughout the discussion of much of this book:

1   Be collaborative

Monitoring will only make a successful contribution to forest management if devised and developed as a genuinely collaborative venture between all interested parties. Open communication and the sharing and dissemination of ideas and knowledge provide an essential foundation for any participatory process, yet run counter to the way in which most academic and commercial forestry institutions operate. Achieving a shift in attitude requires a new and more active approach to learning through experience (Fazey et al, 2006) and a new breed of professionals who are able to break down the culture of insularity that exists both between different scientific disciplines (e.g. field ecologists and statisticians) and between scientists, managers and regulators involved in the management of biodiversity (Field et al, 2007).

2 Be realistic

It is unrealistic to expect that reliable knowledge concerning large-scale and ecologically complex management problems typical of human-modified forest landscapes can be generated within a few years (Taylor et al, 1997; Wilhere, 2002; Hobbs, 2003). Time-lags in the response of biodiversity to human disturbances, the time taken to gather sufficient evidence, and the uptake of any recommendations for management mean that it is important to be strategic in selecting objectives for monitoring, and maximize the amount of 'information return' on any investment made (see Chapter 10). This is particularly true for areas of the world that are under increasing threat from forest conversion and degradation, and where opportunities for conservation are rapidly being lost.

Monitoring should follow a structured timetable that is targeted at informing the management process at time scales that are relevant to real-world decision-making, rather than acting as a passive–feeding source of information. Deciding when strategic reviews of evidence for management should take place is difficult, and is determined by a trade-off between the need to start developing effective management strategies and the need to accumulate more evidence. The fact that evidence relating to different management questions will accumulate at different rates, and that some indicators demand closer and more immediate attention than others (e.g. locally threatened species), means that the entire review process cannot be conducted at standard intervals but needs to encompass multiple spatial and temporal scales.

It is also unrealistic to expect that detailed forest biodiversity monitoring initiatives can be implemented in every landscape. Instead, a multi-tiered approach is necessary, where monitoring for implementation and basic measures of management effectiveness (i.e. stand and landscape structural indicators) takes place across as large a network of sites as possible, while more detailed validation monitoring and research regarding actual changes in biodiversity is limited to a carefully selected subset of the most viable and rewarding locations (e.g. as proposed by Smith et al, 2001, for monitoring in Australian and New Zealand forests).

3   Be adaptive

The entire monitoring and management system is inherently dynamic and adaptive, and learning should be embedded throughout the process. There are no hard and fast rules for success, nor are we likely to ever find absolute answers for many problems in forest management. Sustainability is neither a black and white nor a static phenomenon. Instead, realizing the design of more sustainable forest landscapes should be viewed as a continuous process of enactment in which both the human and ecological systems are in intimate and constant interaction (Haila, 2007).

Within this context, the development of a monitoring programme should not be conceived of as a linear activity; decisions made within individual components both guide and constrain the choices that are made elsewhere. The learning process should be constantly refined and improved through the action of multiple feedback loops that operate both within the monitoring programme per se (e.g. through the correction and revision of field sampling protocols and research hypotheses to describe management impacts, while exerting sufficient caution so as not to jeopardize the value of past work) and between the monitoring programme and wider management system (through revision of management standards).

That said the process of biodiversity monitoring should not be a never-ending story, as this could undermine its credibility and support among decision-makers and the general public (Nichols and Williams, 2006). The scientific community must recognize where it is necessary to draw a line under persistent yet outdated debates that are stifling future progress, and continually strive to identify the real areas of uncertainty that require urgent empirical and theoretical attention.

4   Keep sight of the bigger picture

Biodiversity conservation, whether of forests or any other ecosystem, is ultimately a societal problem that is guided by science, and not (as is often perceived by scientists) a scientific problem that is influenced by society. It is a normative and ethically motivated pursuit that must incorporate the values and voices of non-scientists if it is to meet its purpose (Sheil and Lawrence, 2004). Of course, the conservation and restoration of biodiversity represents just one goal out of many for the managers, policy makers and special interest groups who are involved in shaping the future of forest landscapes around the world. Moreover, other economic, social and policy goals invariably carry more weight in stakeholder negotiations than concerns for the natural environment. The task of simultaneously satisfying these different stakeholder groups and 'closing the loop' of the AM can be extremely challenging (Gibbs et al, 1999; IUCN, 2004; Leslie, 2004; and see Chapter 15).

Like most conservation dilemmas, the problem of managing human-modified forest landscapes is best solved through a process of negotiated compromise that integrates the values and needs of different stakeholders, while also incorporating new scientific knowledge from the monitoring process (Wilhere, 2002; Sayer and Maginnis, 2005a). Consensus building is an important prerequisite to achieving integrated and collaborative management of natural resources (Keough and Blahna, 2006). Techniques for more fairly balancing contributions from multiple stakeholders have seen a significant development in recent years (see Chapter 15) and important lessons can be learnt from past efforts, such as ICDPs (Garnett et al, 2007) and environmental conflict management (Niemelä et al, 2005).

It is easy to be demoralized about the ability of research and monitoring to influence complex multi-stakeholder resource management problems. However, while feasibility will always be a societal question, it remains the responsibility of science to ensure that the relative costs and benefits of different management options are well understood. The challenge of putting biodiversity monitoring to work will ultimately depend, more than anything else, on human behaviour and our capacity to change. As John Meynard Keynes so astutely put it, '*The difficulty lies not so much in developing new ideas as in escaping from old ones.*'

References

Aguilar-Amuchastegui, N. and Henebry, G. M. (2007) 'Assessing sustainability indicators for tropical forests: Spatio-temporal heterogeneity, logging intensity, and dung beetle communities', *Forest Ecology and Management*, vol 253, pp56–67

Akçakaya, H. R., Ferson, S., Burgman, M. A., Keith, D. A., Mace, G. M. and Todd, C. R. (2000) 'Making consistent IUCN classifications under uncertainty', *Conservation Biology*, vol 14, pp1001–1013

Aldrich, M., Belokurov, A., Bowling, J., Dudley, N., Elliot, C., Higgins-Zogib, L., Hurd, J., Lacerda, L., Mansourian, S., McShane, T., Pollard, D., Sayer, J. and Schuyt, K. (2004) *Integrating Forest Protection, Management and Restoration at a Landscape Scale*, WWF International Gland, Switzerland

Allan, C. and Curtis, A. (2005) 'Nipped in the bud: Why regional scale adaptive management is not blooming', *Environmental Management*, vol 36, pp414–425

Allen, G. M. and Gould, E. M. (1986) 'Complexity, wickedness, and public forests', *Journal of Forestry*, vol 84, pp20–23

Allen, R. B., Bellingham, P. J. and Wiser, S. K. (2003) 'Developing a forest biodiversity monitoring approach for New Zealand', *New Zealand Journal of Ecology*, vol 27, pp207–220

Anand, M., Krishnaswamy, J. and Das, A. (2008) 'Proximity to forests drives bird conservation value of coffee plantations: Implications for certification', *Ecological Applications*, vol 18, pp1754–1763

Andelman, S. J. and Fagan, W. F. (2000) 'Umbrellas and flagships: Efficient conservation surrogates or expensive mistakes?', *Proceedings of the National Academy of Sciences of the United States of America*, vol 97, pp5954–5959

Andelman, S., Cochrane, J. F., Gerber, L., Gomez-Priego, P., Groves, C., Haufler, J., Holthausen, R., Lee, D., Maguire, L., Noon, B., Ralls, K. and Regan, H. (2001) *Scientific standards for Conducting Viability Assessments under National Forest Management Act: Report and Recommendations of the NCEAs Working Group* National Centre for Ecological Analysis and Synthesis Santa Barbara, California

Andersen, A. N. and Majer, J. D. (2004) 'Ants show the way down under: Invertebrates as bioindicators in land management', *Frontiers in Ecology and the Environment*, vol 2, pp291–298

Anderson, D. R., Burnham, K. P. and Thompson, W. L. (2000) 'Null hypothesis testing: Problems, prevalence, and an alternative', *Journal of Wildlife Management*, vol 64, pp912–923

Angermeier, P. L. and Karr, J. R. (1994) 'Biological integrity versus biological diversity as policy directives: Protecting biotic resources', *Bioscience*, vol 44, pp690–697

Aragão, L., Malhi, Y., Roman-Cuesta, R. M., Saatchi, S., Anderson, L. O. and Shimabukuro, Y. E. (2007) 'Spatial patterns and fire response of recent Amazonian droughts', *Geophysical Research Letters*, vol 34, article no L07701

Arcese, P. and Sinclair, A. R. E. (1997) 'The role of protected areas as ecological baselines', *Journal of Wildlife Management*, vol 61, pp587–602

Arroyo-Rodriguez, V., Pineda, E., Escobar, F. and Benitez-Malvido, J. (2009) 'Value of small patches in the conservation of plant-species diversity in highly fragmented rainforest', *Conservation Biology*, vol 23, pp729–739

Asner, G. P., Hughes, R. F., Vitousek, P. M., Knapp, D. E., Kennedy-Bowdoin, T., Boardman, J., Martin, R. E., Eastwood, M. and Green, R. O. (2008) 'Invasive plants transform the three-dimensional structure of rain forests', *Proceedings of the National Academy of Sciences of the United States of America*, vol 105, pp4519–4523

Asner, G. P., Knapp, D. E., Broadbent, E. N., Oliveira, P. J. C., Keller, M. and Silva, J. N. (2005) 'Selective logging in the Brazilian Amazon', *Science*, vol 310, pp480–482

Azevedo-Ramos, C., de Carvalho, O. and do Amaral, B. D. (2006) 'Short-term effects of reduced-impact logging on eastern Amazon fauna', *Forest Ecology and Management*, vol 232, pp26–35

Bailey, R. C., Norris, R. H. and Reynoldson, T. B. (2004) *Bioassessment of Freshwater Ecosystems: Using the Reference Condition Approach*, Springer Science, New York

Bailey, R. C., Reynoldson, T. B., Yates, A. G., Bailey, J. and Linke, S. (2007) 'Integrating stream bioassessment and landscape ecology as a tool for land use planning', *Freshwater Biology*, vol 52, pp908–917

Balmford, A. and Bond, W. (2005) 'Trends in the state of nature and their implications for human well-being', *Ecology Letters*, vol 8, pp1218–1234

Balmford, A. and Whitten, T. (2003) 'Who should pay for tropical conservation, and how could the costs be met?', *Oryx*, vol 37, pp238–250

Balmford, A., Bennun, L., ten Brink, B., Cooper, D., Cote, I. M., Crane, P., Dobson, A., Dudley, N., Dutton, I., Green, R. E., Gregory, R. D., Harrison, J., Kennedy, E. T., Kremen, C., Leader-Williams, N., Lovejoy, T. E., Mace, G., May, R., Mayaux, P., Morling, P., Phillips, J., Redford, K., Ricketts, T. H., Rodriguez, J. P., Sanjayan, M., Schei, P. J., van Jaarsveld, A. S. and Walther, B. A. (2005) 'The convention on biological diversity's 2010 target', *Science*, vol 307, pp212–213

Balmford, A., Green, M. J. B. and Murray, M. G. (1996) 'Using higher-taxon richness as a surrogate for species richness: 1. Regional tests', *Proceedings of the Royal Society of London Series B-Biological Sciences*, vol 263, pp1267–1274

Balmford, A., Green, R. E. and Jenkins, M. (2003) 'Measuring the changing state of nature', *Trends in Ecology & Evolution*, vol 18, pp326–330

Bani, L., Massimino, D., Bottoni, L. and Massa, R. (2006) 'A multiscale method for selecting indicator species and priority conservation areas: A case study for broadleaved forests in Lombardy, Italy', *Conservation Biology*, vol 20, pp512–516

Banks-Leite, C. (2009) *Conservação da Comunidade de Aves de Sub-bosque em Paisagens Fragmentadas da Floresta Atlântica*, Universidade de São Paulo, São Paulo

Barbour, M. T., Gerritsen, J., Griffith, G. E., Frydenborg, R., McCarron, E., White, J. S. and Bastian, M. L. (1996) 'A framework for biological criteria for Florida streams using benthic macroinvertebrates', *Journal of the North American Benthological Society*, vol 15, pp185–211

Barlow, J. and Peres, C. A. (2004) 'Ecological responses to el Niño-induced surface fires in central Brazilian Amazonia: Management implications for flammable tropical forests', *Philosophical Transactions of the Royal Society of London Series B-Biological Sciences*, vol 359, pp367–380

Barlow, J., Araujo, I. S., Overal, W. L., Gardner, T. A., da Silva Mendes, F., Lake, I. and Peres, C. A. (2008) 'Diversity and composition of fruit-feeding butterflies in tropical eucalyptus plantations', *Biodiversity and Conservation*, vol 17, pp1089–1104

Barlow, J., Haugaasen, T. and Peres, C. A. (2002) 'Effects of ground fires on understorey bird assemblages in Amazonian forests', *Biological Conservation*, vol 105, pp157–169

Barlow, J., Gardner, T. A., Araujo, I. S., Avila-Pires, T. C. S., Bonaldo, A. B., Costa, J. E., Esposito, M. C., Ferreira, L. V., Hawes, J., Hernandez, M. I. M., Hoogmoed, M., Leite, R. N., Lo-Man-Hung, N. F., Malcolm, J. R., Martins, M. B., Mestre, L. A. M., Miranda-Santos, R., Nunes-Gutjahr, A. L., Overal, W. L., Parry, L. T. W., Peters, S. L., Ribeiro-Junior, M. A., da Silva, M. N. F., da Silva Motta, C. and Peres, C. (2007a) 'Quantifying the biodiversity value of tropical primary, secondary and plantation forests', *Proceedings of the National Academy of Sciences of the United States of America*, vol 104, pp18,555–18,560

Barlow, J., Gardner, T. A., Ferreira, L. V. and Peres, C. A. (2007b) 'Litter fall and decomposition in primary, secondary and plantation forests in the Brazilian Amazon', *Forest Ecology and Management*, vol 247, pp91–97

Barlow, J., Overal, W. L., Araujo, I. S., Gardner, T. A. and Peres, C. A. (2007c) 'The value of primary, secondary and plantation forests for fruit-feeding butterflies in the Brazilian Amazon', *Journal of Applied Ecology*, vol 44, pp1001–1012

Barlow, J., Mestre, L. A. M., Gardner, T. A. and Peres, C. A. (2007d) 'The value of primary, secondary and plantation forests for Amazonian birds', *Biological Conservation*, vol 126, pp212–231

Barlow, J. and Peres, C. A. (2008) 'Fire-mediated dieback and compositional cascade in an Amazonian forest', *Philosophical Transactions of the Royal Society B-Biological Sciences*, vol 363, pp1787–1794

Barrows, C. W., Swartz, M. B., Hodges, W. L., Allen, M. F., Rotenberry, J. T., Li, B. L., Scott, T. A. and Chen, X. W. (2005) 'A framework for monitoring multiple-species conservation plans', *Journal of Wildlife Management*, vol 69, pp1333–1345

Basset, Y., Mavoungou, J. F., Mikissa, J. B., Missa, O., Miller, S. E., Kitching, R. L. and Alonso, A. (2004a) 'Discriminatory power of different arthropod data sets for the biological monitoring of anthropogenic disturbance in tropical forests', *Biodiversity and Conservation*, vol 13, pp709–732

Basset, Y., Missa, O., Alonso, A., Miller, S. E., Curletti, G., De Meyer, M., Eardley, C., Lewis, O. T., Mansell, M. W., Novotny, V. and Wagner, T. (2008a) 'Changes in arthropod assemblages along a wide gradient of disturbance in Gabon', *Conservation Biology*, vol 22, issue 6, pp1552–1563, DOI: 10.1111/j.1523-1739.2008.01017.x

Basset, Y., Missa, O., Alonso, A., Miller, S. E., Curletti, G., De Meyer, M., Eardley, C. D., Mansell, M. W., Novotny, V. and Wagner, T. (2008b) 'Faunal turnover of arthropod assemblages along a wide gradient of disturbance in Gabon', *African Entomology*, vol 16, pp47–59

Basset, Y., Missa, O., Alonso, A., Miller, S. E., Curletti, G., De Meyer, M., Eardley, C. L., Lewis, O. T., Mansell, M. W., Novotny, V. and Wagner, T. (2008c) 'Choice of metrics for studying arthropod responses to habitat disturbance: One example from Gabon', *Insect Conservation and Diversity*, vol 1, pp55–66

Basset, Y., Novotny, V., Miller, S. E., Weiblen, G. D., Missa, O. and Stewart, A. J. A. (2004b) 'Conservation and biological monitoring of tropical forests: The role of parataxonomists', *Journal of Applied Ecology*, vol 41, pp163–174

Bawa, K. S. and Seidler, R. (1998) 'Natural forest management and conservation of biodiversity in tropical forests', *Conservation Biology*, vol 12, pp46–55

Bawa, K. S., Kress, W. J., Nadkarni, N. M., Lele, S., Raven, P. H., Janzen, D. H., Lugo, A. E., Ashton, P. S. and Lovejoy, T. E. (2004) 'Tropical Ecosystems into the 21st century', *Science*, vol 306, pp227–228

Becker, C. G., Fonseca, C. R., Haddad, C. F. B., Batista, R. F. and Prado, P. I. (2007) 'Habitat split and the global decline of amphibians', *Science*, vol 318, pp1775–1777

Bengtsson, J., Nilsson, S. G., Franc, A. and Menozzi, P. (2000) 'Biodiversity, disturbances, ecosystem function and management of european forests', *Forest Ecology and Management*, vol 132, pp39–50

Bennett, A. and Radford, J. Q. (2007) 'Emergent Properties of Land Mosaics: Implications for Land Management and Biodiversity', in B. D. Lindenmayer and N. T. Hobbs (eds) *Managing and Designing Landscapes for Conservation*, pp201–214, Blackwell, Oxford

Bennett, A. F., Radford, J. Q. and Haslem, A. (2006) 'Properties of land mosaics: Implications for nature conservation in agricultural environments', *Biological Conservation*, vol 133, pp250–264

Bennett, E. L. (2000) 'Timber certification: Where is the voice of the biologist?', *Conservation Biology*, vol 14, pp921–923

Bergeron, Y., Harvey, B., Leduc, A. and Gauthier, S. (1999) 'Forest management guidelines based on natural disturbance dynamics: Stand- and forest-level considerations', *Forestry Chronicle*, vol 75, pp49–54

Beukema, H., Danielsen, F., Vincent, G., Hardiwinoto, S. and van Andel, J. (2007) 'Plant and bird diversity in rubber agroforests in the lowlands of Sumatra, Indonesia', *Agroforestry Systems*, vol 70, pp217–242

Bhagwat, S. A., Willis, K. J., Birks, H. J. B. and Whittaker, R. J. (2008) 'Agroforestry: A refuge for tropical biodiversity?', *Trends in Ecology & Evolution*, vol 23, pp261–267

Bibby, C. J. (1999) 'Making the most of birds as environmental indicators', *Ostrich*, vol 70, pp81–88

Bockstaller, C. and Girardin, P. (2003) 'How to validate environmental indicators', *Agricultural Systems*, vol 76, pp639–653

Bodmer, R. E., Fang, T. G., Moya, L. and Gill, R. (1994) 'Managing wildlife to conserve Amazonian forests: Population biology and economic-considerations of game hunting', *Biological Conservation*, vol 67, pp29–35

Bolker, B. (2008) *Ecological Models and Data in R*, Princeton University Press, Princeton

Bolker, B. M., Brooks, M. E., Clark, C. J., Geange, S. W., Poulsen, J. R., Stevens, M. H. H. and White, J. S. S. (2009) 'Generalized linear mixed models: A practical guide for ecology and evolution', *Trends in Ecology & Evolution*, vol 24, pp127–135

Bonn, A., Rodrigues, A. S. L. and Gaston, K. J. (2002) 'Threatened and endemic species: Are they good indicators of patterns of biodiversity on a national scale?', *Ecology Letters*, vol 5, pp733–741

Borcard, D., Legendre, P., Avois-Jacquet, C. and Tuomisto, H. (2004) 'Dissecting the spatial structure of ecological data at multiple scales', *Ecology*, vol 85, pp1826–1832

Bormann, B. T., Haynes, R. W. and Martin, J. R. (2007) 'Adaptive management of forest ecosystems: Did some rubber hit the road?', *Bioscience*, vol 57, pp186–191

Bowen, M. E., McAlpine, C. A., House, A. P. N. and Smith, G. C. (2007) 'Regrowth forests on abandoned agricultural land: A review of their habitat values for recovering forest fauna', *Biological Conservation*, vol 140, pp273–296

Bradshaw, C. J. A., Sodhi, N. S. and Brook, B. W. (2008) 'Tropical turmoil: A biodiversity tragedy in progress', *Frontiers in Ecology and the Environment*, vol 7, pp79–87 DOI: 10.1890/070193

Bradshaw, C. J. A., Warkentin, I. G. and Sodhi, N. S. (in press) 'Urgent preservation of boreal carbon stocks and biodiversity', *Trends in Ecology & Evolution*

Brashares, J. S. and Sam, M. K. (2005) 'How much is enough? Estimating the minimum sampling required for effective monitoring of African reserves', *Biodiversity and Conservation*, vol 14, pp2709–2722

Broadbent, E. N., Asner, G. P., Keller, M., Knapp, D. E., Oliviera, P. J. C. and Silva, J. N. M. (2008) 'Forest fragmentation and edge effects from deforestation and selective logging in the Brazilian Amazon', *Biological Conservation*, vol 141, pp1745–1757

Brockerhoff, E. G., Jactel, H., Parrotta, J. A., Quine, C. P. and Sayer, J. (2008) 'Plantation forests and biodiversity: Oxymoron or opportunity?', *Biodiversity and Conservation*, vol 17, pp925–951

Brockerhoff, E. G., Liebhold, A. M. and Jactel, H. (2006) 'The ecology of forest insect invasions and advances in their management', *Canadian Journal of Forest Research – Revue Canadienne De Recherche Forestiere*, vol 36, pp263–268

Brook, B. W., Bradshaw, C. J. A., Koh, L. P. and Sodhi, N. S. (2006) 'Momentum drives the crash: Mass extinction in the tropics', *Biotropica*, vol 38, pp302–305

Brook, B. W., Sodhi, N. S. and Bradshaw, C. J. A. (2008) 'Synergies among extinction drivers under global change', *Trends in Ecology & Evolution*, vol 23, pp453–460

Brooks, R. P., O'Connell, T. J., Wardrop, D. H. and Jackson, L. E. (1998) 'Towards a regional index of biological integrity: The example of forested riparian ecosystems', *Environmental Monitoring and Assessment*, vol 51, pp131–143

Brooks, T., da Fonseca, G. A. B. and Rodrigues, A. S. L. (2004a) 'Species, data, and conservation planning', *Conservation Biology*, vol 18, pp1682–1688

Brooks, T. M., da Fonseca, G. A. B. and Rodrigues, A. S. L. (2004b) 'Protected areas and species', *Conservation Biology*, vol 18, pp616–618

Brown Jr, K. S. (1997) 'Diversity, disturbance, and sustainable use of neotropical forests: Insects as indicators for conservation monitoring', *Journal of Insect Conservation*, vol 1, pp25–42

Brown, N. R., Noss, R. F., Diamond, D. D. and Myers, M. N. (2001) 'Conservation biology and forest certification: Working together toward ecological sustainability', *Journal of Forestry*, vol 99, pp18–25

Bryce, S. A. (2006) 'Development of a bird integrity index: Measuring avian response to disturbance in the Blue Mountains of Oregon, USA', *Environmental Management*, vol 38, pp470–486

Buck, A. and Rametsteiner, E. (2003) 'Monitoring, Assessment and Reporting' (IUFRO Occasional Paper No 15), in A. Buck, J. Parrotta and G. Wolfrum (eds) *Science and Technology – Building the Future of the World's Forests: Planted Forests and Biodiversity*, pp21–24, International Union of Forest Research Organizations (IUFRO), Geneva, Switzerland

Buck, A., Parrotta, J. and Wolfrum, G. (2003) *Science and Technology – Building the Future of the World's Forests: Planted Forests and Biodiversity*, Contributions to the Third Session of the United Nations Forum on Forests, International Union of Forest Research Organizations (IUFRO), Geneva, Switzerland

Bunnell, F. L. and Huggard, D. J. (1999) 'Biodiversity across spatial and temporal scales: Problems and opportunities', *Forest Ecology and Management*, vol 115, pp113–126

Burgman, M. (2005) *Risks and Decisions for Conservation and Environmental Management*, Cambridge University Press, Cambridge

Burgman, M. A., Lindenmayer, D. B. and Elith, J. (2005) 'Managing landscapes for conservation under uncertainty', *Ecology*, vol 86, pp2007–2017

Burnham, K. P. and Anderson, D. (2002) *Model Selection and Multi-model Inference*, Springer, London

Cantarello, E. and Newton, A. C. (2008) 'Identifying cost-effective indicators to assess the conservation status of forested habitats in Natura 2000 sites', *Forest Ecology and Management*, vol 256, pp815–826

Canterbury, G. E., Martin, T. E., Petit, D. R., Petit, L. J. and Bradford, D. F. (2000) 'Bird communities and habitat as ecological indicators of forest condition in regional monitoring', *Conservation Biology*, vol 14, pp544–558

Cao, Y., Williams, D. D. and Larsen, D. P. (2002) 'Comparison of ecological communities: The problem of sample representativeness', *Ecological Monographs*, vol 72, pp41–56

Cardoso, P., Borges, P. A. V. and Gaspar, C. (2007) 'Biotic integrity of the arthropod communities in the natural forests of Azores', *Biodiversity and Conservation*, vol 16, pp2883–2901

Cardoso, P., Silva, I., de Oliveira, N. G. and Serrano, A. R. M. (2004) 'Indicator taxa of spider (araneae) diversity and their efficiency in conservation', *Biological Conservation*, vol 120, pp517–524

Carignan, V. and Villard, M. A. (2002) 'Selecting indicator species to monitor ecological integrity: A review', *Environmental Monitoring and Assessment*, vol 78, pp45–61

Carnus, J. M., Parrotta, J., Brockerhoff, E., Arbez, M., Jactel, H., Kremer, A., Lamb, D., O'Hara, K. and Walters, B. (2006) 'Planted forests and biodiversity', *Journal of Forestry*, vol 104, pp65–77

Caro, T. (2010) *Surrogate species: Umbrellas, flagships, keystones, indicators and other conservation short cuts*, Island Press

Caro, T., Engilis, A., Fitzherbert, E. and Gardner, T. (2004) 'Preliminary assessment of the flagship species concept at a small scale', *Animal Conservation*, vol 7, pp63–70

Caro, T. M. and O'Doherty, G. (1999) 'On the use of surrogate species in conservation biology', *Conservation Biology*, vol 13, pp805–814

Cashore, B., Gale, F., Meidinger, E. and Newsom, D. (eds) (2006) *Confronting Sustainability: Forest Certification in Developing and Transition Countries*, Yale School of Forestry & Environmental Studies, New Haven

Castelletta, M., Sodhi, N. S. and Subaraj, R. (2000) 'Heavy extinctions of forest avifauna in Singapore: Lessons for biodiversity conservation in Southeast Asia', *Conservation Biology*, vol 14, pp1870–1880

Cauley, H. A., Peters, C. M., Donovan, R. Z. and O'Connor, J. M. (2001) 'Forest Stewardship council certification', *Conservation Biology*, vol 15, pp311–312

CBD (1996) *The Biodiversity Agenda: Decisions From the Third Meeting of the Conference of the Parties to the Convention on Biological Diversity*, United Nations, New York

CBD (2004) *Report of the Seventh Meeting of the Conference of the Parties to the Convention on Biological Diversity*, UNEP/CBD/COP/7/11(Decision VII/11), United Nations, New York

Chamberlain, T. C. (1890) 'The method of multiple working hypotheses', *Science*, vol 148, pp754–759

Chambers, J. Q., Asner, G. P., Morton, D. C., Anderson, L. O., Saatch, S. S., Espirito-Santo, F. D. B., Palace, M. and Souza, C. (2007) 'Regional ecosystem structure and function: Ecological insights from remote sensing of tropical forests', *Trends in Ecology & Evolution*, vol 22, pp414–423

Chapron, G. and Arlettaz, R. (2008) 'Conservation: Academics should "conserve or perish"', *Nature*, vol 451, pp127

Chazdon, R. L. (2003) 'Tropical forest recovery: Legacies of human impact and natural disturbances', *Perspectives in Plant Ecology Evolution and Systematics*, vol 6, pp51–71

Chazdon, R. L. (2008) 'Beyond deforestation: Restoring forests and ecosystem services on degraded lands', *Science*, vol 320 pp1458–1460

Chazdon, R. L., Harvey, C. A., Komar, O., Griffith, D. M., Ferguson, B. G., Martinez-Ramos, M., Morales, H., Nigh, R., Soto-Pinto, L., van Breugel, M. and Philpott, S. M. (2009) 'Beyond reserves: A research agenda for conserving biodiversity in human-modified tropical landscapes', *Biotropica*, vol 41, pp142–153

Chazdon, R. L., Peres, C. A., Dent, A., Sheil, D., Lugo, A. E., Lamb, D., Strork, N. E. and Miller, S. (in press) 'The potential for species conservation in tropical secondary forests', *Conservation Biology*

CICI (2003) *International Conference on the Contribution of Criteria and Indicators for Sustainable Forest Management: The Way Forward (cici 2003)*, Food and Agriculture Organization (FAO), Caracas, Venezuela

CIFOR (1999) *The CIFOR Criteria and Indicators Generic Template*, Centre for International Forestry Research (CIFOR), Jakarta, Indonesia

Clarke, K. R. and Warwick, R. M. (2001) *Change in Marine Communities: An Approach to Statistical Analysis and Intepretation*, Primer-E Ltd, Plymouth

Cleary, D. (2006) 'The questionable effectiveness of science spending by in international conservation organisations in the tropics', *Conservation Biology*, vol 20, pp733–738

Cleary, D. F. R. and Mooers, A. O. (2006) 'Burning and logging differentially affect endemic versus widely distributed butterfly species in Borneo', *Diversity and Distributions*, vol 12 pp409–416

Colwell, R. K., Mao, C. X. and Chang, J. (2004) 'Interpolating, extrapolating, and comparing incidence-based species accumulation curves', *Ecology*, vol 85, pp2717–2727

Connor, R. and Dovers, S. (2004) *Institutional Change for Sustainable Development*, Edward Elgar, Cheltenham

Constantino, P. D. L., Fortini, L. B., Kaxinawa, F. R. S., Kaxinawa, A. M., Kaxinawa, E. S., Kaxinawa, A. P., Kaxinawa, L. S., Kaxinawa, J. M. and Kaxinawa, J. P. (2008) 'Indigenous collaborative research for wildlife management in Amazonia: The case of the Kaxinawa, Acre, Brazil', *Biological Conservation*, vol 141, pp2718–2729

Crawley, M. J. (2005) *Statistics: An Introduction Using R*, John Wiley, New York

Crawley, M. J. (2007) *The R Book*, John Wiley, New York

Crist, T. O. and Veech, J. A. (2006) 'Additive partitioning of rarefaction curves and species-area relationships: Unifying alpha-, beta- and gamma-diversity with sample size and habitat area', *Ecology Letters*, vol 9, pp923–932

Cundill, G. and Fabricius, C. (2009) 'Monitoring in adaptive co-management: Toward a learning based approach', *Journal of Environmental Management*, vol 90, pp3205–3211

Curran, M. P., Miller, R. E., Howes, S. W., Maynard, D. G., Terry, T. A., Heninger, R. L., Niemann, T., van Rees, K., Powers, R. F. and Schoenholtz, S. H. (2005) 'Progress towards more uniform assessment and reporting of soil disturbance for operations, research, and sustainability protocols', *Forest Ecology and Management*, vol 220, pp17–30

Cushman, S. A. and McGarigal, K. (2002) 'Hierarchical, multi-scale decomposition of species-environment relationships', *Landscape Ecology*, vol 17, pp637–646

Cushman, S. A. and McGarigal, K. (2004) 'Hierarchical analysis of forest bird species-environment relationships in the Oregon Coast Range', *Ecological Applications*, vol 14, pp1090–1105

da Mata, R., McGeoch, M. and Tidon, R. (2008) 'Drosophilid assemblages as a bioindicator system of human disturbance in the Brazilian savanna', *Biodiversity and Conservation*, vol 17, pp2899–2916

Daily, G. C. (2001) 'Ecological forecasts', *Nature*, vol 411, pp245–245

Dale, V. H. and Beyeler, S. C. (2001) 'Challenges in the development and use of ecological indicators', *Ecological Indicators*, vol 1 pp3–10

Danielsen, F., Burgess, N. D. and Balmford, A. (2005a) 'Monitoring matters: Examining the potential of locally-based approaches', *Biodiversity and Conservation*, vol 14, pp2507–2542

Danielsen, F., Burgess, N. D., Balmford, A., Donald, P. F., Funder, M., Jones, J. P. G., Alviola, P., Balete, D. S., Blomley, T., Brashares, J., Child, B., Enghoff, M., Fjeldsa, J., Holt, S., Hubertz, H., Jensen, A. E., Jensen, P. M., Massao, J., Mendoza, M. M., Ngaga, Y., Poulsen, M. K., Rueda, R., Sam, M., Skielboe, T., Stuart-Hill, G., Topp-Jorgensen, E. and Yonten, D. (2009) 'Local participation in natural resource monitoring: A characterization of approaches', *Conservation Biology*, vol 23, pp31–42

Danielsen, F., Jensen, A. E., Alviola, P. A., Balete, D. S., Mendoza, M., Tagtag, A., Custodio, C. and Enghoff, M. (2005b) 'Does monitoring matter? A quantitative assessment of management decisions from locally-based monitoring of protected areas', *Biodiversity and Conservation*, vol 14, pp2633–2652

Danielsen, F., Mendoza, M. M., Alviola, P., Balete, D. S., Enghoff, M., Poulsen, M. K. and Jensen, A. E. (2003) 'Biodiversity monitoring in developing countries: What are we trying to achieve?', *Oryx*, vol 37, pp407–409

Danielsen, F., Mendoza, M. M., Tagtag, A., Alviola, P. A., Balete, D. S., Jensen, A. E., Enghoff, M. and Poulsen, M. K. (2007) 'Increasing conservation management action by involving local people in natural resource monitoring', *Ambio*, vol 36, pp566–570

Davis, A. J., Holloway, J. D., Huijbregts, H., Krikken, J., Kirk-Spriggs, A. H. and Sutton, S. L. (2001a) 'Dung beetles as indicators of change in the forests of northern Borneo', *Journal of Applied Ecology*, vol 38, pp593–616

Davis, L., Johnson, K. N., Bettingger, P. S. and Howard, T. E. (2001b) *Forest Management to Sustain Ecological, Economic and Social Values*, 4th edn, McGraw-Hill, New York

Dawkins, H. C. and Philip, M. S. (1998) *Tropical Moist Forest Silviculture and Management: A History of Failure and Success*, CAB International, Wallingford

DeFries, R. (2008) 'Terrestrial vegetation in the coupled human-earth system: Contributions of remote sensing ', *Annual Review of Environment and Resources*, vol 33, pp369–90

DeFries, R., Foley, J. A. and Asner, G. P. (2004) 'Land-use choices: Balancing human needs and ecosystem function', *Frontiers in Ecology and the Environment*, vol 2, pp249–257

DeFries, R., Rovero, F., Wright, P., Ahumada, J., Andelman, S., Brandon, K., Dempewolf, J., Hansen, A., Hewson, J. and Liu, J. (2009) 'From plot to landscape scale: Linking tropical biodiversity measurements across spatial scales', *Frontiers in Ecology and Evolution*, DOI: 10.1890/080104

Delgado-Acevedo, J. and Restrepo, C. (2008) 'The contribution of habitat loss to changes in body size, allometry, and bilateral asymmetry in two eleutherodactylus frogs from Puerto Rico', *Conservation Biology*, vol 22, pp773–782

DeWalt, S. J., Maliakal, S. K. and Denslow, J. S. (2003) 'Changes in vegetation structure and composition along a tropical forest chronosequence: Implications for wildlife', *Forest Ecology and Management*, vol 182, pp139–151

Di Stefano, J. (2001) 'Power analysis and sustainable forest management', *Forest Ecology and Management*, vol 154, pp141–153

Diamond, J. M. (1986) 'Overview: Laboratory Experiments, Field Experiments, and Natural Experiments', in J. M. Diamond and T. J. Case (eds) *Community Ecology*, pp3–22, Harper & Row, New York

Dickinson, J. C., Forgach, J. M. and Wilson, T. E. (2004) 'The Business of Certification', in D. J. Zarin, J. R. R. Alavalapati, F. E. Putz and M. Schmink (eds) *Working Forests in the Neotropics: Conservation Through Sustainable Management?*, pp97–118, Colombia University, New York

Didham, R. K., Ghazoul, J., Stork, N. E. and Davis, A. J. (1996) 'Insects in fragmented forests: A functional approach', *Trends in Ecology & Evolution*, vol 11, pp255–260

Didham, R. K., Tylianakis, J. M., Gernmell, N. J., Rand, T. A. and Ewers, R. M. (2007) 'Interactive effects of habitat modification and species invasion on native species decline', *Trends in Ecology & Evolution*, vol 22, pp489–496

Diffendorfer, J. E., Fleming, G. M., Duggan, J. M., Chapman, R. E., Rahn, M. E., Mitrovich, M. J. and Fisher, R. N. (2007) 'Developing terrestrial, multi-taxon indices of biological integrity: An example from coastal sage scrub', *Biological Conservation*, vol 140, pp130–141

Dixon, P. M., Olsen, A. R. and Kahn, B. M. (1998) 'Measuring trends in ecological resources', *Ecological Applications*, vol 8, pp225–227

Dovers, S. (2005) 'Clarifying the imperative of integration research for sustainable environmental management', *Journal of Research Practice*, vol 1, no 2, pp1–19

Dudley, N., Baldock, D., Nasi, R. and Stolton, S. (2005) 'Measuring biodiversity and sustainable management in forests and agricultural landscapes', *Philosophical Transactions of the Royal Society B-Biological Sciences*, vol 360, pp457–470

Dufrêne, M. and Legendre, P. (1997) 'Species assemblages and indicator species: The need for a flexible asymmetrical approach', *Ecological Monographs*, vol 67, pp345–366

Dumbrell, A. J., Clark, E. J., Frost, G. A., Randell, T. E., Pitchford, J. W. and Hill, J. K. (2008) 'Changes in species diversity following habitat disturbance are dependent on spatial scale: Theoretical and empirical evidence', *Journal of Applied Ecology*, vol 45 pp1531–1539

Dunn, R. R. (2004) Recovery of faunal communities during tropical forest regeneration', *Conservation Biology*, vol 18, pp302–309

Dunn, R. R. and Romdal, T. S. (2005) 'Mean latitudinal range sizes of bird assemblages in six neotropical forest chronosequences', *Global Ecology and Biogeography*, vol 14, pp359–366

Dunning, J. B., Danielson, B. J. and Pulliam, H. R. (1992) 'Ecological processes that affect populations in complex landscapes', *Oikos*, vol 65, pp169–175

Eberhardt, L. L. (2003) 'What should we do about hypothesis testing?', *Journal of Wildlife Management*, vol 67, pp241–247

Edenius, L. and Mikusinski, G. (2006) 'Utility of habitat suitability models as biodiversity assessment tools in forest management', *Scandinavian Journal of Forest Research*, vol 21, pp62–72

Elith, J., Graham, C. H., Anderson, R. P., Dudik, M., Ferrier, S., Guisan, A., Hijmans, R. J., Huettmann, F., Leathwick, J. R., Lehmann, A., Li, J., Lohmann, L. G., Loiselle, B. A., Manion, G., Moritz, C., Nakamura, M., Nakazawa, Y., Overton, J. M., Peterson, A. T., Phillips, S. J., Richardson, K., Scachetti-Pereira, R., Schapire, R. E., Soberon, J., Williams, S., Wisz, M. S. and Zimmermann, N. E. (2006) 'Novel methods improve prediction of species' distributions from occurrence data', *Ecography*, vol 29, pp129–151

Elliot, C. (2003) *'WWF vision for planted forests'*, Forests for Life, WWF International, delivered at UNFF Intersessional Experts Meeting on the Role of Planted Forests in Sustainable Forest Management, 24–30 March 2003, Wellington, New Zealand

Elliott, L. P. and Brook, B. W. (2007) 'Revisiting Chamberlain: Multiple working hypotheses for the 21st century', *Bioscience*, vol 57, pp608–614

Ellis, E. C. and Ramankutty, N. (2008) 'Putting people in the map: Anthropogenic biomes of the world', *Frontiers in Ecology and the Environment*, vol 6, pp439–447

Escobar, F., Halffter, G., Solis, A., Halffter, V. and Navarrete, D. (2008) 'Temporal shifts in dung beetle community structure within a protected area of tropical wet forest: A 35-year study and its implications for long-term conservation', *Journal of Applied Ecology*, vol 45, pp1584–1592

Evans, K. and Gauriguata, M. R. (2008) *Participatory Monitoring in Tropical Forest Management: A Review of Tools, Concepts and Lessons Learned*, Center for International Forestry Research, Bogor, Indonesia

Ewers, R. M. and Didham, R. K. (2006) 'Confounding factors in the detection of species responses to habitat fragmentation', *Biological Reviews*, vol 81, pp117–142

Ewers, R. M., Kapos, V., Coomes, D. A., Lafortezza, R. and Didham, (in press) 'Mapping community change in modified landscapes', *Biological Conservation*

Ewers, R. M., Thorpe, S. and Didham, R. K. (2007) 'Synergistic interactions between edge and area effects in a heavily fragmented landscape', *Ecology*, vol 88, pp96–106

Eycott, A. E., Watkinson, A. R. and Dolman, P. M. (2006) 'Ecological patterns of plant diversity in a plantation forest managed by clearfelling', *Journal of Applied Ecology*, vol 43, pp1160–1171

Failing, L. and Gregory, R. (2003) 'Ten common mistakes in designing biodiversity indicators for forest policy', *Journal of Environmental Management*, vol 68, pp121–132

FAO (2006) *Global Forest Resources Assessment 2005: Progress Towards Sustainable Forest Management*, Food and Agriculture Organization of the United Nations, Rome

FAO (2007) *State of the World's Forests 2007*, Food and Agriculture Organization of the United Nations, Rome

Faria, D., Paciencia, M. L. B., Dixo, M., Laps, R. R. and Baumgarten, J. (2007) 'Ferns, frogs, lizards, birds and bats in forest fragments and shade cacao plantations in two contrasting landscapes in the Atlantic Forest, Brazil, *Biodiversity and Conservation*, vol 16, pp2335–2357

Fazey, I., Fazey, J. A. and Fazey, D. M. A. (2005) 'Learning more effectively from experience', *Ecology and Society*, vol 10, article 4

Fazey, I., Fazey, J. A., Salisbury, J. G., Lindenmayer, D. B. and Dovers, S. (2006) 'The nature and role of experiential knowledge for environmental conservation', *Environmental Conservation*, vol 33, pp1–10

Feeley, K. J. and Terborgh, J. W. (2008) 'Direct versus indirect effects of habitat reduction on the loss of avian species from tropical forest fragments', *Animal Conservation*, vol 11, pp353–360

Feinsinger, P. (2001) *Designing Field Studies for Biodiversity Conservation*, The Nature Conservancy and Island Press, Washington DC

Felton, A. M., Engstrom, L. M., Felton, A. and Knott, C. D. (2003) 'Orangutan population density, forest structure and fruit availability in hand-logged and unlogged peat swamp forests in West Kalimantan, Indonesia', *Biological Conservation*, vol 114, pp91–101

Ferraz, G., Nichols, J. D., Hines, J. E., Stouffer, P. C., Bierregaard, R. O. and Lovejoy, T. E. (2007) 'A large-scale deforestation experiment: Effects of patch area and isolation on Amazon birds', *Science*, vol 315, pp238–241

Ferrier, S. (2002) 'Mapping spatial pattern in biodiversity for regional conservation planning: Where to from here?', *Systematic Biology*, vol 51, pp331–363

Ferrier, S. and Guisan, A. (2006) 'Spatial modelling of biodiversity at the community level', *Journal of Applied Ecology*, vol 43, pp393–404

Ferrier, S., Manion, G., Elith, J. and Richardson, K. (2007) 'Using generalized dissimilarity modelling to analyse and predict patterns of beta diversity in regional biodiversity assessment', *Diversity and Distributions*, vol 13, pp252–264

Field, S. A., O'Connor, P. J., Tyre, A. J. and Possingham, H. P. (2007) 'Making monitoring meaningful', *Austral Ecology*, vol 32, pp485–491

Field, S. A., Tyre, A. J. and Possingham, H. P. (2005) 'Optimizing allocation of monitoring effort under economic and observational constraints', *Journal of Wildlife Management*, vol 69, pp473–482

Fimbel, A. F., Grajal, A. and Robinson, J. (eds) (2001) *The Cutting Edge: Conserving Wildlife in Logged Tropical Forests*, Columbia University Press, New York

Finegan, B. (2005) 'Global Standards and Locally Adapted Forestry: The Problems of Biodiversity Indicators', in J. Sayer and S. Maginnis (eds) *Forests in Landscapes: Ecosystem Approaches to Sustainability*, pp47–58, Earthscan, London

Finegan, B., Hayes, J. P., Delgado, D. and Gretzinger, S. (2004) *Ecological Monitoring of Forest Management in the Humid Tropics: A Guide for Forest Operators and Certifiers with Emphasis on High Conservation Forests*, WWF Central America, available at www.webcatie.ac.cr/information/RFCA/rev42/page29–42.pdf (in Spanish)

Fischer, J. and Lindenmayer, D. B. (2006) 'Beyond fragmentation: The continuum model for fauna research and conservation in human-modified landscapes', *Oikos*, vol 112, pp473–480

Fischer, J. and Lindenmayer, B. D. (2007) 'Landscape modification and habitat fragmentation: A synthesis', *Global Ecology and Biogeography*, vol 16, pp265–280

Fischer, J., Lindenmayer, D. B. and Manning, A. D. (2006) 'Biodiversity, ecosystem function, and resilience: Ten guiding principles for commodity production landscapes', *Frontiers in Ecology and the Environment*, vol 4, pp80–86

Fischer, J., Manning, A. D., Steffen, W., Rose, D. B., Daniell, K., Felton, A., Garnett, S., Gilna, B., Heinsohn, R., Lindenmayer, D. B., MacDonald, B., Mills, F., Newell, B., Reid, J., Robin, L., Sherren, K. and Wade, A. (2007) 'Mind the sustainability gap', *Trends in Ecology & Evolution*, vol 22, pp621–624

Fischer, J., Peterson, G. D., Gardner, T. A., Gordon, L. J., Fazey, I., Elmqvist, T., Felton, A., Folke, C. and Dovers, S. (2009) 'Integrating resilience thinking and optimisation for conservation', *Trends in Ecology & Evolution*, vol 24, pp549–554

Fisher, B. L. (1999) 'Improving inventory efficiency: A case study of leaf-litter ant diversity in Madagascar', *Ecological Applications*, vol 9, pp714–731

Fitzherbert, E. B., Struebig, M. J., Morel, A., Danielson, F., Bruhl, C. A., Donald, P. F. and Phalan, B. (2008) 'How will oil palm expansion affect biodiversity?', *Trends in Ecology & Evolution*, vol 23, pp538–545

Folke, C., Carpenter, S., Walker, B., Scheffer, M., Elmqvist, T., Gunderson, L. and Holling, C. S. (2004) 'Regime shifts, resilience, and biodiversity in ecosystem management', *Annual Review of Ecology Evolution and Systematics*, vol 35, pp557–581

Franco, A. M. A., Palmeirim, J. M. and Sutherland, W. J. (2007) 'A method for comparing effectiveness of research techniques in conservation and applied ecology', *Biological Conservation*, vol 134, pp96–105

Franklin, J. F. (1993) 'Preserving biodiversity: Species, ecosystems, or landscapes', *Ecological Applications*, vol 3, pp202–205

Franklin, J. F. (1995) 'Scientists in wonderland', *Bioscience*, ppS74–S78

Franklin, J. F. and Lindenmayer, D. B. (2009) 'Importance of matrix habitats in maintaining biological diversity', *Proceedings of the National Academy of Sciences of the United States of America*, vol 106, pp349–350

Franklin, J. F. and Swanson, M. E. (2007) 'Forest Landscape Structure, Degradation and Condition: Some Commentary and Fundamental Principles', in B. D. Lindenmayer and R. Hobbs (eds) *Managing and Designing Landscapes for Conservation*, pp131–145, Blackwell, London

Freer-Smith, P. and Carnus, J. M. (2008) 'The sustainable management and protection of forests: Analysis of the current position globally', *Ambio*, vol 37, pp254–262

FSC (2002) *FSC Principles and Criteria for Forest Stewardship*, Forest Stewardship Council, Bonn, Germany, available at www.fsc.org

FSC (2004) *Structure and Content of Forest Stewardship Standards*, Forest Stewardship Council, Bonn, Germany, available at www.fsc.org

FSC Brazil (2004) *Padroes de Certificãcao do FSC – Forest Stewardship Council – Para Manejo de Plantações no Brasil*, Conselho Brasileiro de Manejo Florestal (FSC Brasil), Brasilia

FSC Canada (2004) *National Boreal Standard*, Forest Stewardship Council Canada Working Group

Garcia, C. A. and Lescuyer, G. (2008) 'Monitoring, indicators and community based forest management in the tropics: Pretexts or red herrings?', *Biodiversity and Conservation*, vol 17, pp1303–1317

Garden, J. G., McAlpine, C. A., Possingham, H. P. and Jones, D. N. (2007) 'Using multiple survey methods to detect terrestrial reptiles and mammals: What are the most successful and cost-efficient combinations?', *Wildlife Research*, vol 34, pp218–227

Gardner, M. J. and Altman, D. G. (2000) 'Confidence Intervals Rather Than P Values', in D. G. Altman, D. Machin, T. N. Bryant and M. J. Gardner (eds) *Statistics With Confidence: Confidence Intervals and Statistical Guidelines*, 2nd edn, pp15–27, BMJ Books, London

Gardner, T. A., Barlow, J., Araujo, I. S., Avila-Pires, T. C. S., Bonaldo, A. B., Costa, J. E., Esposito, M. C., Ferreira, L. V., Hawes, J., Hernandez, M. I. M., Hoogmoed, M., Leite, R. N., Lo-Man-Hung, N. F., Malcolm, J. R., Martins, M. B., Mestre, L. A. M., Miranda-Santos, R., Nunes-Gutjahr, A. L., Overal, W. L., Parry, L. T. W., Peters, S. L., Ribeiro-Junior, M. A., da Silva, M. N. F., da Silva Motta, C. and Peres, C. (2008a) 'The cost-effectiveness of biodiversity surveys in tropical forests', *Ecology Letters*, vol 11, pp139–150

Gardner, T. A., Barlow, J., Chazdon, R. L., Ewers, R., Harvey, C. A., Peres, C. A. and Sodhi, N. S. (2009) 'Prospects for tropical forest biodiversity in a human-modified world', *Ecology Letters*, vol 12, pp561–582

Gardner, T. A., Barlow, J., Parry, L. T. W. and Peres, C. A. (2007a) 'Predicting the uncertain future of tropical forest species in a data vacuum', *Biotropica*, vol 39, pp25–30

Gardner, T. A., Barlow, J. and Peres, C. A. (2007b) 'Paradox, presumption and pitfalls in conservation biology: Consequences of habitat change for amphibians and reptiles', *Biological Conservation*, vol 138, pp166–179

Gardner, T. A., Côté, I. M., Gill, J. A., Grant, A. and Watkinson, A. R. (2003) 'Long-term region wide declines in Caribbean coral reefs', *Science*, vol 301, pp958–960

Gardner, T. A., Hernández, M. M. I., Barlow, J. and Peres, C. (2008b) 'Understanding the biodiversity consequences of habitat change: The value of secondary and plantation forests for neotropical dung beetles', *Journal of Applied Ecology*, vol 45, pp883–893

Gardner, T. A., Ribeiro Jr, M. A., Barlow, J., Ávila-Pires, T. A. S., Hoogmoed, M. and Peres, C. A. (2007c) 'The value of primary, secondary and plantation forests for a neotropical herpetofauna', *Conservation Biology*, vol 21, pp775–787

Garnett, S. T., Sayer, J. and du Toit, J. (2007) 'Improving the effectiveness of interventions to balance conservation and development: A conceptual framework', *Ecology and Society*, vol 12, article 2

Gaston, K. (2009) 'Biodiversity', in N. Sodhi and P. Ehrlich (eds) *Conservation Biology for All*, pp27–44, Oxford University Press, Oxford

Gaston, K. J. and Fuller, R. A. (2007) 'Biodiversity and extinction: Losing the common and the widespread', *Progress in Physical Geography*, vol 31, pp213–225

Gaston, K. J. and Spicer, J. I. (2004) *Biodiversity: An Introduction*, Blackwell Publishing, Oxford

Ghazoul, J. (2001) 'Barriers to biodiversity conservation in forest certification', *Conservation Biology*, vol 15, pp315–317

Ghazoul, J. and Hellier, A. (2000) 'Setting critical limits to ecological indicators of sustainble tropical forestry', *International Forestry Review*, vol 2, pp243–253

Ghazoul, J. and McAllister, M. (2003) 'Communicating complexity and uncertainty in decision making contexts: Bayesian approaches to forest research', *International Forestry Review*, vol 5, pp9–19

Gibbs, J. P., Snell, H. L. and Causton, C. E. (1999) 'Effective monitoring for adaptive wildlife management: Lessons from the Galapagos Islands', *Journal of Wildlife Management*, vol 63, pp1055–1065

Gillison, A. (2002) 'A generic, computer-assisted method for rapid vegetation classification and survey: Tropical and temperate case studies', *Conservation Ecology*, vol 6, article 3

Gillison, A. and Liswanti, N. (2004) 'Assessing biodiversity at landscape level in northern Thailand and Sumatra (Indonesia): The importance of environmental context', *Agriculture Ecosystems & Environment*, vol 104, pp75–86

Gillman, L. N. (2008) 'Assessment of sustainable forest management in New Zealand indigenous forest', *New Zealand Geographer*, vol 64, pp57–67

Glennon, M. J. and Porter, W. F. (2005) 'Effects of land use management on biotic integrity: An investigation of bird communities', *Biological Conservation*, vol 126, pp499–511

Godoy, R. A. and Bawa, K. S. (1993) 'The economic value and sustainable harvest of plants and animals from the tropical forest: Assumptions, hypotheses and methods', *Economic Botany*, vol 47, pp215–219

Gotelli, N. J. and Colwell, R. K. (2001) 'Quantifying biodiversity: Procedures and pitfalls in the measurement and comparison of species richness', *Ecology Letters*, vol 4, pp379–391

Grace, J. B. (2006) *Structural Equation Modelling and Natural Systems*, Cambridge University Press, Cambridge

Grafen, A. and Hails, R. (2002) *Modern Statistics for the Life Sciences*, Oxford University Press, Oxford

Green, R. E., Balmford, A., Crane, P. R., Mace, G. M., Reynolds, J. D. and Turner, R. K. (2005) 'A framework for improved monitoring of biodiversity: Responses to the world summit on sustainable development', *Conservation Biology*, vol 19, pp56–65

Greenpeace (2008) *Holding the Line with FSC*, Greenpeace International Amsterdam, Netherlands

Gregory, R. D., van Strien, A., Vorisek, P., Meyling, A. W. G., Noble, D. G., Foppen, R. P. B. and Gibbons, D. W. (2005) 'Developing indicators for European birds', *Philosophical Transactions of the Royal Society B-Biological Sciences*, vol 360, pp269–288

Grenyer, R., Orme, C. D. L., Jackson, S. F., Thomas, G. H., Davies, R. G., Davies, T. J., Jones, K. E., Olson, V. A., Ridgely, R. S., Rasmussen, P. C., Ding, T. S., Bennett, P. M., Blackburn, T. M., Gaston, K. J., Gittleman, J. L. and Owens, I. P. F. (2006) 'Global distribution and conservation of rare and threatened vertebrates', *Nature*, vol 444, pp93–96

Grieser Johns, A. (2001) 'Natural Forest Management and Biodiversity Conservation: Field Study Design and Integration at the Operational Level', in R. A. Fimbel, A. Grajal and J. G. Robinson (eds) *The Cutting Edge: Conserving Wildlife in Logged Tropical Forests*, pp405–422, Colombia University, New York

Groves, C. R., Jensen, D. B., Valutis, L. L., Redford, K. H., Shaffer, M. L., Scott, J. M., Baumgartner, J. V., Higgins, J. V., Beck, M. W. and Anderson, M. G. (2002) 'Planning for biodiversity conservation: Putting conservation science into practice', *Bioscience*, vol 52, pp499–512

Grumbine, R. E. (1994) 'What is ecosystem management', *Conservation Biology*, vol 8, pp27–38

Grundel, R. and Pavlovic, N. B. (2008) 'Using conservation value to assess land restoration and management alternatives across a degraded oak savanna landscape', *Journal of Applied Ecology*, vol 45, pp315–324

Guisan, A. and Thuiller, W. (2005) 'Predicting species distribution: Offering more than simple habitat models', *Ecology Letters*, vol 8, pp993–1009

Guisan, A. and Zimmermann, N. E. (2000) 'Predictive habitat distribution models in ecology', *Ecological Modelling*, vol 135, pp147–186

Gullison, R. E. (2003) 'Does forest certification conserve biodiversity?', *Oryx*, vol 37, pp153–165

Guynn, D. C., Guynn, S. T., Layton, P. A. and Wigley, T. B. (2004) 'Biodiversity metrics in sustainable forestry certification programs', *Journal of Forestry*, vol 102, pp46–52

Hagan, J. M. and Whitman, A. A. (2006) 'Biodiversity indicators for sustainable forestry: Simplifying complexity', *Journal of Forestry*, vol 104, pp203–210

Haila, Y. (2002) 'A conceptual genealogy of fragmentation research: From island biogeography to landscape ecology ', *Ecological Applications*, vol 12, pp321–334

Haila, Y. (2007) 'Enacting Landscape Design: From Specific Cases to General Principles', in B. D. Lindenmayer and R. J. Hobbs (eds) *Managing and Designing Landscapes for Conservation: Moving from Perspectives to Principles*, pp22–34, Blackwell, London

Halffter, G. and Favila, M. E. (1993) 'The Scarabaeinae (insecta: Coleoptera), an animal group for analyzing, inventorying and monitoring biodiversity in tropical rain forest and modified landscapes', *Biology International*, vol 27, pp15–21

Halffter, G. and Matthews, E. G. (1966) 'The natural history of dung beetles in the subfamily scarabaeidae (coleoptera, scarabaeidae)', *Folia Entomoliogica Mexicana*, vol 12–14, pp1–312

Hall, C. M., Rhind, S. M. and Wilson, M. J. (2009) 'The potential for use of gastropod molluscs as bioindicators of endocrine disrupting compounds in the terrestrial environment', *Journal of Environmental Monitoring*, vol 11, pp491–497

Hall, P. and Bawa, K. (1993) 'Methods to assess the impact of extraction of non-timber tropical forest products on plant-populations', *Economic Botany*, vol 47, pp234–247

Hamer, K. C. and Hill, J. K. (2000) 'Scale-dependent effects of habitat disturbance on species richness in tropical forests', *Conservation Biology*, vol 14, pp1435–1440

Hammond, A. L. (1995) *Environmental Indicators: A Systematic Approach to Measuring and Reporting on Environmental Policy Performance in the Context of Sustainable Development*, World Resources Institute, Washington DC

Hammond, D. S. and Zagt, R. J. (2006) 'Considering background condition effects in tailoring tropical forest management systems for sustainability', *Ecology and Society*, vol 11, available at www.ecologyandsociety.org/vol11/iss1/art37/

Hansen, M. C., Stehman, S. V., Potapov, P. V., Loveland, T. R., Townshend, J. R. G., DeFries, R. S., Pittman, K. W., Arunarwati, B., Stolle, F., Steininger, M. K., Carroll, M. and DiMiceli, C. (2008) 'Humid tropical forest clearing from 2000 to 2005 quantified by using multitemporal and multiresolution remotely sensed data', *Proceedings of the National Academy of Sciences of the United States of America*, vol 105, pp9439–9444

Hanski, I. (2005) *The Shrinking World: Ecological Consequences of Habitat Loss*, International Ecology Institute, Oldendorf/Luhe

Hanski, I. and Cambefort, Y. (eds) (1991) *Dung Beetle Ecology*, xiii + 481pp, Princeton University Press, Princeton, NJ

Hanski, I. and Gilpin, M. E. (1991) 'Metapopulation dynamics: Brief history and conceptual domain', *Biological Journal of the Linnean Society*, vol 42, pp3–16

Hanski, I., Koivulehto, H., Cameron, A. and Rahagalala, P. (2007) 'Deforestation and apparent extinctions of endemic forest beetles in Madagascar', *Biology Letters*, vol 3, pp344–347

Hartley, M. J. (2002) 'Rationale and methods for conserving biodiversity in plantation forests', *Forest Ecology and Management*, vol 155, pp81–95

Hartley, S. and Kunin, W. E. (2003) 'Scale dependency of rarity, extinction risk, and conservation priority, *Conservation Biology*, vol 17, pp1559–1570

Harvey, C. A., Medina, A., Sanchez, D. M., Vilchez, S., Hernandez, B., Saenz, J. C., Maes, J. M., Casanoves, F. and Sinclair, F. L. (2006) 'Patterns of animal diversity in different forms of tree cover in agricultural landscapes', *Ecological Applications*, vol 16, pp1986–1999

Hausner, V. H., Yoccoz, N. G. and Ims, R. A. (2003) 'Selecting indicator traits for monitoring land use impacts: Birds in northern coastal birch forests', *Ecological Applications*, vol 13, pp999–1012

Hawes, J., Barlow, J., Gardner, T. A. and Peres, C. (2008) 'The value of forest strips for understorey birds in an Amazonian plantation landscape', *Biological Conservation*, vol 141, pp2262–2278

Heckenberger, M. J., Kuikuro, A., Kuikuro, U. T., Russell, J. C., Schmidt, M., Fausto, C. and Franchetto, B. (2003) 'Amazonia 1492: Pristine forest or cultural parkland?', *Science*, vol 301, pp1710–1714

Heikkinen, R. K., Luoto, M., Kuussaari, M. and Poyry, J. (2005) 'New insights into butterfly–environment relationships using partitioning methods', *Proceedings of the Royal Society B-Biological Sciences*, vol 272, pp2203–2210

Henle, K., Davies, K. F., Kleyer, M., Margules, C. and Settele, J. (2004) 'Predictors of species sensitivity to fragmentation', *Biodiversity and Conservation*, vol 13, pp207–251

Henry, P. Y., Lengyel, S., Nowicki, P., Julliard, R., Clobert, J., Celik, T., Gruber, B., Schmeller, D., Babij, V. and Henle, K. (2008) 'Integrating ongoing biodiversity monitoring: Potential benefits and methods', *Biodiversity and Conservation*, vol 17, pp3357–3382

Hess, G. R., Bartel, R. A., Leidner, A. K., Rosenfeld, K. M., Rubino, M. J., Snider, S. B. and Ricketts, T. H. (2006) 'Effectiveness of biodiversity indicators varies with extent, grain, and region', *Biological Conservation*, vol 132, pp448–457

Heyer, W. R., Donnelly, M. A., McDiarmid, R. W., Hayek, L. C. and Foster, M. S. (eds) (1994) *Measuring and Monitoring Biological Diversity: Standard Methods for Amphibians*, 364pp, Smithsonian Institution Press, Washington DC

Hickey, G. M. (2004) 'Regulatory approaches to monitoring sustainable forest management', *International Forestry Review*, vol 6, pp89–98

Hickey, G. M. and Innes, J. L. (2008) 'Indicators for demonstrating sustainable forest management in British Columbia, Canada: An international review', *Ecological Indicators*, vol 8, pp131–140

Hickey, G. M., Innes, J. L., Kozak, R. A., Bull, G. Q. and Vertinsky, I. (2005) 'Monitoring and information reporting for sustainable forest management: An international multiple case study analysis', *Forest Ecology and Management*, vol 209, pp237–259

Higman, S., Mayers, J., Bass, S., Judd, N. and Nussbaum, R. (2005) *The Sustainable Forestry Handbook*, 2nd edn, Earthscan, London

Hilborn, R. (1992) 'Can fisheries agencies learn from experience', *Fisheries*, vol 17, pp6–14

Hilborn, R. and Mangel, M. (1997) *The Ecological Detective: Confronting Models with Data*, Princeton University Press, Princeton

Hill, D., Fasham, M., Tucker, G., Shewry, M. and Shaw, P. (eds) (2005) *Handbook of Biodiversity Methods: Survey, Evaluation and Monitoring*, Cambridge University Press, Cambridge

Hill, J. K. and Hamer, K. C. (2004) 'Determining impacts of habitat modification on diversity of tropical forest fauna: The importance of spatial scale', *Journal of Applied Ecology*, vol 41, pp744–754

Hillers, A., Veith, M. and Rodelt, M. O. (2008) 'Effects of forest fragmentation and habitat degradation on West African leaf-litter frogs', *Conservation Biology*, vol 22, pp762–772

Hilty, J. and Merenlender, A. (2000) 'Faunal indicator taxa selection for monitoring ecosystem health', *Biological Conservation*, vol 92, pp185–197

Hobbs, N. T. (2003) 'Challenges and opportunities in integrating ecological knowledge across scales', *Forest Ecology and Management*, vol 181, pp223–238

Hobbs, N. T. and Hilborn, R. (2006) 'Alternatives to statistical hypothesis testing in ecology: A guide to self teaching', *Ecological Applications*, vol 16, pp5–19

Hobbs, R. J. and Huenneke, L. F. (1992) 'Disturbance, diversity and invasion: Implications for conservation', *Conservation Biology*, vol 6, pp324–337

Hobbs, R. J., Arico, S., Aronson, J., Baron, J. S., Bridgewater, P., Cramer, V. A., Epstein, P. R., Ewel, J. J., Klink, C. A., Lugo, A. E., Norton, D., Ojima, D., Richardson, D. M., Sanderson, E. W., Valladares, F., Vila, M., Zamora, R. and Zobel, M. (2006) 'Novel ecosystems: Theoretical and management aspects of the new ecological world order', *Global Ecology and Biogeography*, vol 15, pp1–7

Holck, M. H. (2008) 'Participatory forest monitoring: An assessment of the accuracy of simple cost-effective methods', *Biodiversity and Conservation*, vol 17, pp2023–2036

Holling, C. S. (1978) *Adaptive Environmental Assessment and Management*, Wiley, Chichester

Holling, C. S. (1998) 'Two cultures of ecology', *Ecology and Society*, vol 2, available at www.ecologyandsociety.org/vol2/iss2/art4/

Holling, C. S. and Meffe, G. K. (1996) 'Command and control and the pathology of natural resource management', *Conservation Biology*, vol 10, pp328–337

Holt, A. R., Gaston, K. J. and He, F. L. (2002) 'Occupancy–abundance relationships and spatial distribution: A review', *Basic and Applied Ecology*, vol 3, pp1–13

Holvoet, B. and Muys, B. (2004) 'Sustainable forest management worldwide: A comparative assessment of standards', *International Forestry Review*, vol 6, pp99–122

Hopkins, G. W. and Freckleton, R. P. (2002) 'Declines in the numbers of amateur and professional taxonomists: Implications for conservation', *Animal Conservation*, vol 5, pp245–249

Howe, R. W., Regal, R. R., Niemi, G. J., Danz, N. P. and Hanowski, J. M. (2007) 'Probability-based indicator of ecological condition', *Ecological Indicators*, vol 7, pp793–806

Hughes, R. M., Kaufmann, P. R., Herlihy, A. T., Kincaid, T. M., Reynolds, L. and Larsen, D. P. (1998) 'A process for developing and evaluating indices of fish assemblage integrity', *Canadian Journal of Fisheries and Aquatic Sciences*, vol 55, pp1618–1631

Hulbert, S. H. (1971) 'The nonconcept of species diversity: A critique and alternative parameters', *Ecology*, vol 52, pp577–586

Hulbert, S. H. (1984) 'Pseudoreplication and the design of ecological field experiments', *Ecological Monographs*, vol 54, pp187–211

Hunter, M. L. (1990) *Wildlife, Forests and Forestry. Principles of Managing Forests for Biological Diversity*, Prentice Hall, New Jersey

Hunter, M. L. (1996) 'Benchmarks for managing ecosystems: Are human activities natural?', *Conservation Biology*, vol 10, pp695–697

Hunter, M. L. (1999) *Maintaining Biodiversity in Forest Ecosystems*, Cambridge University Press, Cambridge

Huston, M. A. (1994) *Biological Diversity: The Coexistence of Species in Changing Species*, Cambridge University Press, Cambridge

Hutto, R. L. and Gallo, S. M. (2006) 'The effects of postfire salvage logging on cavity-nesting birds', *Condor*, vol 108, pp817–831

Irwin, L. L. and Wigley, T. B. (1993) 'Toward an experimental basis for protecting forest wildlife', *Ecological Applications*, vol 3, pp213–217

Isaac, N. J. B., Mallet, J. and Mace, G. M. (2004) 'Taxonomic inflation: Its influence on macroecology and conservation', *Trends in Ecology & Evolution*, vol 19, pp464–469

ITTO (2006) *Status of Tropical Forest Management 2005*, International Tropical Timber Organization, Yokohama, Japan

ITTO (2007) *Annual Review and Assessment of the World Timber Situation 2007*, International Tropical Timber Organization, Yokohama, Japan

IUCN (2004) 'Ecosystem approaches and sustainable forest management', a discussion paper for the UNFF Secretariat prepared by the Forest Conservation Program (IUCN), Program on Forests (PROFOR), and the World Bank, Gland, Switzerland

James, A. N., Gaston, K. J. and Balmford, A. (1999) 'Balancing the Earth's accounts', *Nature*, vol 401, pp323–324

Janzen, D. H. (2004) 'Setting up tropical biodiversity for conservation through non-damaging use: Participation by parataxonomists', *Journal of Applied Ecology*, vol 41, pp181–187

Jenkins, M., Green, R. E. and Madden, J. (2003) 'The challenge of measuring global change in wild nature: Are things getting better or worse?', *Conservation Biology*, vol 17, pp20–23

Jennings, S., Nussbaum, R., Judd, N. and Evans, T. (2003) *The High Conservation Value Forest Toolkit*, ProForest, Oxford

Johnson, J. B. and Omland, K. S. (2004) 'Model selection in ecology and evolution', *Trends in Ecology & Evolution*, vol 19, pp101–108

Jones, C. G., Lawton, J. H. and Shachak, M. (1994) 'Organisms as ecosystem engineers', *Oikos*, vol 69, pp373–386

Jonsson, B. G. and Jonsell, M. (1999) 'Exploring potential biodiversity indicators in boreal forests', *Biodiversity and Conservation*, vol 8, pp1417–1433

Joseph, L. N., Wintle, B., Ensbey, M., Tyre, D., O'Connor, P. and Possingham, H. (in press) 'Optimal monitoring for conservation', *Conservation Biology*

Kainer, K. A., DiGiano, M. L., Duchelle, A. E., Wadt, L. H. O., Bruna, E. and Dain, J. L. (2009) 'Partnering for greater success: Local stakeholders and research in tropical biology and conservation', *Biotropica*, vol 41, pp555–562

Kanowski, J., Catterall, C. P. and Wardell-Johnson, G. W. (2005) 'Consequences of broadscale timber plantations for biodiversity in cleared rainforest landscapes of tropical and subtropical Australia', *Forest Ecology and Management*, vol 208, pp359–372

Kareiva, P., Watts, S., McDonald, R. and Boucher, T. (2007) 'Domesticated nature: Shaping landscapes and ecosystems for human welfare', *Science*, vol 316, pp1866–1869

Karr, J. R. (1991) 'Biological integrity: A long-neglected aspect of water-resource management', *Ecological Applications*, vol 1, pp66–84

Karr, J. R. (1993) 'Measuring Biological Integrity: Lessons From Streams', in S. Woodley, J. Kay and G. Francis (eds) *Ecological Integrity and the Management of Ecosystems*, pp83–104, St. Lucie Press, Ottawa

Karr, J. R. and Chu, E. W. (1999) *Restoring Life in Running Waters: Better Biological Monitoring*, Island Press, Washington DC

Karr, J. R. and Dudley, D. R. (1981) 'Ecological perspective on water-quality goals', *Environmental Management*, vol 5, pp55–68

Karsenty, A. and Gourlet-Fleury, S. (2006) 'Assessing sustainability of logging practices in the Congo Basin's managed forests: The issue of commercial species recovery', *Ecology and Society*, vol 1, article 26

Kattan, G. H., Alvarezlopez, H. and Giraldo, M. (1994) 'Forest fragmentation and bird extinctions: San Antonio 80 years later', *Conservation Biology*, vol 8, pp138–146

Keough, H. L. and Blahna, D. J. (2006) 'Achieving integrative, collaborative ecosystem management', *Conservation Biology*, vol 20, pp1373–1382

Kessler, M., Abrahamczyk, S., Bos, M., Buchori, D., Putra, D. D., Gradstein, S. R., Hohn, P., Kluge, J., Orend, F., Pitopang, R., Saleh, S., Schulze, C. H., Sporn, S. G., Steffan-Dewenter, I., Tjitrosoedirdjo, S. S. and Tscharntke, T. (in press) 'Alpha and beta-diversity of plants and animals along a tropical land-use gradient', *Ecological Applications*

Kiker, G. A., Bridges, T. S., Varghese, A., Seager, P. T. P. and Linkov, I. (2005) 'Application of multicriteria decision analysis in environmental decision making', Integrated Environmental Assessment and Management, vol 1, pp95–108

King, E. G. and Hobbs, R. J. (2006) 'Identifying linkages among conceptual models of ecosystem degradation and restoration: Towards an integrative framework', *Restoration Ecology*, vol 14, pp369–378

Klein, A. M., Steffan-Dewenter, I., Buchori, D. and Tscharntke, T. (2002) 'Effects of land-use intensity in tropical agroforestry systems on coffee flower-visiting and trap-nesting bees and wasps', *Conservation Biology*, vol 16, pp1003–1014

Klein, B. C. (1989) 'Effects of forest fragmentation on dung and carrion beetle communities in central Amazonia', *Ecology*, vol 70, pp1715–1725

Kneeshaw, D. D., Leduc, A., Drapeau, P., Gauthier, S., Pare, D., Carignan, R., Doucet, R., Bouthillier, L. and Messier, C. (2000) 'Development of integrated ecological standards of sustainable forest management at an operational scale', *Forestry Chronicle*, vol 76, pp481–493

Knight, A. T., Cowling, R. M. and Campbell, B. M. (2006) 'An operational model for implementing conservation action', *Conservation Biology*, vol 20, pp408–419

Koh, L. P. and Gardner, T. A. (2009) 'Conservation in Human Modified Landscapes', in N. S. Sodhi and P. E. Ehrlich (eds) *Conservation Biology*, Oxford University Press, Oxford

Koh, L. P., Dunn, R. R., Sodhi, N. S., Colwell, R. K., Proctor, H. C. and Smith, V. S. (2004) 'Species coextinctions and the biodiversity crisis', *Science*, vol 305, pp1632–1634

Koleff, P., Gaston, K. J. and Lennon, J. J. (2003) 'Measuring beta diversity for presence-absence data', *Journal of Animal Ecology*, vol 72, pp367–382

Koper, N., Schmiegelow, F. K. A. and Merrill, E. H. (2007) 'Residuals cannot distinguish between ecological effects of habitat amount and fragmentation: Implications for the debate', *Landscape Ecology*, vol 22, pp811–820

Krell, F. T. (2004) 'Parataxonomy vs. taxonomy in biodiversity studies: Pitfalls and applicability of "morphospecies" sorting', *Biodiversity and Conservation*, vol 13, pp795–812

Kremen, C. (1992) 'Assessing the indicator properties of species assemblages for natural areas monitoring', *Ecological Applications*, vol 2, pp203–217

Kremen, C. (1994) 'Biological inventory using target taxa: A case-study of the butterflies of Madagascar', *Ecological Applications*, vol 4, pp407–422

Kremen, C., Colwell, R. K., Erwin, T. L., Murphy, D. D., Noss, R. F. and Sanjayan, M. A. (1993) 'Terrestrial arthropod assemblages: Their use in conservation planning', *Conservation Biology*, vol 7, pp796–808

Kremen, C., Merenlender, A. M. and Murphy, D. D. (1994) 'Ecological monitoring: A vital need for integrated conservation and development programs in the tropics', *Conservation Biology*, vol 8, pp388–397

Kunzmann, K. (2008) 'The non-legally binding instrument on sustainable management of all types of forests: Towards a legal regime for sustainable forest management?', *German Law Journal*, vol 9, pp981–1006

Kupfer, J. A., Malanson, G. P. and Franklin, S. B. (2006) 'Not seeing the ocean for the islands: The mediating influence of matrix-based processes on forest fragmentation effects', *Global Ecology and Biogeography*, vol 15, pp8–20

Lamb, D., Erskine, P. D. and Parrotta, J. A. (2005) 'Restoration of degraded tropical forest landscapes', *Science*, vol 310, pp1628–1632

Lambeck, R. J. (1997) 'Focal species: A multi-species umbrella for nature conservation', *Conservation Biology*, vol 11, pp849–856

Lambeck, R. J. (2002) 'Focal species and restoration ecology: Response to Lindenmayer et al', *Conservation Biology*, vol 16, pp549–551

Lande, R., DeVries, P. J. and Walla, T. R. (2000) 'When species accumulation curves intersect: Implications for ranking diversity using small samples', *Oikos*, vol 89, pp601–605

Landres, P. B. (1983) 'Use of the guild concept in environmental-impact assessment', *Environmental Management*, vol 7, pp393–397

Landres, P. B., Morgan, P. and Swanson, F. J. (1999) 'Overview of the use of natural variability concepts in managing ecological systems', *Ecological Applications*, vol 9, pp1179–1188

Landres, P. B., Verner, J. and Thomas, J. W. (1988) 'Ecological uses of vertebrate indicator species: A critique', *Conservation Biology*, vol 2, pp316–328

Larsen, T. H., Lopera, A. and Forsyth, A. (2008) 'Understanding trait-dependent community disassembly: Dung beetles, density functions, and forest fragmentation', *Conservation Biology*, vol 22, pp1288–1298

Larsen, T. H., Williams, N. M. and Kremen, C. (2005) 'Extinction order and altered community structure rapidly disrupt ecosystem functioning', *Ecology Letters*, vol 8, pp538–547

Larson, M. A., Thompson, F. R., Millspaugh, J. J., Dijak, W. D. and Shifley, S. R. (2004) 'Linking population viability, habitat suitability, and landscape simulation models for conservation planning', *Ecological Modelling*, vol 180, pp103–118

Laurance, W. F. (2002) 'Hyperdynamism in fragmented habitats', *Journal of Vegetation Science*, vol 13, pp595–602

Laurance, W. F. (2007) 'Have we overstated the tropical biodiversity crisis?', *Trends in Ecology & Evolution*, vol 22, pp65–70

Laurance, W. F. (2008) 'Theory meets reality: How habitat fragmentation research has transcended island biogeographic theory', *Biological Conservation*, vol 141, pp1731–1744

Laurance, W. F., Lovejoy, T. E., Vasconcelos, H. L., Bruna, E. M., Didham, R. K., Stouffer, P. C., Gascon, C., Bierregaard, R. O., Laurance, S. G. and Sampaio, E. (2002) 'Ecosystem decay of Amazonian forest fragments: A 22-year investigation', *Conservation Biology*, vol 16, pp605–618

Lawrence, D., D'Odorico, P., Diekmann, L., DeLonge, M., Das, R. and Eaton, J. (2007) 'Ecological feedbacks following deforestation create the potential for a catastrophic ecosystem shift in tropical dry forest', *Proceedings of the National Academy of Sciences of the United States of America*, vol 104, pp20,696–20,701

Lawton, J. H. and Gaston, K. J. (2001) 'Indicator Species', in S. Levin (ed) *Encyclopedia of Biodiversity*, vol 3, pp437–450, Academic Press, San Diego

Lawton, J. H., Bignell, D. E., Bolton, B., Bloemers, G. F., Eggleton, P., Hammond, P. M., Hodda, M., Holt, R. D., Larsen, T. B., Mawdsley, N. A., Stork, N. E., Srivastava, D. S. and Watt, A. D. (1998) 'Biodiversity inventories, indicator taxa and effects of habitat modification in tropical forest', *Nature*, vol 391, pp72–76

Leary, R. F. and Allendorf, F. W. (1989) 'Fluctuating asymmetry as an indicator of stress: Implications for conservation biology', *Trends in Ecology & Evolution*, vol 4, pp214–217

Lee, K. (1993) *Compass and Gyroscope: Integrating Science and Politics for the Environment*, Island Press, Washington DC

Lees, A. C. and Peres, C. A. (2006) 'Rapid avifaunal collapse along the Amazonian deforestation frontier', *Biological Conservation*, vol 133, pp198–211

Lees, A. C. and Peres, C. A. (2008a) 'Avian life-history determinants of local extinction risk in a hyper-fragmented neotropical forest landscape', *Animal Conservation*, vol 11, pp128–137

Lees, A. C. and Peres, C. A. (2008b) 'Conservation value of remnant riparian forest corridors of varying quality for Amazonian birds and mammals', *Conservation Biology*, vol 22, pp439–449

Legendre, P. and Legendre, L. (1998) *Numerical Ecology*, Elsevier Science, Amsterdam

Legg, C. J. and Nagy, L. (2006) 'Why most conservation monitoring is, but need not be, a waste of time', *Journal of Environmental Management*, vol 78, pp194–199

Leslie, A. D. (2004) 'The impacts and mechanics of certification', *International Forestry Review*, vol 6, pp30–39

Lewis, O. T. (2001) 'Effect of experimental selective logging on tropical butterflies', *Conservation Biology*, vol 15, pp389–400

Liebsch, D., Marques, M. C. M. and Goldenberg, R. (2008) 'How long does the Atlantic rain forest take to recover after a disturbance? Changes in species composition and ecological features during secondary succession', *Biological Conservation*, vol 141, pp1717–1725

Lindbladh, M., Niklasson, M. and Nilsson, S. G. (2003) 'Long-time record of fire and open canopy in a high biodiversity forest in southeast Sweden', *Biological Conservation*, vol 114, pp231–243

Lindblom, C. (1959) 'The science of muddling through', *Public Administration Review*, vol 19, pp79–88

Lindenmayer, D. B. (1999) 'Future directions for biodiversity conservation in managed forests: Indicator species, impact studies and monitoring programs', *Forest Ecology and Management*, vol 115, pp277–287

Lindenmayer, D. B. (2003) 'Integrating Wildlife Conservation and Wood Production in Victorian Mountain Ash Forests', in D. B. Lindenmayer and J. Franklin (eds) *Towards Forest Sustainability*, CSIRO Publishing, Collingwood, Victoria

Lindenmayer, D. B. (2007) *The Variable Retention Harvest System and its Implications for Biodiversity in the Mountain Ash Forests of the Central Highlands of Victoria*, Australian National University, Canberra

Lindenmayer, D. B. (2009) *Large Scale Landscape Experiments: Lessons from Tumut*, Cambridge University Press, Cambridge

Lindenmayer, D. B. and Burgman, M. (2005) *Practical Conservation Biology*, CSIRO Publishing, Collingwood, Victoria

Lindenmayer, D. B. and Fischer, J. (2003) 'Sound science or social hook: A response to Brooker's application of the focal species approach', *Landscape and Urban Planning*, vol 62, pp149–158

Lindenmayer, D. B. and Fischer, J. (2006) *Habitat Fragmentation and Landscape Change: An Ecological and Conservation Synthesis*, Island Press, Washington DC

Lindenmayer, D. B. and Franklin, J. F. (2002) *Conserving Biodiversity: A Comprehensive Multiscaled Approach*, Island Press, Washington DC

Lindenmayer, D. B. and Franklin, J. F. (eds) (2003) *Towards Forest Sustainability*, CSIRO, Collingwood, Victoria

Lindenmayer, D. B. and Hobbs, R. J. (2004) 'Fauna conservation in Australian plantation forests: A review', *Biological Conservation*, vol 119, pp151–168

Lindenmayer, D. B. and Hobbs, R. J. (eds) (2007) *Managing and Designing Landscapes for Conservation*, Blackwell, London

Lindenmayer, D. B. and Likens, G. E. (2009) 'Adaptive monitoring: A new paradigm for long-term research and monitoring', *Trends in Ecology & Evolution*, vol 24, pp482–486

Lindenmayer, D. B., Cunningham, R. B., Donnelly, C. F. and Lesslie, R. (2002a) 'On the use of landscape surrogates as ecological indicators in fragmented forests', *Forest Ecology and Management*, vol 159, pp203–216

Lindenmayer, D. B., Cunningham, R. B., MacGregor, C., Crane, M., Michael, D., Fischer, J., Montague-Drake, R., Felton, A. and Manning, A. (2008b) 'Temporal changes in vertebrates during landscape transformation: A large-scale "natural experiment"', *Ecological Monographs*, vol 78, pp567–590

Lindenmayer, D. B., Franklin, J. F. and Fischer, J. (2006) 'General management principles and a checklist of strategies to guide forest biodiversity conservation', *Biological Conservation*, vol 131, pp433–445

Lindenmayer, D. B., Fischer, J., Felton, A., Montague-Drake, R., Manning, A. D., Simberloff, D., Youngentob, K., Saunders, D., Wilson, D., Felton, A. M., Blackmore, C., Lowe, A., Bond, S., Munro, N. and Elliott, C. P. (2007) 'The complementarity of single-species and ecosystem-oriented research in conservation research', *Oikos*, vol 116, pp1220–1226

Lindenmayer, D. B., Hobbs, R. J., Montague-Drake, R., Alexandra, J., Bennett, A., Burgman, M., Cale, P., Calhoun, A., Cramer, V., Cullen, P., Driscoll, D., Fahrig, L., Fischer, J., Franklin, J., Haila, Y., Hunter, M., Gibbons, P., Lake, S., Luck, G., MacGregor, C., McIntyre, S., Mac Nally, R., Manning, A., Miller, J., Mooney, H., Noss, R., Possingham, H., Saunders, D., Schmiegelow, F., Scott, M., Simberloff, D., Sisk, T., Tabor, G., Walker, B., Wiens, J., Woinarski, J. and Zavaleta, E. (2008a) 'A checklist for ecological management of landscapes for conservation', *Ecology Letters*, vol 11, pp78–91

Lindenmayer, D. B., Manning, A. D., Smith, P. L., Possingham, H. P., Fischer, J., Oliver, I. and McCarthy, M. A. (2002b) 'The focal-species approach and landscape restoration: A critique', *Conservation Biology*, vol 16, pp338–345

Lindenmayer, D. B., Margules, C. R. and Botkin, D. B. (2000) 'Indicators of biodiversity for ecologically sustainable forest management', *Conservation Biology*, vol 14, pp941–950

Lindenmayer, D. B., McIntyre, S. and Fischer, J. (2003b) 'Birds in eucalypt and pine forests: Landscape alteration and its implications for research models of faunal habitat use', *Biological Conservation*, vol 110, pp45–53

Linke, S. and Norris, R. (2003) 'Biodiversity: Bridging the gap between condition and conservation', *Hydrobiologia*, vol 500, pp203–211

Linke, S., Pressey, R. L., Bailey, R. C. and Norris, R. H. (2007) 'Management options for river conservation planning: Condition and conservation re-visited', *Freshwater Biology*, vol 52, pp918–938

Liu, J., Dietz, T., Carpenter, S. R., Alberti, M., Folke, C., Moran, E., Pell, A. N., Deadman, P., Kratz, T., Lubchenco, J., Ostrom, E., Ouyang, Z. Y., Provencher, W., Redman, C. L., Schneider, S. H. and Taylor, W. W. (2007) 'Complexity of coupled human and natural systems', *Science*, vol 317, pp1513–1516

Longino, J. T. and Colwell, R. K. (1997) 'Biodiversity assessment using structured inventory: Capturing the ant fauna of a tropical rain forest', *Ecological Applications*, vol 7, pp1263–1277

Lookingbill, T. R., Gardner, R. H., Townsend, P. A. and Carter, S. L. (2007) 'Conceptual models as hypotheses in monitoring urban landscapes', *Environmental Management*, vol 40, pp171–182

Lovett, G. M., Burns, D. A., Driscoll, C. T., Jenkins, J. C., Mitchell, M. J., Rustad, L., Shanley, J. B., Likens, G. E. and Haeuber, R. (2007) 'Who needs environmental monitoring?', *Frontiers in Ecology and the Environment*, vol 5, pp253–260

Ludwig, D. (2001) 'The era of management is over', *Ecosystems*, vol 4, pp758–764

Ludwig, D., Hilborn, R. and Walters, C. (1993) 'Uncertainty, resource exploitation, and conservation: Lessons from history', *Science*, vol 260, pp17–18

Lugo, A. E. and Helmer, E. (2004) 'Emerging forests on abandoned land: Puerto Rico's new forests', *Forest Ecology and Management*, vol 190, pp145–161

Mac Nally, R. (1997) 'Monitoring forest bird communities for impact assessment: The influence of sampling intensity and spatial scale', *Biological Conservation*, vol 82, pp355–367

Mac Nally, R. and Fleishman, E. (2002) 'Using "indicator" species to model species richness: Model development and predictions', *Ecological Applications*, vol 12, pp79–92

Mac Nally, R. and Fleishman, E. (2004) 'A successful predictive model of species richness based on indicator species', *Conservation Biology*, vol 18, pp646–654

Mac Nally, R., Parkinson, A., Horrocks, G., Conole, L. and Tzaros, C. (2001) 'Relationships between terrestrial vertebrate diversity, abundance and availability of coarse woody debris on south-eastern Australian floodplains', *Biological Conservation*, vol 99, pp191–205

Mace, G., Masundire, H. and Baillie, J. (2005) 'Biodiversity', in R. Hassan, R. Scholes and N. Ash (eds) *Ecosystems and Human Well-being: Current State and Trends: Findings of the Condition and Trends Working Group*, .vol 1, pp77–122, Island Press, Washington DC

Mace, G. M. (2004) 'The role of taxonomy in species conservation', *Philosophical Transactions of the Royal Society of London Series B-Biological Sciences*, vol 359, pp711–719

Mace, G. M., Collar, N. J., Gaston, K. J., Hilton-Taylor, C., Akcakaya, H. R., Leader-Williams, N., Milner-Gulland, E. J. and Stuart, S. N. (2008) 'Quantification of extinction risk: IUCN's system for classifying threatened species', *Conservation Biology*, vol 22, pp1424–1442

MacKenzie, D. I. and Kendall, W. L. (2002) 'How should detection probability be incorporated into estimates of relative abundance?', *Ecology*, vol 83, pp2387–2393

MacKenzie, D. I., Nichols, J. D., Royle, J. A., Pollock, K. H., Bailey, L. L. and Hines, J. E. (2006) *Occupancy Estimation and Modelling*, Elsevier, London

Maginnis, S. and Jackson, W. (2005) 'Balancing restoration and development', *ITTO Tropical Forest Update*, vol 5, no 2, pp4–6

Maginnis, S., Jackson, W. and Dudley, N. (2004) 'Conservation Landscapes: Whose Landscapes? Whose Trade-offs?', in T. D. McShane and M. P. Wells (eds) *Getting Biodiversity Projects to Work*, pp321–339, Columbia University Press, New York

Magnusson, W. E. and Mourão, G. M. (in press) *Estatística sem Matemática*, 2nd edn, Editora Planta, Londrina

Magnusson, W. E., Costa, F., Lima, A., Baccaro, F., Braga-Neto, R., Romero, R. L., Menin, M., Penha, J., Hero, J. M. and Lawson, B. E. (2008) 'A program for monitoring biological diversity in the Amazon: An alternative perspective to threat-based monitoring', *Biotropica*, vol 40, pp409–411

Magurran, A. E. (2004) *Measuring Biological Diversity*, Blackwell Science, Oxford

Malcolm, J. R., Campbell, B. D., Kuttner, B. G. and Sugar, A. (2004) 'Potential indicators of the impacts of forest management on wildlife habitat in northeastern Ontario: A multivariate application of wildlife habitat suitability matrices', *Forestry Chronicle*, vol 80, pp91–106

Maness, T. (2007) 'Trade-off analysis for decision-making in natural resources: Where we are and where we are headed', *BC Journal of Ecosystems and Management*, vol 8, pp1–16

Manley, P. N., Zielinski, W. J., Schlesinger, M. D. and Mori, S. R. (2004) 'Evaluation of a multiple-species approach to monitoring species at the ecoregional scale', *Ecological Applications*, vol 14, pp296–310

Manly, B. F. J., McDonald, L., Thomas, D. L., McDonald, T. L. and Erickson, W. P. (2002) *Resource Selection by Animals: Statistical Design and Analysis for Field Studies*, 2nd edn, Kluwer Academic Publishers, Dordrecht, The Netherlands

Manning, A. D., Lindenmayer, D. B. and Fischer, J. (2006) 'Stretch goals and backcasting: Approaches for overcoming barriers to large-scale ecological restoration', *Restoration Ecology*, vol 14, pp487–492

Manning, A. D., Lindenmayer, D. B. and Nix, H. A. (2004) 'Continua and Umwelt: Novel perspectives on viewing landscapes', *Oikos*, vol 104, pp621–628

Margoluis, R. and Salafsky, N. (1998) *Measures of Success: Designing, Managing, and Monitoring Conservation and Development Projects*, Island Press, Washington DC

Margoluis, R., Stem, C., Salafsky, N. and Brown, M. (2009) 'Using conceptual models as a planning and evaluation tool in conservation', *Evaluation and Program Planning*, vol 32, pp138–147

Margules, C. R. and Austin, M. P. (eds) (1991) *Nature Conservation: Cost Effective Biological Surveys and Data Analysis*, CSIRO, Melbourne

Margules, C. R. and Sarkar, S. (2007) *Systematic Conservation Planning*, Cambridge University Press, Cambridge

Margules, C. R., Pressey, R. L. and Williams, P. H. (2002) 'Representing biodiversity: Data and procedures for identifying priority areas for conservation', *Journal of Biosciences*, vol 27, pp309–326

Mason, D. (1996) 'Responses of venezuelan understory birds to selective logging, enrichment strips, and vine cutting', *Biotropica*, vol 28, pp296–309

McAfee, B. J., Malouin, C. and Fletcher, N. (2006) 'Achieving forest biodiversity outcomes across scales, jurisdictions and sectors with cycles of adaptive management integrated through criteria and indicators', *Forestry Chronicle*, vol 82, pp321–334

McAlpine, C. A. and Eyre, T. J. (2002) 'Testing landscape metrics as indicators of habitat loss and fragmentation in continuous eucalypt forests (Queensland, Australia)', *Landscape Ecology*, vol 17, pp711–728

McAlpine, C. A., Spies, T. A., Norman, P. and Peterson, A. (2007) 'Conserving forest biodiversity across multiple land ownerships: Lessons from the northwest forest plan and the southeast Queensland regional forests agreement (Australia)', *Biological Conservation*, vol 134, pp580–592

McCarthy, M. A. (2007) *Bayesian Methods for Ecology*, Cambridge University Press, Cambridge

McCarthy, M. A., Parris, K., van der Ree, R., McDonnell, M. J., Burgman, M. A., Williams, N. S. G., McLean, N., Harper, M. J., Meyer, R., Hahs, A. and Coates, T. (2004) 'The habitat hectares approach to vegetation assessment: An evaluation and suggestions for improvement', *Ecological Management and Restoration*, vol 5, pp24–27

McComb, B. C., Spies, T. A. and Olsen, K. A. (2007) 'Sustaining biodiversity in the Oregon Coast Range: Potential effects of forest policies in a multi-ownership province', *Ecology and Society*, vol 12, article 29

McDonald, G. T. and Lane, M. B. (2004) 'Converging global indicators for sustainable forest management', *Forest Policy and Economics*, vol 6, pp63–70

McElhinny, C., Gibbons, P., Brack, C. and Bauhus, J. (2005) 'Forest and woodland stand structural complexity: Its definition and measurement', *Forest Ecology and Management*, vol 218, pp1–24

McGeoch, M. A. (1998) 'The selection, testing and application of terrestrial insects as bioindicators', *Biological Reviews*, vol 73, pp181–201

McGeoch, M. A. (2007) 'Insects and Bioindication: Theory and Progress', in A. J. A. Stewart, T. R. New and O. T. Lewis (eds) *Insect Conservation Biology*, pp144–174, CABI Publishing, Wallingford

McGeoch, M. A. and Chown, S. L. (1998) 'Scaling up the value of bioindicators', *Trends in Ecology & Evolution*, vol 13, pp46–47

McGeoch, M. A., Van Rensburg, B. J. and Botes, A. (2002) 'The verification and application of bioindicators: A case study of dung beetles in a savanna ecosystem', *Journal of Applied Ecology*, vol 39, pp661–672

McGill, B. J., Etienne, R. S., Gray, J. S., Alonso, D., Anderson, M. J., Benecha, H. K., Dornelas, M., Enquist, B. J., Green, J. L., He, F. L., Hurlbert, A. H., Magurran, A. E., Marquet, P. A., Maurer, B. A., Ostling, A., Soykan, C. U., Ugland, K. I. and White, E. P. (2007) 'Species abundance distributions: Moving beyond single prediction theories to integration within an ecological framework', *Ecology Letters*, vol 10, pp995–1015

McGinley, K. and Finegan, B. (2003) 'The ecological sustainability of tropical forest management: Evaluation of the national forest management standards of Costa Rica and Nicaragua, with emphasis on the need for adaptive management', *Forest Policy and Economics*, vol 5, pp421–431

McNeely, J. A. (2004) 'Nature vs. nurture: Managing relationships between forests, agroforestry and wild biodiversity', *Agroforestry Systems*, vol 61, pp155–165

McNeely, J. A. and Scherr, S. J. (2003) *Ecoagriculture: Strategies to Feed the World*, Island Press, Washington DC

Meijaard, E. and Sheil, D. (2008) 'The persistence and conservation of Borneo's mammals in lowland rain forests managed for timber: Observations, overviews and opportunities', *Ecological Research*, vol 23, pp21–34

Meijaard, E., Sheil, D., Nasi, R. and Stanley, S. A. (2006) 'Wildlife conservation in Bornean timber concessions', *Ecology and Society*, vol 11, article 47

Meijerink, G. (2008) 'The Role of Measurement Problems and Monitoring in PES Schemes', in R. B. Dellink and A. Ruijs (eds) *Economics of Poverty, Environment and Natural Resource Use*, pp61–85, Springer, Dordrecht

Mendoza, G. A. and Martins, H. (2006) 'Multi-criteria decision analysis in natural resource management: A critical review of methods and new modelling paradigms', *Forest Ecology and Management*, vol 230, pp1–22

Metzger, J. P. (2009) 'Conservation issues in the Brazilian Atlantic Forest', *Biological Conservation*, vol 142, pp1138–1140

Metzger, J. P., Martensen, A. C., Dixo, M., Bernacci, L. C., Ribeiro, M. C., Teixeira, A. M. G. and Pardini, R. (2009) 'Time-lag in biological responses to landscape changes in a highly dynamic Atlantic Forest region', *Biological Conservation*, vol 142, pp1166–1177

Miles, L. and Kapos, V. (2008) 'Reducing greenhouse gas emissions from deforestation and forest degradation: Global land-use implications', *Science*, vol 320, pp1454–1455

Mills, L. S., Soule, M. E. and Doak, D. F. (1993) 'The keystone species concept in ecology and conservation', *Bioscience*, vol 43, pp219–224

Milner-Gulland, E. J. and Rowcliffe, J. M. (2007) *Conservation and Sustainable Use: A Handbook of Techniques*, Oxford University Press, Oxford

Moilanen, A., Wilson, K. A. and Possingham, H. (2009) *Spatial Conservation Prioritization: Quantitative Methods and Computational Tools*, Oxford University Press, Oxford

Moreira, F. M. D., Huising, E. J. and Bignell, D. E. (eds) (2008) *A Handbook of Tropical Soil Biology*, Earthscan, London

Mortensen, H. S., Dupont, Y. L. and Olesen, J. M. (2008) 'A snake in paradise: Disturbance of plant reproduction following extirpation of bird flower-visitors on Guam', *Biological Conservation*

Müller, J. and Brandl, R. (2009) 'Relationships between habitat parameters derived by airborne lidar and forest beetle assemblages', *Journal of Applied Ecology*

Muller, J., Bussler, H., Gossner, M., Rettelbach, T. and Duelli, P. (2008) 'The European spruce bark beetle Ips typographus in a national park: From pest to keystone species', *Biodiversity and Conservation*, vol 17, pp2979–3001

Murilo, S. D., Magnusson, W. E. and Zuanon, J. (2009) 'Effects of reduced impact logging on fish assemblages in central amazonia', *Conservation Biology*

Murphy, D. D. and Noon, B. D. (1991) 'Coping with uncertainty in wildlife biology', *Journal of Wildlife Management*, vol 55, pp773–782

Naidoo, R., Balmford, A., Ferraro, P. J., Polasky, S., Ricketts, T. and Rouget, M. (2006) 'Integrating economic costs into conservation planning', *Trends in Ecology & Evolution*, vol 21, pp681–687

Nature (2007) 'The great divide', *Nature*, vol 450, pp135–136

Neel, M. C., McGarigal, K. and Cushman, S. A. (2004) 'Behavior of class-level landscape metrics across gradients of class aggregation and area ', *Landscape Ecology*, vol 19, pp435–455

New, T. R. (1998) *Invertebrate Surveys for Conservation*, Oxford University Press, Oxford

Newton, A. (2007) *Forest Ecology and Conservation: A Handbook of Techniques*, Oxford University Press, Oxford

Newton, A. (2008) *Forest Ecology and Conservation: A Handbook of Techniques*, Oxford University Press, Oxford

Newton, A. C. and Kapos, V. (2002) 'Biodiversity indicators in national forest inventories', *Unaslyva*, vol 53, pp56–64

Newton, A. C. and Oldfield, S. (2005) 'Forest Policy, the Precautionary Principle and Sustainable Forest Management', (eds) *Earthscan*, London

Nichols, E., Gardner, T. A., Peres, C. A. and Spector, S. (2009) 'Co-declining mammals and dung beetles: An impending ecological cascade', *Oikos*, vol 118, pp481–487

Nichols, E., Larsen, T. B., Spector, S., Davis, A. L. V., Escobar, F., Favila, M., Vulinec, K. and Network., T. S. R. (2007) 'Global dung beetle response to tropical forest modification and fragmentation: A quantitative literature review and meta-analysis', *Biological Conservation*, vol 137, pp1–19

Nichols, J. D. and Williams, B. K. (2006) 'Monitoring for conservation', *Trends in Ecology & Evolution*, vol 21, pp668–673

Nichols, L. A., Spector, S., Louzada, J. N. C., Larsen, T., Amezquita, S. and Favila, M. E. (2008) 'Ecological functions and ecosystem services provided by Scarabaeinae dung beetles', *Biological Conservation*, vol 141, pp1461–1474

Nielsen, S. E., Bayne, E. M., Schieck, J., Herbers, J. and Boutin, S. (2007) 'A new method to estimate species and biodiversity intactness using empirically derived reference conditions', *Biological Conservation*, vol 137, pp403–414

Nielsen, S. E., Johnson, C. J., Heard, D. C. and Boyce, M. S. (2005) 'Can models of presence-absence be used to scale abundance? Two case studies considering extremes in life history', *Ecography*, vol 28, pp197–208

Niemeijer, D. and de Groot, R. S. (2008) 'A conceptual framework for selecting environmental indicator sets', *Ecological Indicators*, vol 8, pp14-25

Niemelä, J. (2000) 'Biodiversity monitoring for decision-making', *Annales Zoologici Fennici*, vol 37, pp307–317

Niemelä, J., Young, J., Alard, D., Askasibar, M., Henle, K., Johnson, R., Kurttila, M., Larsson, T. B., Matouch, S., Nowicki, P., Paiva, R., Portoghesi, L., Smulders, R., Stevenson, A., Tartes, U. and Watt, A. (2005) 'Identifying, managing and monitoring conflicts between forest biodiversity conservation and other human interests in Europe', *Forest Policy and Economics*, vol 7, pp877–890

Niemi, G. J. and McDonald, M. E. (2004) Application of ecological indicators', *Annual Review of Ecology Evolution and Systematics*, vol 35, pp89–111

Nilsson, S. G., Hedin, J. and Niklasson, M. (2001) 'Biodiversity and its assessment in boreal and nemoral forests', *Scandinavian Journal of Forest Research*, vol 16, pp10–26

Nilsson, S. V. (2009) 'Selecting Biodiversity Indicators to Set Conservation Targets: Species, Structures, or Processes?', in M.-A. Villard and B. G. Jonsson (eds) *Setting Conservation Targets for Managed Forest Landscapes*, Cambridge University Press, Cambridge

Norris, K. (2008) 'Agriculture and biodiversity conservation: Opportunity knocks', *Conservation Letters*, vol 1, pp2–11

Noss, R. F. (1990) 'Indicators for monitoring biodiversity: A hierarchical approach', *Conservation Biology*, vol 4, pp355–364

Noss, R. F. (1996) 'Ecosystems as conservation targets', *Trends in Ecology & Evolution*, vol 11, pp351–351

Noss, R. F. (1999) 'Assessing and monitoring forest biodiversity: A suggested framework and indicators', *Forest Ecology and Management*, vol 115, pp135–146

Noss, R. F. (2004) 'Some suggestions for keeping national wildlife refuges healthy and whole', *Natural Resources Journal*, vol 44, pp1093–1111

Noss, R. F. and Cooperrider, A. Y. (1994) *Saving Nature's Legacy: Protecting and Restoring Biodiversity*, Island Press, Washington DC

Nussbaum, R. and Simula, M. (2005) *The Forest Certification Handbook*, 2nd edn, Earthscan, London

Nussbaum, R., Jennings, S. and Garforth, M. (2002) *Assessing Forest Certification Schemes: A Practical Guide*, ProForest, available at www.proforest.net/, Oxford

O'Connell, T. J., Jackson, L. E. and Brooks, R. P. (2000) 'Bird guilds as indicators of ecological condition in the central Appalachians', *Ecological Applications*, vol 10, pp1706–1721

O'Hara, R. B. (2005) 'Species richness estimators: How many species can dance on the head of a pin?', *Journal of Animal Ecology*, vol 74, pp375–386

Oliver, I. and Beattie, A. J. (1996) 'Designing a cost-effective invertebrate survey: A test of methods for rapid assessment of biodiversity', *Ecological Applications*, vol 6, pp594–607

Oliver, I., Beattie, A. J. and York, A. (1998) 'Spatial fidelity of plant, vertebrate, and invertebrate assemblages in multiple-use forest in eastern Australia', *Conservation Biology*, vol 12, pp822–835

Oliver, I., Jones, H. and Schmoldt, D. L. (2007) 'Expert panel assessment of attributes for natural variability benchmarks for biodiversity ', *Austral Ecology*, vol 32, pp453–475

Oliver, I., Mac Nally, R. and York, A. (2000) 'Identifying performance indicators of the effects of forest management on ground-active arthropod biodiversity using hierarchical partitioning and partial canonical correspondence analysis', *Forest Ecology and Management*, vol 139, pp21–40

Overton, J. M., Stephens, R. T. T., Leathwick, J. R. and Lehmann, A. (2002) 'Information pyramids for informed biodiversity conservation', *Biodiversity and Conservation*, vol 11, pp2093–2116

Ozinga, S. (2001) *Behind the Logo. An Environmental and Social Assessment of Forest Certification Schemes*, FERN, Moreton-in-Marsh

Paine, R. T. (1966) 'Food complexity and species diversity', *American Naturalist*, vol 100, pp65–75

Palmer, M., Bernhardt, E., Chornesky, E., Collins, S., Dobson, A., Duke, C., Gold, B., Jacobson, R., Kingsland, S., Kranz, R., Mappin, M., Martinez, M. L., Micheli, F., Morse, J., Pace, M., Pascual, M., Palumbi, S., Reichman, O. J., Simons, A., Townsend, A. and Turner, M. (2004) 'Ecology for a crowded planet', *Science*, vol 304, pp1251–1252

Pardini, R., Faria, D., Accacio, G. M., Laps, R. R., Mariano-Neto, E., Paciencia, M. L. B., Dixo, M. and Baumgarten, J. (2009) 'The challenge of maintaining Atlantic Forest biodiversity: A multi-taxa conservation assessment of specialist and generalist species in an agro-forestry mosaic in southern Bahia', *Biological Conservation*, vol 142, pp1178–1190

Parkes, D., Newell, G. and Cheal, D. (2003) 'Assessing the quality of native vegetation: The "habitat hectares" approach', *Ecological Management and Restoration*, vol 4, ppS29–S38

Parrish, J. D., Braun, D. P. and Unnasch, R. S. (2003) 'Are we conserving what we say we are? Measuring ecological integrity within protected areas', *Bioscience*, vol 53, pp851–860

Parry, L., Barlow, J. and Peres, C. A. (2009) 'Allocation of hunting effort by Amazonian smallholders: Implications for conserving wildlife in mixed-use landscapes', *Biological Conservation*, vol 142, pp1777–1786

Pawar, S. (2003) 'Taxonomic chauvinism and the methodologically challenged', *Bioscience*, vol 53, pp861–864

Pawar, S. S., Birand, A. C., Ahmed, M. F., Sengupta, S. and Shankar Raman, T. R. (2006) 'Conservation biogeography in north-east India: Hierarchical analysis of cross-taxon distributional congruence', *Diversity and Distributions*, vol 13, pp53–65 DOI: 10.1111/j.1472-4642.2006.00298.x

Pearce, J. and Venier, L. (2005) 'Small mammals as bioindicators of sustainable boreal forest management', *Forest Ecology and Management*, vol 208, pp153–175

Pearce, J. L. and Venier, L. A. (2006) 'The use of ground beetles (coleoptera : Carabidae) and spiders (araneae) as bioindicators of sustainable forest management: A review', *Ecological Indicators*, vol 6, pp780–793

Pearman, P. B. (1997) 'Correlates of amphibian diversity in an altered landscape of Amazonian Ecuador', *Conservation Biology*, vol 11, pp1211–1225

Pearman, P. B. and Weber, D. (2007) 'Common species determine richness patterns in biodiversity indicator taxa', *Biological Conservation*, vol 138, pp109–119

Pearson, D. L. (1994) 'Selecting indicator taxa for the quantitative assessment of biodiversity', *Philosophical Transactions of the Royal Society of London Series B-Biological Sciences*, vol 345, pp75–79

Pearson, D. L. and Cassola, F. (1992) 'Worldwide species richness patterns of tiger beetles (coleoptera, Cicindelidae) – Indicator taxon for biodiversity and conservation studies', *Conservation Biology*, vol 6, pp376–391

Peck, S. B. and Forsyth, A. (1982) 'Composition, structure, and competitive behavior in a guild of Ecuadorian rain-forest dung beetles (coleoptera, Scarabaeidae)', *Canadian Journal of Zoology – Revue Canadienne De Zoologie*, vol 60, pp1624–1634

Pereira, H. M. and Cooper, H. D. (2006) 'Towards the global monitoring of biodiversity change', *Trends in Ecology & Evolution*, vol 21, pp123–129

Peres, C. A. (2000) 'Effects of subsistence hunting on vertebrate community structure in Amazonian forests', *Conservation Biology*, vol 14, pp240–253

Peres, C. A. and Lake, I. R. (2003) 'Extent of nontimber resource extraction in tropical forests: Accessibility to game vertebrates by hunters in the Amazon basin', *Conservation Biology*, vol 17, pp521–535

Peres, C. A. and Michalski, F. (2006) 'Synergistic Effects of Habitat Disturbance and Hunting in Amazonian Forest Fragments', in W. F. Laurance and C. A. Peres (eds) *Emerging Threats to Tropical Forests*, pp105–126, University of Chicago Press, Chicago

Perfecto, I. and Vandermeer, J. (2008) 'Biodiversity conservation in tropical agroecosystems', *Annals of the New York Academy of Science*, vol 1134, pp173–200

Perfecto, I., Vandermeer, J., Mas, A. and Pinto, L. S. (2005) 'Biodiversity, yield, and shade coffee certification', *Ecological Economics*, vol 54, pp435–446

Peters, R. H. (1991) *A Critique for Ecology*, Cambridge University Press, Cambridge

Peters, S. L., Malcolm, J. R. and Zimmerman, B. L. (2006) 'Effects of selective logging on bat communities in the southeastern Amazon', *Conservation Biology*, vol 20, pp1410–1421

Phillips, O. L., Baker, T. R., Arroyo, L., Higuchi, N., Killeen, T. J., Laurance, W. F., Lewis, S. L., Lloyd, J., Malhi, Y., Monteagudo, A., Neill, D. A., Vargas, P. N., Silva, J. N. M., Terborgh, J., Martinez, R. V., Alexiades, M., Almcida, S., Brown, S., Chave, J., Comiskey, J. A., Czimczik, C. I., Di Fiore, A., Erwin, T., Kuebler, C., Laurance, S. G., Nascimento, H. E. M., Olivier, J., Palacios, W., Patino, S., Pitman, N. C. A., Quesada, C. A., Salidas, M., Lezama, A. T. and Vinceti, B. (2004) 'Pattern and process in Amazon tree turnover, 1976–2001', *Philosophical Transactions of the Royal Society of London Series B-Biological Sciences*, vol 359, pp381–407

Philpott, S. M., Arendt, W. J., Armbrecht, I., Bichier, P., Diestch, T. V., Gordon, C., Greenberg, R., Perfecto, I., Reynoso-Santos, R., Soto-Pinto, L., Tejeda-Cruz, C., Williams-Linera, G., Valenzuela, J. and Manuel Zolotoff, J. (2008) 'Biodiversity loss in Latin American coffee landscapes: Review of evidence on ants, birds, and trees', *Conservation Biology*, vol 22, pp1093–1105

Pimentel, D., Stachow, U., Takacs, D. A., Brubaker, H. W., Dumas, A. R., Meaney, J. J., Oneil, J. A. S., Onsi, D. E. and Corzilius, D. B. (1992) 'Conserving biological diversity in agricultural forestry systems: Most biological diversity exists in human-managed ecosystems', *Bioscience*, vol 42, pp354–362

Pineda, E., Moreno, C., Escobar, F. and Halffter, G. (2005) 'Frog, bat, and dung beetle diversity in the cloud forest and coffee agroecosystems of Veracruz, Mexico', *Conservation Biology*, vol 19, pp400–410

Pinheiro, J. C. and Bates, D. (2000) *Mixed-effects Models in S and S-plus*, Springer-Verlag, New York

Platt, J. R. (1964) 'Strong inference', *Science*, vol 146, pp347–353

Pohl, G. R., Langor, D. W. and Spence, J. R. (2007) 'Rove beetles and ground beetles (coleoptera: Staphylinidae, carabidae) as indicators of harvest and regeneration practices in western Canadian foothills forests', *Biological Conservation*, vol 137, pp294–307

Pokorny, B. and Adams, M. (2003) 'What do criteria and indicators assess? An analysis of five C&I sets relevant for forest management in the Brazilian Amazon', *International Forestry Review*, vol 5, pp20–28

Polasky, S., Nelson, E., Camm, J., Csuti, B., Fackler, P., Lonsdorf, E., Montgomery, C. A., White, D., Arthur, J. L., Garber-Yonts, B., Haight, R., Kagan, J., Starfield, A. and Tobalske, C. (2008) 'Where to put things? Spatial land management to sustain biodiversity and economic returns', *Biological Conservation*, vol 141, pp1505–1524

Poore, D. (2003) *Changing Landscapes*, Earthscan, London

Posa, M. R. C. and Sodhi, N. S. (2006) 'Effects of anthropogenic land use on forest birds and butterflies in Subic Bay, Philippines', *Biological Conservation*, vol 129, pp256–270

Possingham, H.P. (2001) *The Business of Biodiversity: Applying Decision Theory Principles to Nature Conservation*, Earthwatch Institute, Melbourne

Possingham, H. P. and Nicholson, E. (2007) 'Principles of Landscape Design that Emerge from a Formal Problem Solving Approach', in B. D. Lindenmayer and N. T. Hobbs (eds) *Managing and Designing Landscapes for Conservation*, Blackwell Publishing, London

Possingham, H.P., Andelman, S., Burgman, M., Medellin, R. A., Master, L. L. and Keith, D. A. (2002) 'Limits to the use of threatened species lists', *Trends in Ecology & Evolution*, vol 17, pp503–507

Possingham, H. P., Andelman, S. J., Noon, B. R., Trombulak, S. and Pulliam, H. R. (2001) 'Making Smart Conservation Decisions', in M. E. Soulé and G. H. Orians (eds) *Conservation Biology: Research Priorities for the Next Decade*, pp225–244, Island Press, Washington DC

Poulson, J., Appelgate, G. and Raymond, D. (2001) *Linking C&I to a Code of Practice for Industrial Tropical Tree Plantations*, Center for International Forestry Research (CIFOR), Jakarta, Indonesia

Pounds, J. A., Bustarnante, M. R., Coloma, L. A., Consuegra, J. A., Fogden, M. P. L., Foster, R. N., Marca, E. L., Master, K. L., Merino-Viteri, A., Puschendorf, R., Santiago, R. R., Sanchez-Azofeifa, A., Still, C. J. and Young, B. E. (2006) 'Widespread amphibian extinctions from epidemic disease driven by global warming', *Nature*, vol 439, pp161–167

Prabhu, R., Colfer, C. J. P. and Dudley, R. G. (1999) *Guidelines for Developing, Testing and Selecting Criteria and Indicators for Sustainable Forest Management*, Criteria and Indicators Toolbox Series 1, CIFOR, Bogor, Indonesia

Prabhu, R., Colfer, C. J. P., Venkateswarlu, P., Tan, L. C., Soekmadi, R. and Wollenberg, E. (1996) *Testing Criteria and Indicators for Sustainable Management of Forest, Phase 1*, Final Report, CIFOR Special Publications, Bogor, Indonesia

Prabhu, R., Ruitenbeek, H. J., Boyle, T. J. B. and Colfer, C. J. P. (2001) 'Between Voodoo Science and Adaptive Management: The Role and Research Needs for Indicators of Sustainable Forest Management', in J. Raison, A. Brown and D. Flinn (eds) *Criteria and Indicators for Sustainable Forest Management*, pp39–66, IUFRO Research Series No. 7, CABI Publishing, Wallingford

Prato, T. (2005) 'Bayesian adaptive management of ecosystems', *Ecological Modelling*, vol 183, pp147–156

Prendergast, J. R. and Eversham, B. C. (1997) 'Species richness covariance in higher taxa: Empirical tests of the biodiversity indicator concept', *Ecography*, vol 20, pp210–216

Prendergast, J. R., Quinn, R. M., Lawton, J. H., Eversham, B. C. and Gibbons, D. W. (1993) 'Rare species, the coincidence of diversity hotspots and conservation strategies', *Nature*, vol 365, pp335–337

Presley, S. J., Willig, M. R., Wunderle, J. M. and Saldanha, L. N. (2008) 'Effects of reduced-impact logging and forest physiognomy on bat populations of lowland Amazonian forest', *Journal of Applied Ecology*, vol 45, pp14–25

Proulx, R. and Parrott, L. (2008) 'Measures of structural complexity in digital images for monitoring the ecological significance of an old growth forest ecosystem ', *Ecological Indicators*, vol 8, pp270–284

Prugh, L. R., Hodges, K. E., Sinclair, A. R. E. and Brashares, J. S. (2008) 'Effect of habitat area and isolation on fragmented animal populations', *Proceedings of the National Academy of Sciences of the United States of America*, vol 105, pp20,770–20,775

Pullin, A. S. and Stewart, G. B. (2006) 'Guidelines for systematic review in conservation and environmental management', *Conservation Biology*, vol 20, pp1647–1656

Putz, F. E. and Romero, C. (2001) 'Biologists and timber certification', *Conservation Biology*, vol 15, pp313–314

Putz, F. E., Blate, G. M., Redford, K. H., Fimbel, R. and Robinson, J. (2001) 'Tropical forest management and conservation of biodiversity: An overview', *Conservation Biology*, vol 15, pp7–20

Putz, F. E., Redford, K. H., Robinson, J., Fimbel, R. and Blate, G. M. (2000) *Biodiversity Conservation in the Context of Tropical Forest Management*, World Bank Environment Department, Washington DC

Putz, F. E., Sisk, P., Fredericksen, T. and Dykstra, D. (2008) 'Reduced-impact logging: Challenges and opportunities', *Forest Ecology and Management*, vol 256, pp1427–1433

Quinn, J. P. and Keough, M. J. (2002) *Experimental Design and Data Analysis for Biologists*, Cambridge University Press, Cambridge

Quintero, I. and Roslin, T. (2005) 'Rapid recovery of dung beetle communities following habitat fragmentation in central Amazonia', *Ecology*, vol 12, pp3303–3311

Raffa, K. F., Aukema, B. H., Bentz, B. J., Carroll, A. L., Hicke, J. A., Turner, M. G. and Romme, W. H. (2008) 'Cross-scale drivers of natural disturbances prone to anthropogenic amplification: The dynamics of bark beetle eruptions', *Bioscience*, vol 58, pp501–517

Rainio, J. and Niemela, J. (2003) 'Ground beetles (coleoptera: Carabidae) as bioindicators', *Biodiversity and Conservation*, vol 12, pp487–506

Raison, J. and Rabb, M. A. (2001) 'Guiding Concepts for the Applications of Indicators to Interpret Change in Soil Properties and Processes in Forests', in J. Raison, A. G. Brown and D. W. Flinn (eds) *Criteria and Indicators for Sustainable Forest Management*, pp231–258, CABI, Wallingford

Raison, J., Brown, A., Brown, A. G., Franc, A. and Flinn, D. (eds) (2001) *Criteria and Indicators for Sustainable Forest Management*, CABI, Wallingford

Rametsteiner, E. and Simula, M. (2003) 'Forest certification: An instrument to promote sustainable forest management?', *Journal of Environmental Management*, vol 67, pp87–98

Rapp, V. (2008) *Northwest Forest Plan: The First 10 Years (1994–2003): First-decade Results of the Northwest Forest Plan*, US Department of Agriculture, Forest Service, Pacifc Northwest Research Station

Redford, K. H. (1992) 'The empty forest', *Bioscience*, vol 42, pp412–422

Regan, H. M., Colyvan, M. and Burgman, M. A. (2002) 'A taxonomy and treatment of uncertainty for ecology and conservation biology', *Ecological Applications*, vol 12, pp618–628

Regan, H. M., Hierl, L. A., Franklin, J., Deutschman, D. H., Schmalbach, H. L., Winchell, C. S. and Johnson, B. S. (2008) 'Species prioritization for monitoring and management in regional multiple species conservation plans', *Diversity and Distributions*, vol 14, pp462–471

Rempel, R. S. and Kaufmann, C. K. (2003) 'Spatial modeling of harvest constraints on wood supply versus wildlife habitat objectives', *Environmental Management*, vol 32, pp334–347

Rempel, R. S., Andison, D. W. and Hannon, S. J. (2004) 'Guiding principles for developing an indicator and monitoring framework', *Forestry Chronicle*, vol 80, pp82–90

Rhodes, J. R., Callaghan, J. G., McAlpine, C. A., De Jong, C., Bowen, M. E., Mitchell, D. L., Lunney, D. and Possingham, H. P. (2008) 'Regional variation in habitat-occupancy thresholds: A warning for conservation planning', *Journal of Applied Ecology*, vol 45, pp549–557

Ribeiro Jr, M. A., Gardner, T. A. and Avila-Pires, T. C. S. (2008) 'Evaluating the effectiveness of herpetofaunal sampling techniques across a habitat change gradient in a neotropical forest landscape', *Journal of Herpetology*, vol 42, pp733–749

Ribeiro, M. C., Metzger, J. P., Martensen, A. C., Ponzoni, F. J. and Hirota, M. M. (2009) 'The Brazilian Atlantic Forest: How much is left, and how is the remaining forest distributed? Implications for conservation', *Biological Conservation*, vol 142, pp1141–1153

Richter, B. D. and Redford, K. H. (1999) 'The art (and science) of brokering deals between conservation and use', *Conservation Biology*, vol 13, pp1235–1237

Rickenbach, M. and Overdevest, C. (2006) 'More than markets: Assessing Forest Stewardship Council (FSC) certification as a policy tool', *Journal of Forestry*, vol 104, pp143–147

Roberge, J. M. and Angelstam, P. (2004) 'Usefulness of the umbrella species concept as a conservation tool', *Conservation Biology*, vol 18, pp76–85

Roberts, J. (2001) 'Catchment and Process Studies in Forest Hydrology: Implications for Indicators of Sustainable Forest Management', in J. Raison, A. G. Brown and D. W. Flinn (eds) *Criteria and Indicators for Sustainable Forest Management*, pp259–310, CABI, Wallingford

Robinson, J. G. (1993) 'The limits to caring: Sustainable living and the loss of biodiversity', *Conservation Biology*, vol 7, pp20–28

Robinson, W. D. (1999) 'Long-term changes in the avifauna of Barro Colorado Island, Panama, a tropical forest isolate', *Conservation Biology*, vol 13, pp85–97

Rockstrom, J., Steffen, W., Noone, K., Persson, A., Chapin, F. S., Lambin, E. F., Lenton, T. M., Scheffer, M., Folke, C., Schellnhuber, H. J., Nykvist, B., de Wit, C. A., Hughes, T., van der Leeuw, S., Rodhe, H., Sorlin, S., Snyder, P. K., Costanza, R., Svedin, U., Falkenmark, M., Karlberg, L., Corell, R. W., Fabry, V. J., Hansen, J., Walker, B., Liverman, D., Richardson, K., Crutzen, P. and Foley, J. A. (2009) 'A safe operating space for humanity', *Nature*, vol 461, pp472–475

Rodrigues, A. S. L., Andelman, S. J., Bakarr, M. I., Boitani, L., Brooks, T. M., Cowling, R. M., Fishpool, L. D. C., da Fonseca, G. A. B., Gaston, K. J., Hoffmann, M., Long, J. S., Marquet, P. A., Pilgrim, J. D., Pressey, R. L., Schipper, J., Sechrest, W., Stuart, S. N., Underhill, L. G., Waller, R. W., Watts, M. E. J. and Yan, X. (2004) 'Effectiveness of the global protected area network in representing species diversity', *Nature*, vol 428, pp640–643

Rohr, J. R., Mahan, C. G. and Kim, K. C. (2007) 'Developing a monitoring program for invertebrates: Guidelines and a case study', *Conservation Biology*, vol 21, pp422–433

Romesburg, H. C. (1981) 'Wildlife science: Gaining reliable knowledge', *Journal of Wildlife Management*, vol 45, pp293–313

Rondon, X. J., Gorchov, D. L. and Cornejo, F. (2009) 'Tree species richness and composition 15 years after strip clear-cutting in the Peruvian Amazon', *Plant Ecology*, vol 201, pp23–37

Rosenzweig, M. L. (2003) *Win–Win Ecology. How Earth's Species Can Survive in the Midst of Human Enterprise*, Oxford University Press, New York

Routledge, R. D. (1979) 'Diversity indexes: Which ones are admissible', *Journal of Theoretical Biology*, vol 76, pp503–515

Sala, O. E. (ed) (2005) 'Biodiversity across scenarios'. *Millenium Ecosystem Assessment*, Available online: www.maweb.org

Salafsky, N. and Margoluis, R. (1999) *Greater Than the Sum of Their Parts: Designing Conservation and Development Programs to Maximise Results and Learning*, Biodiversity Support Program, Washington DC

Salafsky, N. and Margoluis, R. (2003) 'What conservation can learn from other fields about monitoring and evaluation', *Bioscience*, vol 53, pp120–122

Salafsky, N. and Margoluis, R. (2004) 'Using Adaptive Management to Improve ICDPs', in T. O. McShane and M. P. Wells (eds) *Getting Biodiversity Projects to Work: Toward More Effective Conservational Development*, pp372–394, Colombia University Press, New York

Salafsky, N. and Salzer, D. (2005) 'The unglamorous essential foundation of conservation science', *Oryx*, vol 39, pp235–236

Salafsky, N., Margoluis, R., Redford, K. H. and Robinson, J. G. (2002) 'Improving the practice of conservation: A conceptual framework and research agenda for conservation science', *Conservation Biology*, vol 16, pp1469–1479

Salafsky, N., Salzer, D., Stattersfield, A. J., Hilton-Taylor, C., Neugarten, R., Butchart, S. H. M., Collen, B., Cox, N., Master, L. L., O'Connor, S. and Wilkie, D. (2008) 'A standard lexicon for biodiversity conservation: Unified classifications of threats and actions', *Conservation Biology*, vol 22, pp897–911

Sanderson, E. W., Redford, K. H., Vedder, A., Coppolillo, P. B. and Ward, S. E. (2002) 'A conceptual model for conservation planning based on landscape species requirements', *Landscape and Urban Planning*, vol 58, pp41–56

Santos, E. M. R., Franklin, E. and Magnusson, W. E. (2008) 'Cost-efficiency of subsampling protocols to evaluate oribatid-mite communities in an Amazonian savanna', *Biotropica*, vol 40, pp728–735

Sayer, J. and Maginnis, S. (eds) (2005a) *Forests in Landscapes: Ecosystem Approaches to Sustainability*, Earthscan, London

Sayer, J. and Maginnis, S. (2005b) 'Forests in Landscapes: Expanding Horizons for Ecosystem Forestry', in J. Sayer and S. Maginnis (eds) *Forests in Landscapes: Ecosystem Approaches to Sustainability*, pp177–192, Earthscan, London

Sayer, J. and Maginnis, S. (2005c) 'New Challenges for Forest Management, in J. Sayer and S. Maginnis (eds) *Forests in Landscapes: Ecosystem Approaches to Sustainability*, pp1–16, Earthscan, London

Sayer, J. A. and Campbell, B. M. (2004) *The Science of Sustainable Development: Local Livelihoods and the Global Environment*, Cambridge University Press, Cambridge

Scales, B. R. and Marsden, S. J. (2008) 'Biodiversity in small-scale tropical agroforests: A review of species richness and abundance shifts and the factors influencing them', *Environmental Conservation*, vol 35, pp160–172

Scheiner, S. M. (2001) 'Theories, Hypotheses and Statistics', in S. M. Scheiner and J. Gurevitch (eds) *Design and Analysis of Ecological Experiments*, Oxford University Press, Oxford

Scheiner, S. M. and Gurevitch, J. (eds) (2001) *Design and Analysis of Ecological Experiments*, Oxford University Press, Oxford

Scherr, S. J. and McNeely, J. A. (eds) (2007) *Farming with Nature: The Science and Practice of Ecoagriculture*, Island Press, Washington DC

Schiller, A., Hunsaker, C. T., Kane, M. A., Wolfe, A. K., Dale, V. H., Suter, G. W., Russell, C. S., Pion, G., Jensen, M. H. and Konar, V. C. (2001) 'Communicating ecological indicators to decision makers and the public', *Conservation Ecology*, vol 5, article 19

Schlaepfer, R., Gorgerat, V. and Butler, R. (2004) *A Comparative Analysis between Sustainable Forest Management (SFM) and the Ecosystem Approach (EA)*, Swiss Agency for Environment, Forests and Landscape, Swiss Forest Agency, Laboratory for Ecosystem Management, Swiss Federal Institute of Technology, Lousanne

Schmeller, D. S., Henry, P. Y., Julliard, R., Gruber, B., Clobert, J., Dziock, F., Lengyel, S., Nowicki, P., Deri, E., Budrys, E., Kull, T., Tali, K., Bauch, B., Settele, J., Van Swaay, C., Kobler, A., Babij, V., Papastergiadou, E. and Henle, K. (2009) 'Advantages of volunteer-based biodiversity monitoring in Europe', *Conservation Biology*, vol 23, pp307–316

Schmidt, B. R. (2003) 'Count data, detection probabilities, and the demography, dynamics, distribution, and decline of amphibians', *Comptes Rendus Biologies*, vol 326, ppS119–S124

Schmidt, B. R. and Meyer, A. H. (2008) 'On the analysis of monitoring data: Testing for no trend in population size', *Journal for Nature Conservation*, vol 16, pp157–163

Schmitt, C. B., Belokurov, A., Besançon, C., Boisrobert, L., Burgess, N. D., Campbell, A., Coad, L., Fish, L., Gliddon, D., Humphries, K., Kapos, V., Loucks, C., Lysenko, I., Miles, L., Mills, C., Minnemeyer, S., Pistorius, T., Ravilious, C., Steininger, M. and Winkel, G. (2008) *Global Ecological Forest Classification and Forest Protected Area Gap Analysis; Analyses and recommendations in view of the 10% target for forest protection under the Convention on Biological Diversity* (CBD), University of Freiburg, Freiburg

Schroth, G. and Harvey, C. A. (2007) 'Biodiversity conservation in cocoa production landscapes: An overview', *Biodiversity and Conservation*, vol 16, pp2237–2244

Schroth, G., Da Fonseca, G. A. B., Harvey, C. A., Gascon, C., Vasconcelos, D. and Izac, A.-M., N. (eds) (2004) *Agroforestry and Biodiversity Conservation in Tropical Landscapes*, Island Press, Washington DC

Schulze, C. H., Waltert, M., Kessler, P. J. A., Pitopang, R., Shahabuddin, Veddeler, D., Muhlenberg, M., Gradstein, S. R., Leuschner, C., Steffan-Dewenter, I. and Tscharntke, T. (2004) 'Biodiversity indicator groups of tropical land-use systems: Comparing plants, birds, and insects', *Ecological Applications*, vol 14, pp1321–1333

Scott, M. J., Heglund, P. and Morrison, M. L. (eds) (2002) *Predicting Species Occurrences: Issues of Accuracy and Scale*, Island Press, London

Sekercioglu, C. H. (2002) 'Effects of forestry practices on vegetation structure and bird community of Kibale National Park, Uganda', *Biological Conservation*, vol 107, pp229–240

Sekercioglu, C. H. (2006) 'Increasing awareness of avian ecological function', *Trends in Ecology & Evolution*, vol 21, pp464–471

Sekercioglu, C. H., Ehrlich, P. R., Daily, G. C., Aygen, D., Goehring, D. and Sandi, R. F. (2002) 'Disappearance of insectivorous birds from tropical forest fragments', *Proceedings of the National Academy of Sciences of the United States of America*, vol 99, pp263–267

Sekercioglu, C. H., Loarie, S. R., Brenes, F. O., Ehrlich, P. R. and Daily, G. C. (2007) 'Persistence of forest birds in the Costa Rican agricultural countryside', *Conservation Biology*, vol 21, pp482–494

Sergio, F., Caro, T., Brown, D., Clucas, B., Hunter, J., Ketchum, J., McHugh, K. and Hiraldo, F. (2008) 'Top predators as conservation tools: Ecological rationale, assumptions, and efficacy', *Annual Review of Ecology Evolution and Systematics*, vol 39, pp1–19

Sheil, D. (2001) 'Conservation and biodiversity monitoring in the tropics: Realities, priorities and distractions', *Conservation Biology*, vol 15, pp1179–1182

Sheil, D. (2002) 'Why doesn't biodiversity monitoring support conservation priorities in the tropics?', *Unaslyva*, vol 53, pp50–54

Sheil, D. and Lawrence, A. (2004) 'Tropical biologists, local people and conservation: New opportunities for collaboration', *Trends in Ecology & Evolution*, vol 19, pp634–638

Sheil, D., Nasi, R. and Johnson, B. (2004) 'Ecological criteria and indicators for tropical forest landscapes: Challenges in the search for progress', *Ecology and Society*, vol 9, article 7

Sigel, B. J., Sherry, T. W. and Young, B. E. (2006) 'Avian community response to lowland tropical rainforest isolation: 40 years of change at La Selva Biological Station, Costa Rica', *Conservation Biology*, vol 20, pp111–121

Simberloff, D. (1998) 'Flagships, umbrellas, and keystones: Is single-species management passe in the landscape era?', *Biological Conservation*, vol 83, pp247–257

Simberloff, D. and Dayan, T. (1991) 'The guild concept and the structure of ecological communities', *Annual Review of Ecology and Systematics*, vol 22, pp115–143

Similä, M., Kouki, J., Monkkonen, M., Sippola, A. L. and Huhta, E. (2006) 'Co-variation and indicators, of species diversity: Can richness of forest-dwelling species be predicted in northern boreal forests?', *Ecological Indicators*, vol 6, pp686–700

Sist, P., Sheil, D., Kartawinata, K. and Priyadi, H. (2003) 'Reduced-impact logging in Indonesian Borneo: Some results confirming the need for new silvicultural prescriptions', *Forest Ecology and Management*, vol 179, pp415–427

Slik, J. W. F., Bernard, C. S., Breman, F. C., Van Beek, M., Salim, A. and Sheil, D. (2008) 'Wood density as a conservation tool: Quantification of disturbance and identification of conservation-priority areas in tropical forests', *Conservation Biology*, vol 22, pp1299–1308

Smith, C. T., Gordon, A. D., Payn, T. W., Richardson, B., Schoenholtz, S. H., Skinner, M. F., Snowdon, P. and West, G. G. (2001) 'Indicators for Sustained Productive Capacity of New Zealand and Australian Plantation Forests', in R. J. Raison, A. G. Brown and D. W. Flinn (eds) *Criteria and Indicators for Sustainable Forest Management*, pp183–198, CABI, Wallingford

Smith, G. F., Gittings, T., Wilson, M., French, L., Oxbrough, A., O'Donoghue, S., O'Halloran, J., Kelly, D. L., Mitchell, F. J. G., Kelly, T., Iremonger, S., McKee, A. M. and Giller, P. (2008) 'Identifying practical indicators of biodiversity for stand-level management of plantation forests', *Biodiversity and Conservation*, vol 17, pp991–1015

Smyth, A., Mac Nally, R. and Lamb, D. (2002) 'Comparative influence of forest management and habitat structural factors on the abundances of hollow-nesting bird species in subtropical Australian eucalypt forest', *Environmental Management*, vol 30, pp547–559

Soberón, J., Rodriguez, P. and Vazquez-Dominguez, E. (2000) 'Implications of the hierarchical structure of biodiversity for the development of ecological indicators of sustainable use', *Ambio*, vol 29, pp136–142

Sodhi, N. S., Lee, T. M., Koh, L. P. and Prawiradilaga, D. M. (2006) 'Long-term avifaunal impoverishment in an isolated tropical woodlot', *Conservation Biology*, vol 20, pp772–779

Soulé, M. E., Estes, J. A., Miller, B. and Honnold, D. L. (2005) 'Strongly interacting species. Conservation policy, management, and ethics', *Bioscience*, vol 55, pp168–176

Souza, C. M., Roberts, D. A. and Cochrane, M. A. (2005) 'Combining spectral and spatial information to map canopy damage from selective logging and forest fires', *Remote Sensing of Environment*, vol 98, pp329–343

Spector, S. (2006) 'Scarabaeine dung beetles (Coleoptera: Scarabaeidae: Scarabaeinae): An invertebrate focal taxon for biodiversity research and conservation', *Coleopterists Bulletin*, vol 60, pp71–83

Spector, S. and Forsyth, A. B. (1998) 'Indicator Taxa in the Vanishing Tropics', in A. Balmford and G. Mace (eds) *Conservation in a Changing World*, pp181–210, Cambridge University Press, Cambridge

Spellerberg, I. F. and Sawyer, J. W. D. (1996) 'Standards for biodiversity: A proposal based on biodiversity standards for forest plantations', *Biodiversity and Conservation*, vol 5, pp447–459

Spies, T. A., Johnson, K. N., Burnett, K. M., Ohmann, J. L., McComb, B. C., Reeves, G. H., Bettinger, P., Kline, J. D. and Garber-Yonts, B. (2007) 'Cumulative ecological and socioeconomic effects of forest policies in coastal Oregon', *Ecological Applications*, vol 17, pp5–17

Srbek-Araujo, A. C. and Chiarello, A. G. (2008) 'Domestic dogs in Atlantic Forest preserves of south-eastern Brazil: A camera-trapping study on patterns of entrance and site occupancy rates', *Brazilian Journal of Biology*, vol 68, pp771–779

Stadt, J. J., Schieck, J. and Stelfox, H. A. (2006) 'Alberta biodiversity monitoring program: Monitoring effectiveness of sustainable forest management planning', *Environmental Monitoring and Assessment*, vol 121, pp33–46

Stankey, G. H., Bormann, B. T., Ryan, C., Shindler, B., Sturtevant, V., Clark, R. N. and Philpot, C. (2003) 'Adaptive management and the northwest forest plan: Rhetoric and reality', *Journal of Forestry*, vol 101, pp40–46

Stattersfield, A. J. and Capper, D. R. (eds) (2000) *Threatened Birds of the World*, Birdlife International, Cambridge

Stauffer, D. F. (2002) 'Linking Populations and Habitats: Where Have We Been? Where Are We Going?', in J. M. Scott, P. J. Heglund, M. L. Morrison, J. B. Haufler, M. G. Raphael, W. A. Wall and F. B. Samson (eds) *Predicting Species Occurences: Issues of Accuracy and Scale*, pp53–61, Island Press, Washington DC

Steffan-Dewenter, I., Munzenberg, U., Burger, C., Thies, C. and Tscharntke, T. (2002) 'Scale-dependent effects of landscape context on three pollinator guilds', *Ecology*, vol 83, pp1421–1432

Stem, C., Margoluis, R., Salafsky, N. and Brown, M. (2005) 'Monitoring and evaluation in conservation: A review of trends and approaches', *Conservation Biology*, vol 19, pp295–309

Stephens, P. A., Buskirk, S. W., Hayward, G. D. and Del Rio, C. M. (2005) 'Information theory and hypothesis testing: A call for pluralism', *Journal of Applied Ecology*, vol 42, pp4–12

Stephens, P. A., Buskirk, S. W., Hayward, G. D. and Martinez del Rio, C. (2007) 'A call for statistical pluralism answered', *Journal of Applied Ecology*, vol 44, pp461–463

Stoddard, J. L., Larsen, D. P., Hawkins, C. P., Johnson, R. K. and Norris, R. H. (2006) 'Setting expectations for the ecological condition of streams: The concept of reference condition', *Ecological Applications*, vol 16, pp1267–1276

Storch, I. and Bissonette, J. A. (2003) 'The Problem of Linking Scales in the Use of Indicator Species in Conservation Biology', in J. A. Bissonette and I. Storch (eds) *Landscape Theory and Resource Management: Linking Theory to Practice*, Island Press, Covelo, California

Stork, N. E., Boyle, T. J., Dale, V., Eeley, H., Finegan, B., Lawes, M., Manokaran, N., Prabhu, R. and Soberon, J. (1997) 'Criteria and indicators for assessing the sustainability of forest management: Conservation of biodiversity', Working Paper No. 17, Center for International Forestry Research, Bogor, Indonesia

Su, J. C., Debinski, D. M., Jakubauskas, M. E. and Kindscher, K. (2004) 'Beyond species richness: Community similarity as a measure of cross-taxon congruence for coarse-filter conservation', *Conservation Biology*, vol 18, pp167–173

Summerville, K. S. and Crist, T. O. (2002) 'Effects of timber harvest on forest Lepidoptera: Community, guild, and species responses', *Ecological Applications*, vol 12, pp820–835

Sutherland, W. J. (ed) (1996) *Ecological Census Techniques: A Handbook*, Cambridge University Press, Cambridge

Sutherland, W. J. (2006) 'Predicting the ecological consequences of environmental change: A review of the methods', *Journal of Applied Ecology*, vol 43, pp599–616

Sutherland, W. J., Newton, I. and Green, R. E. (2004a) *Bird Ecology and Conservation: A Handbook of Techniques*, Oxford University Press, Oxford

Sutherland, W. J., Pullin, A. S., Dolman, P. M. and Knight, T. M. (2004b) 'The need for evidenced-based conservation', *Trends in Ecology & Evolution*, vol 19, pp305–308

Suzuki, N. and Olson, D. H. (2008) 'Options for biodiversity conservation in managed forest landscapes of multiple ownerships in Oregon and Washington, USA', *Biodiversity and Conservation*, vol 17, pp1017–1039

Szaro, R. C. (1986) 'Guild management: An evaluation of avian guilds as a predictive tool', *Environmental Management*, vol 10, pp681–688

Tabarelli, M., Da Silva, M. J. C. and Gascon, C. (2004) 'Forest fragmentation, synergisms and the impoverishment of neotropical forests', *Biodiversity and Conservation*, vol 13, pp1419–1425

Taper, M. L. and Lele, S. R. (2004) *The Nature of Scientific Evidence: Statistical, Philosophical and Empirical Considerations*, Chicago University Press, Chicago

Taylor, B., Kremsater, L. and Ellis, R. (1997) *Adaptive Management of Forests in British Columbia*, British Columbia Ministry of Forests. Forest Practices Branch, Victoria, BC

Tear, T. H., Kareiva, P., Angermeier, P. L., Comer, P., Czech, B., Kautz, R., Landon, L., Mehlman, D., Murphy, K., Ruckelshaus, M., Scott, J. M. and Wilhere, G. (2005) 'How much is enough? The recurrent problem of setting measurable objectives in conservation', *Bioscience*, vol 55, pp835–849

Teder, T., Moora, M., Roosaluste, E., Zobel, K., Partel, M., Koljalg, U. and Zobel, M. (2007) 'Monitoring of biological diversity: A common-ground approach', *Conservation Biology*, vol 21, pp313–317

Terborgh, J. (1974) 'Preservation of natural diversity: Problem of extinction prone species', *Bioscience*, vol 24, pp715–722

Terborgh, J., Feeley, K., Silman, M., Nunez, P. and Balukjian, B. (2006) 'Vegetation dynamics of predator-free land-bridge islands', *Journal of Ecology*, vol 94, pp253–263

Terborgh, J., Lopez, L., Nunez, P., Rao, M., Shahabuddin, G., Orihuela, G., Riveros, M., Ascanio, R., Adler, G. H., Lambert, T. D. and Balbas, L. (2001) 'Ecological meltdown in predator-free forest fragments', *Science*, vol 294, pp1923–1926

Terborgh, J., Nunez-Iturri, G., Pitman, N. C. A., Valverde, F. H. C., Alvarcz, P., Swamy, V., Pringle, E. G. and Paine, C. E. T. (2008) 'Tree recruitment in an empty forest', *Ecology*, vol 89, pp1757–1768

Tews, J., Brose, U., Grimm, V., Tielborger, K., Wichmann, M. C., Schwager, M. and Jeltsch, F. (2004) 'Animal species diversity driven by habitat heterogeneity/diversity: The importance of keystone structures', *Journal of Biogeography*, vol 31, pp79–92

Thiollay, J. M. (1997) 'Disturbance, selective logging and bird diversity: A neotropical forest study', *Biodiversity and Conservation*, vol 6, pp1155–1173

Thompson, S. K. and Seber, G. A. F. (1996) *Adaptive Sampling*, John Wiley & Sons, New York

Thomson, J. R., Fleishman, E., Mac Nally, R. M. and Dobkin, D. S. (2005) 'Influence of the temporal resolution of data on the success of indicator species models of species richness across multiple taxonomic groups', *Biological Conservation*, vol 124, pp503–518

Thomson, J. R., Mac Nally, R., Fleishman, E. and Horrocks, G. (2007) 'Predicting bird species distributions in reconstructed landscapes', *Conservation Biology*, vol 21, pp752–766

Ticktin, T. (2004) 'The ecological implications of harvesting non-timber forest products', *Journal of Applied Ecology*, vol 41, pp11–21

Trainor, C. R. (2007) 'Changes in bird species composition on a remote and well-forested Wallacean island, south-east Asia', *Biological Conservation*, vol 140, pp373–385

Tscharntke, T., Klein, A. M., Kruess, A., Steffan-Dewenter, I. and Thies, C. (2005) 'Landscape perspectives on agricultural intensification and biodiversity: Ecosystem service management', *Ecology Letters*, vol 8, pp857–874

Turner, M. G., Gardner, R. H. and O'Neill, R. V. (2001) *Landscape Ecology in Theory and Practice*, Springer, New York

Turnhout, E., Hisschemoller, M. and Eijsackers, H. (2007) 'Ecological indicators: Between the two fires of science and policy', *Ecological Indicators*, vol 7, pp215–228

Tylianakis, J. M., Didham, R. K., Bascompte, J. and Wardle, D. A. (2008) 'Global change and species interactions in terrestrial ecosystems', *Ecology Letters*, vol 11, pp1351–1363

Tyre, A. J., Tenhumberg, B., Field, S. A., Niejalke, D., Parris, K. and Possingham, H. P. (2003) 'Improving precision and reducing bias in biological surveys: Estimating false–negative error rates', *Ecological Applications*, vol 13, pp1790–1801

Uehara-Prado, M., Fernandes, J. D., Bello, A. D., Machado, G., Santos, A. J., Vaz-de-Mello, F. Z. and Freitas, A. V. L. (2009) 'Selecting terrestrial arthropods as indicators of small-scale disturbance: A first approach in the Brazilian Atlantic Forest', *Biological Conservation*, vol 142, pp1220–1228

Uezu, A., Beyer, D. D. and Metzger, J. M. (2008) 'Can agroforest woodlots work as stepping stones for birds in the Atlantic Forest region?', *Biodiversity and Conservation*, vol 17, pp1907–1922 DOI: 10.1007/s10531-008-9329-0

Ulrich, W. and Ollik, M. (2005) 'Limits to the estimation of species richness: The use of relative abundance distributions', *Diversity and Distributions*, vol 11, pp265–273

Underwood, A. J. (1997) *Experiments in Ecology: Their Logical Design and Interpretation Using Analysis of Variance*, Cambridge University Press, Cambridge

UNEP (2000) *Development of Indicators of Biological Diversity*, United Nations Environment Program, Washington, DC, UNEP/CBD/SBSTTA/5/12

UNEP (2002) *Report of the Sixth Meeting of the Conference of the Parties to the Convention on Biological Diversity*, United Nations Environment Program, Washington, DC, UNEP/CBD/COP/6/20, available at www.biodiv.org/doc/meetings/cop/cop-06/official/cop-06-20-en.pdf

UNFF (2004) *Criteria and Indicators of Sustainable Forest Management*, Report of the Secretary-General, E/CN.18/2004/11, United Nations Economic and Social Council, New York, available at www.un.org/esa/forests/documents-unff.html

UNFF (2006) *United Nations Forum on Forests: Report of the Sixth Session*, Economic and Social Council, UN, New York

Uriarte, M., Ewing, H. A., Eviner, V. T. and Weathers, K. C. (2007) 'Constructing a broader and more inclusive value system in science', *Bioscience*, vol 57, pp71–78

Urquhart, N. S., Paulsen, S. G. and Larsen, D. P. (1998) 'Monitoring for policy-relevant regional trends over time', *Ecological Applications*, vol 8, pp246–257

VanDerWal, J., Shoo, L. P., Johnson, C. N. and Williams, S. E. (2009) 'Abundance and the environmental niche: Environmental suitability estimated from niche models predicts the upper limit of local abundance', *American Naturalist*, vol 174, pp282–291

Vasconcelos, H. L., Vilhena, J. M. S. and Caliri, G. J. A. (2000) 'Responses of ants to selective logging of a central Amazonian forest', *Journal of Applied Ecology*, vol 37, pp508–514

Vaughan, C., Ramirez, O., Herrera, G. and Guries, R. (2007) 'Spatial ecology and conservation of two sloth species in a cacao landscape in Limón, Costa Rica', *Biodiversity and Conservation*, vol 16, pp2293–2310

Veech, J. A., Summerville, K. S., Crist, T. O. and Gering, J. C. (2002) 'The additive partitioning of species diversity: Recent revival of an old idea', *Oikos*, vol 99, pp3–9

Venier, L. A. and Pearce, J. L. (2004) 'Birds as indicators of sustainable forest management', *Forestry Chronicle*, vol 80, pp61–66

Villard, M.-A. and Jonsson, B. G. (eds) (2009) *Setting Conservation Targets for Managed Forest Landscapes*, Cambridge University Press, Cambridge

Vitousek, P. M. (1984) 'Litterfall, nutrient cycling, and nutrient limitation in tropical forests', *Ecology*, vol 65, pp285–298

Vos, P., Meelis, E. and Ter Keurs, W. J. (2000) 'A framework for the design of ecological monitoring programs as a tool for environmental and nature management', *Environmental Monitoring and Assessment*, vol 61, pp317–344

Walker, B., Hollin, C. S., Carpenter, S. R. and Kinzig, A. (2004) 'Resilience, adaptability and transformability in social-ecological systems', *Ecology and Society*, vol 9, article 5

Walpole, M. J., Almond, R. E. A., Besançon, C., Butchart, S. H. M., Campbell-Lendrum, D., Carr, G. M., Collen, B., Collette, L., Davidson, N. C., Dulloo, E., Fazel, A. F., Galloway, J. N., Gill, M., Goverse, T., Hockings, M., Leaman, D. J., Morgan, D. H. W., Revenga, C., Richwood, C. J., Schutyser, F., Simons, S., Stattersfield, A., Tyrrell, T. D., Vié, J.-C. and Zimsky, M. (2009) 'Tracking progress toward the 2010 biodiversity target and beyond', *Science*, vol 325, pp1503–1504

Walshe, T., Wintle, B., Fidler, F. and Burgman, M. (2007) Use of confidence intervals to demonstrate performance against forest management standards', *Forest Ecology and Management*, vol 247, pp237–245

Walters, C. J. and Green, R. (1997) 'Valuation of experimental management options for ecological systems', *Journal of Wildlife Management*, vol 61, pp987–1006

Walters, C. J. and Hilborn, R. (1978) 'Ecological optimization and adaptive management', *Annual Review of Ecology and Systematics*, vol 9, pp157–188

Walters, C. J. and Holling, C. S. (1990) 'Large-scale management experiments and learning by doing', *Ecology*, vol 71, pp2060–2068

Walters, L., Balint, P., Desai, A. and Stewart, R. (2003) *Risk and Uncertainty in Management of the Sierra Nevada National Forests*, USDA Forest Service, Pacific Southwest Region, George Mason University, Fairfax, Virginia

Watt, A. D. (1998) 'Measuring disturbance in tropical forests: A critique of the use of species-abundance models and indicator measures in general', *Journal of Applied Ecology*, vol 35, pp467–469

WCFSD (1999) *Summary Report: World Commission on Forests and Sustainable Development*, World Commission on Forests and Sustainable Development, Winipeg, Canada

Wenger, E., McDermott, R. and Synder, W. M. (2002) *Cultivating Communities of Practice: A Guide to Managing Knowledge*, Harvard Business School Press, Boston, MA

Whitfield, S. M., Bell, K. E., Philippi, T., Sasa, M., Bolanos, F., Chaves, G., Savage, J. M. and Donnelly, M. A. (2007) 'Amphibian and reptile declines over 35 years at La Selva, Costa Rica', *Proceedings of the National Academy of Sciences of the United States of America*, vol 104, pp8352–8356

Whittaker, R. J., Araujo, M. B., Paul, J., Ladle, R. J., Watson, J. E. M. and Willis, K. J. (2005) 'Conservation biogeography: Assessment and prospect', *Diversity and Distributions*, vol 11, pp3–23

Wijewardana, D. (2008) 'Criteria and indicators for sustainable forest management: The road travelled and the way ahead', *Ecological Indicators*, vol 8, pp115–122

Wikelski, M. and Cooke, S. J. (2006) 'Conservation physiology', *Trends in Ecology & Evolution*, vol 21, pp38–46

Wilhere, G. F. (2002) 'Adaptive management in habitat conservation plans', *Conservation Biology*, vol 16, pp20–29

Wilkie, M. L., Holmgren, P. and Castaneda, F. (2003) *Sustainable Forest Management and the Ecosystem Approach: Two Concepts One Goal*, Forest Resources Development Service, FAO, Rome, Italy

Williams, B. K. (1997) 'Logic and science in wildlife biology, *Journal of Wildlife Management*, vol 61, pp1007–1015

Williams, B. K., Nichols, J. D. and Conroy, M. J. (2001) *Analysis and Management of Wildlife Populations: Modelling, Estimating and Decision Making*, Academic Press, London

Willis, K. J. and Birks, H. J. B. (2006) 'What is natural? The need for a long-term perspective in biodiversity conservation', *Science*, vol 314, pp1261–1265

Willis, K. J., Araujo, M. B., Bennett, K. D., Figueroa-Rangel, B., Froyd, C. A. and Myers, N. (2007) 'How can a knowledge of the past help to conserve the future? Biodiversity conservation and the relevance of long-term ecological studies', *Philosophical Transactions of the Royal Society B-Biological Sciences*, vol 362, pp175–186

Willis, K. J., Gillson, L. and Brncic, T. M. (2004) 'How "virgin" is virgin rainforest?', *Science*, vol 304, pp402–403

Wilson, D. E., Cole, R. F., Nichols, J. D., Rudran, R. and Foster, M. S. (1996) *Measuring and Monitoring Biological Diversity: Standard Methods for Mammals*, Smithsonian, Washington DC

Wilson, K. A., Carwardine, J. and Possingham, H. P. (2009) 'Setting conservation priorities', *Annals of the New York Academy of Science*, vol 1162, pp237–264

Winter, S., Chirici, G., McRoberts, R. E., Hauk, E. and Tomppo, E. (2008) 'Possibilities for harmonizing national forest inventory data for use in forest biodiversity assessments', *Forestry*, vol 81, pp33–44

Wintle, B. A. and Lindenmayer, D. B. (2008) 'Adaptive risk management for certifiably sustainable forestry', *Forest Ecology and Management*, vol 256, pp1311–1319

Wintle, B. A., Bekessy, S. A., Venier, L. A., Pearce, J. L. and Chisholm, R. A. (2005) 'Utility of dynamic-landscape metapopulation models for sustainable forest management', *Conservation Biology*, vol 19, pp1930–1943

Wittemyer, G., Elsen, P., Bean, W. T., Burton, C. O. and Brashares, J. S. (2008) 'Accelerated human population growth at protected area edges', *Science*, vol 321, pp123–126

Woinarski, J. C. Z. (2007) 'Nature's Infinite Variety: Conservation Choice and Management for Dynamic Ecological Systems', in B. D. Lindenmayer and R. Hobbs (eds) *Managing and Designing Landscapes for Conservation*, pp101–110, Blackwell, London

Wolters, V., Bengtsson, J. and Zaitsev, A. S. (2006) 'Relationship among species richness of different taxa', *Ecology*, vol 87, pp1886–1895

Wright, S. J. (2005) 'Tropical forests in a changing environment', *Trends in Ecology & Evolution*, vol 20, pp553–560

Wright, S. J. and Muller-Landau, H. C. (2006) 'The future of tropical forest species', *Biotropica*, vol 38, pp207–301

Wunder, S. (2007) 'The efficiency of payments for environmental services in tropical conservation', *Conservation Biology*, vol 21, pp48–58

Wunder, S., Engel, S. and Pagiola, S. (2008) 'Taking stock: A comparative analysis of payments for environmental services programs in developed and developing countries', *Ecological Economics*, vol 65, pp834–852

Yamasaki, S. H., Kneeshaw, D. D., Munson, A. D. and Dorion, F. (2002) 'Bridging boundaries among disciplines and institutions for effective implementation of criteria and indicators', *Forestry Chronicle*, vol 78, pp487–491

Yoccoz, N. G., Nichols, J. D. and Boulinier, T. (2001) 'Monitoring of biological diversity in space and time', *Trends in Ecology & Evolution*, vol 16, pp446–453

Zarin, D. J., Schulze, M. D., Vidal, E. and Lentini, M. (2007) 'Beyond reaping the first harvest: Management objectives for timber production in the Brazilian Amazon', *Conservation Biology*, vol 21, pp916–925

Zuur, A. F., Ieno, E. N. and Smith, G. M. (2007) *Analysing Ecological Data*, Springer, New York

Printed and bound by CPI Group (UK) Ltd, Croydon, CR0 4YY

01/11/2024

01782610-0001